# Statistics for Biology and Health

*Series Editors*
K. Dietz, M. Gail, K. Krickeberg, B. Singer

**Springer**
*New York*
*Berlin*
*Heidelberg*
*Barcelona*
*Budapest*
*Hong Kong*
*London*
*Milan*
*Paris*
*Santa Clara*
*Singapore*
*Tokyo*

# Statistics for Biology and Health

*Klein/Moeschberger:* Survival Analysis: Techniques for Censored and Truncated Data.
*Kleinbaum:* Logistic Regression: A Self-Learning Text.
*Kleinbaum:* Survival Analysis: A Self-Learning Text.
*Lange:* Mathematical and Statistical Methods for Genetic Analysis.
*Manton/Singer/Suzman:* Forecasting the Health of Elderly Populations.
*Salsburg:* The Use of Restricted Significance Tests in Clinical Trials.

Kenneth Lange

# Mathematical and Statistical Methods for Genetic Analysis

Springer

Kenneth Lange
Department of Biostatistics
and Mathematics
University of Michigan
Ann Arbor, MI 48109-2029
USA

*Series Editors*
K. Dietz
Institut für Medizinische Biometrie
Universität Tübingen
Westbahnhofstr. 55
D 72070 Tübingen
Germany

M. Gail
National Cancer Institute
Rockville, MD 20892
USA

K. Krickeberg
3 Rue de L'Estrapade
75005 Paris
France

B. Singer
Office of Population Research
Princeton University
Princeton, NJ 08544
USA

With 30 illustrations.

Library of Congress Cataloging-in-Publication Data
Lange, Kenneth.
    Mathematical and statistical methods for genetic analysis /
Kenneth Lange.
        p.    cm. — (Statistics for biology and health)
    Includes bibliographical references and index.
    ISBN 0-387-94909-7 (hc : alk. paper)
    1. Genetics—Mathematics.   2. Genetics—Statistical methods.
I. Title.   II. Series.
QH438.4.M33L36   1997
576.5'01'51—dc21                                   96-49533

Printed on acid-free paper.

Production managed by Timothy Taylor; manufacturing supervised by Johanna Tschebull.
Camera-ready copy prepared from the author's LaTeX files.
Printed and bound by Maple-Vail Book Manufacturing Group, York, PA.
Printed in the United States of America.

9 8 7 6 5 4 3 2 1

ISBN 0-387-94909-7 Springer-Verlag New York Berlin Heidelberg   SPIN 10557499

*To Genie*

# Preface

When I was a postdoctoral fellow at UCLA more than two decades ago, I learned genetic modeling from the delightful texts of Elandt-Johnson [2] and Cavalli-Sforza and Bodmer [1]. In teaching my own genetics course over the past few years, first at UCLA and later at the University of Michigan, I longed for an updated version of these books. Neither appeared and I was left to my own devices. As my hastily assembled notes gradually acquired more polish, it occurred to me that they might fill a useful niche. Research in mathematical and statistical genetics has been proceeding at such a breathless pace that the best minds in the field would rather create new theories than take time to codify the old. It is also far more profitable to write another grant proposal. Needless to say, this state of affairs is not ideal for students, who are forced to learn by wading unguided into the confusing swamp of the current scientific literature.

Having set the stage for nobly rescuing a generation of students, let me inject a note of honesty. This book is not the monumental synthesis of population genetics and genetic epidemiology achieved by Cavalli-Sforza and Bodmer. It is also not the sustained integration of statistics and genetics achieved by Elandt-Johnson. It is not even a compendium of recommendations for carrying out a genetic study, useful as that may be. My goal is different and more modest. I simply wish to equip students already sophisticated in mathematics and statistics to engage in genetic modeling. These are the individuals capable of creating new models and methods for analyzing genetic data. No amount of expertise in genetics can overcome mathematical and statistical deficits. Conversely, no mathematician or statistician ignorant of the basic principles of genetics can ever hope to identify worthy problems. Collaborations between geneticists on one side and mathematicians and statisticians on the other can work, but it takes patience and a willingness to learn a foreign vocabulary.

So what are my expectations of readers and students? This is a hard question to answer, in part because the level of the mathematics required builds as the book progresses. At a minimum, readers should be familiar with notions of theoretical statistics such as likelihood and Bayes' theorem. Calculus and linear algebra are used throughout. The last few chapters make fairly heavy demands on skills in theoretical probability and combinatorics. For a few subjects such as continuous time Markov chains and Poisson approximation, I sketch enough of the theory to make the exposition of applications self-contained. Exposure to interesting applications should whet students' appetites for self-study of the underlying mathematics. Everything considered, I recommend that instructors cover the chapters in the order indicated and determine the speed of the course by the mathematical sophistication of the students. There is more than ample material here for a full semester, so it is pointless to rush through basic theory if students encounter difficulty early on. Later chapters can be covered at the discretion of the instructor.

The matter of biological requirements is also problematic. Neither the brief review of population genetics in Chapter 1 nor the primer of molecular genetics in the Appendix is a substitute for a rigorous course in modern genetics. Although many of my classroom students have had little prior exposure to genetics, I have always insisted that those intending to do research fill in the gaps in their knowledge. Students in the mathematical sciences occasionally complain to me that learning genetics is hopeless because the field is in such rapid flux. While I am sympathetic to the difficult intellectual hurdles ahead of them, this attitude is a prescription for failure. Although genetics lacks the theoretical coherence of mathematics, there are fundamental principles and crucial facts that will never change. My advice is follow your curiosity and learn as much genetics as you can. In scientific research chance always favors the well prepared.

The incredible flowering of mathematical and statistical genetics over the past two decades makes it impossible to summarize the field in one book. I am acutely aware of my failings in this regard, and it pains me to exclude most of the history of the subject and to leave unmentioned so many important ideas. I apologize to my colleagues. My own work receives too much attention; my only excuse is that I understand it best. Fortunately, the recent book of Michael Waterman delves into many of the important topics in molecular genetics missing here [4].

I have many people to thank for helping me in this endeavor. Carol Newton nurtured my early career in mathematical biology and encouraged me to write a book in the first place. Daniel Weeks and Eric Sobel deserve special credit for their many helpful suggestions for improving the text. My genetics colleagues David Burke, Richard Gatti, and Miriam Meisler read and corrected my first draft of the appendix. David Cox, Richard Gatti, and James Lake kindly contributed data. Janet Sinsheimer and Hongyu Zhao provided numerical examples for Chapters 10 and 12, respectively. Many students at UCLA and Michigan checked the problems and proofread the text. Let me single out Ru-zong Fan, Ethan Lange, Laura Lazzeroni, Eric Schadt, Janet Sinsheimer, Heather Stringham, and Wynn Walker for their diligence. David Hunter kindly prepared the index. Doubtless a few errors remain, and I would be grateful to readers for their corrections. Finally, I thank my wife Genie, to whom I dedicate this book, for her patience and love.

## A Few Words about Software

This text contains several numerical examples that rely on software from the public domain. Readers interested in a copy of the programs MENDEL and FISHER mentioned in Chapters 7 and 8 and the optimization program SEARCH used in Chapter 3 should get in touch with me. Laura Lazzeroni distributes software for testing transmission association and linkage disequilibrium as discussed in Chapter 4. Daniel Weeks is responsible for the software implementing the APM method of linkage analysis featured in Chapter 6. He and Eric Sobel also distribute software for haplotyping and stochastic calculation of location scores as covered in Chapter 9. Readers should contact Eric Schadt or Janet Sinsheimer for the phylogeny software of Chapter 10 and Michael Boehnke for the radiation hybrid software of Chapter 11. Further free software for genetic analysis is listed in the recent book by Ott and Terwilliger [3].

## References

[1] Cavalli-Sforza LL, Bodmer WF (1971) *The Genetics of Human Populations*. Freeman, San Francisco

[2] Elandt-Johnson RC (1971) *Probability Models and Statistical Methods in Genetics*. Wiley, New York

[3] Terwilliger JD, Ott J (1994) *Handbook of Human Genetic Linkage*. Johns Hopkins University Press, Baltimore

[4] Waterman MS (1995) *Introduction to Computational Biology: Maps, Sequences, and Genomes*. Chapman and Hall, London

Acknowledgments

[1] Lange K, Boehnke M (1992) Bayesian methods and optimal experimental design for gene mapping by radiation hybrids. Ann Hum Genet 56:119–144

[2] Goradia TM, Lange K, Miller PL, Nadkarni PM (1992) Fast computation of genetic likelihoods on human pedigree data. Hum Hered 42:42–62

[3] Lange K, Sinsheimer JS (1992) Calculation of genetic identity coefficients. Ann Hum Genet 56:339–346

[4] Sobel E, Lange K (1993) Metropolis sampling in pedigree analysis. Stat Methods Med Res 2:263–282

[5] Lange K (1995) Applications of the Dirichlet distribution to forensic match probabilities. Genetica 96:107–117

[6] Lange K, Boehnke M, Cox DR, Lunetta KL (1995) Statistical methods for polyploid radiation hybrid mapping. Genome Res 5:136–150

[7] Sobel E, Lange K (1996) Descent graphs in pedigree analysis: applications to haplotyping, location scores, and marker sharing statistics. Amer J Hum Genet 58:1323–1337

# Contents

# 1

# Basic Principles of Population Genetics

## 1.1   Introduction

In this chapter we briefly review some elementary results from population genetics [2, 3, 4, 6, 7, 10, 13]. Various genetic definitions are recalled merely to provide a context for this and more advanced mathematical theory. Readers with a limited knowledge of modern genetics are urged to learn molecular genetics by formal course work or informal self-study. The Appendix summarizes a few of the major currents in molecular genetics.

## 1.2   Genetics Background

The classical genetic definitions of interest to us predate the modern molecular era. First, **genes** occur at definite sites, or **loci**, along a **chromosome**. Each locus can be occupied by one of several variant genes called **alleles**. Most human cells contain 46 chromosomes. Two of these are **sex** chromosomes—two paired X's for a female and an X and a Y for a male. The remaining 22 **homologous** pairs of chromosomes are termed **autosomes**. One member of each chromosome pair is maternally derived via an **egg**; the other member is paternally derived via a **sperm**. Except for the sex chromosomes, it follows that there are two genes at every locus. These constitute a person's **genotype** at that locus. If the two alleles are identical, then the person is a **homozygote**; otherwise, he is a **heterozygote**. Typically, one denotes a genotype by two allele symbols separated by a slash /. Genotypes may not be observable. By definition, what is observable is a person's **phenotype**.

TABLE 1.1. Phenotypes at the ABO Locus

| Phenotypes | Genotypes |
|:---:|:---:|
| $A$ | $A/A$, $A/O$ |
| $B$ | $B/B$, $B/O$ |
| $AB$ | $A/B$ |
| $O$ | $O/O$ |

A simple example will serve to illustrate these definitions. The ABO locus resides on the long arm of chromosome 9 at band q34. This locus determines detectable **antigens** on the surface of red blood cells. There are three alleles, $A$, $B$, and $O$, which determine an $A$ antigen, a $B$ antigen, and the absence of either antigen, respectively. Phenotypes are recorded by reacting antibodies for $A$ and $B$ against a blood sample. The four observable phenotypes are $A$ (antigen $A$ alone detected), $B$ (antigen $B$ alone detected), $AB$ (antigens $A$ and $B$ both detected), and $O$ (neither antigen $A$ nor $B$ detected). These correspond to the genotype sets given in Table 1.1.

Note that phenotype $A$ results from either the homozygous genotype $A/A$ or the heterozygous genotype $A/O$; similarly, phenotype $B$ results from either $B/B$ or $B/O$. Alleles $A$ and $B$ both mask the presence of the $O$ allele and are said to be **dominant** to it. Alternatively, $O$ is **recessive** to $A$ and $B$. Relative to one another, alleles $A$ and $B$ are **codominant**.

The six genotypes listed above at the ABO locus are unordered in the sense that maternal and paternal contributions are not distinguished. In some cases it is helpful to deal with **ordered** genotypes. When we do, we will adopt the convention that the maternal allele is listed to the left of the slash and the paternal allele is listed to the right. With three alleles, the ABO locus has nine distinct ordered genotypes.

The **Hardy-Weinberg law** of population genetics permits calculation of genotype frequencies from allele frequencies. In the ABO example above, if the frequency of the $A$ allele is $p_A$ and the frequency of the $B$ allele is $p_B$, then a random individual will have phenotype $AB$ with frequency $2p_A p_B$. The factor of 2 in this frequency reflects the two equally likely ordered genotypes $A/B$ and $B/A$. In essence, Hardy-Weinberg equilibrium corresponds to the random union of two **gametes**, one gamete being an egg and the other being a sperm. A union of two gametes incidentally is called a **zygote**.

In gene mapping studies, several genetic loci on the same chromosome are phenotyped. When these loci are simultaneously followed in a human **pedigree**, the phenomenon of **recombination** can often be observed. This reshuffling of genetic material manifests itself when a parent transmits to a child a chromosome that differs from both of the corresponding homologous parental chromosomes. Recombination takes place during the formation of gametes at **meiosis**. Suppose, for the sake of argument, that in the parent producing the gamete, one member of each chromosome pair is painted black and the other member is painted white.

Instead of inheriting an all-black or an all-white representative of a given pair, a gamete inherits a chromosome that alternates between black and white. The points of exchange are termed **crossovers**. Any given gamete will have just a few randomly positioned crossovers per chromosome. The **recombination fraction** between two loci on the same chromosome is the probability that they end up in regions of different color in a gamete. This event occurs whenever the two loci are separated by an odd number of crossovers along the gamete. Chapter 12 will elaborate on this brief, simplified description of the recombination process.

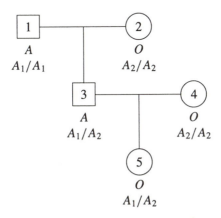

FIGURE 1.1. A Pedigree with ABO and AK1 Phenotypes

As a concrete example, consider the locus AK1 (adenylate kinase 1) in the vicinity of ABO on chromosome 9. With modern biochemical techniques it is possible to identify two codominant alleles, $A_1$ and $A_2$, at this enzyme locus. Figure 1.1 depicts a pedigree with phenotypes listed at the ABO locus and unordered genotypes listed at the AK1 locus. In this pedigree, as in all pedigrees, circles denote females and squares denote males. Individuals 1, 2, and 4 are termed the **founders** of the pedigree. Parents of founders are not included in the pedigree. By convention, each nonfounder or child of the pedigree always has both parents included.

Close examination of the pedigree shows that individual 3 has alleles $A$ and $A_1$ on his paternally derived chromosome 9 and alleles $O$ and $A_2$ on his maternally derived chromosome 9. However, he passes to his child 5 a chromosome with $O$ and $A_1$ alleles. In other words, the gamete passed is recombinant between the loci ABO and AK1. On the basis of many such observations, it is known empirically that doubly heterozygous males like 3 produce recombinant gametes about 12 percent of the time. In females the recombination fraction is about 20 percent.

The pedigree in Figure 1.1 is atypical in several senses. First, it is quite simple graphically. Second, everyone is phenotyped; in larger pedigrees, some people will be dead or otherwise unavailable for typing. Third, it is constructed so that recombination can be unambiguously determined. In most matings, one cannot directly count recombinant and nonrecombinant gametes. This forces geneticists to rely on indirect statistical arguments to overcome the problem of missing information. The experimental situation is analogous to medical imaging, where partial tomographic information is available, but the full details of transmission or emission events must be reconstructed. Part of the missing information in pedigree data has to do with **phase**. Alleles $O$ and $A_2$ are in phase in individual 3 of Figure 1.1. In general, a gamete's sequence of alleles along a chromosome constitutes a **haplotype**. The alleles appearing in the haplotype are said to be in phase. Two such haplotypes together determine a **multilocus** genotype (or simply a genotype when the context is clear).

Recombination or **linkage** studies are conducted with loci called **traits** and **markers**. Trait loci typically determine genetic diseases or interesting biochemical or physiological differences between individuals. Marker loci, which need not be genetic loci in the traditional sense at all, are signposts along the chromosomes. A marker locus is simply a place on a chromosome showing detectable population differences. These differences or alleles permit recombination to be measured between the trait and marker loci. In practice, recombination between two loci can be observed only when the parent contributing a gamete is heterozygous at both loci. In linkage analysis it is therefore advantageous for a locus to have several common alleles. Such loci are said to be **polymorphic**.

The number of haplotypes possible for a given set of loci is the product of the numbers of alleles possible at each locus. In the ABO-AK1 example, there are $k = 3 \times 2 = 6$ possible haplotypes. These can form $k^2$ genotypes based on ordered haplotypes or $k + \frac{k(k-1)}{2} = \frac{k(k+1)}{2}$ genotypes based on unordered haplotypes.

To compute the population frequencies of random haplotypes, one can invoke **linkage equilibrium**. This rule stipulates that a haplotype frequency is the product of the underlying allele frequencies. For instance, the frequency of an $OA_1$ haplotype is $p_O p_{A_1}$, where $p_O$ and $p_{A_1}$ are the population frequencies of the alleles $O$ and $A_1$, respectively. To compute the frequency of a multilocus genotype, one can view it as the union of two random gametes in imitation of the Hardy-Weinberg law. For example, the genotype of person 2 in Figure 1.1 has population frequency $(p_O p_{A_2})^2$, being the union of two $OA_2$ haplotypes. Exceptions to the rule of linkage equilibrium often occur for tightly linked loci.

## 1.3   Hardy-Weinberg Equilibrium

Let us now consider a formal mathematical model for the establishment of Hardy-Weinberg equilibrium. This model relies on the seven following explicit assumptions: (a) infinite population size, (b) discrete generations, (c) random mating, (d) no selection, (e) no migration, (f) no mutation, and (g) equal initial genotype frequencies in the two sexes. Suppose for the sake of simplicity that there are two alleles $A_1$ and $A_2$ at some autosomal locus in this infinite population and that all genotypes are unordered. Consider the result of crossing the genotype $A_1/A_1$ with the genotype $A_1/A_2$. The first genotype produces only $A_1$ gametes, and the second genotype yields gametes $A_1$ and $A_2$ in equal proportion. For the cross under consideration, gametes produced by the genotype $A_1/A_1$ are equally likely to combine with either gamete type issuing from the genotype $A_1/A_2$. Thus, for the cross $A_1/A_1 \times A_1/A_2$, the frequency of offspring obviously is $\frac{1}{2}A_1/A_1$ and $\frac{1}{2}A_1/A_2$. Similarly, the cross $A_1/A_1 \times A_2/A_2$ yields only $A_1/A_2$ offspring. The cross $A_1/A_2 \times A_1/A_2$ produces offspring in the ratio $\frac{1}{4}A_1/A_1, \frac{1}{2}A_1/A_2$, and $\frac{1}{4}A_2/A_2$. These proportions of outcomes for the various possible crosses are known as **segregation ratios**.

TABLE 1.2. Mating Outcomes for Hardy-Weinberg Equilibrium

| Mating Type | Nature of Offspring | Frequency |
|---|---|---|
| $A_1/A_1 \times A_1/A_1$ | $A_1/A_1$ | $u^2$ |
| $A_1/A_1 \times A_1/A_2$ | $\frac{1}{2}A_1/A_1 + \frac{1}{2}A_1/A_2$ | $2uv$ |
| $A_1/A_1 \times A_2/A_2$ | $A_1/A_2$ | $2uw$ |
| $A_1/A_2 \times A_1/A_2$ | $\frac{1}{4}A_1/A_1 + \frac{1}{2}A_1/A_2 + \frac{1}{4}A_2/A_2$ | $v^2$ |
| $A_1/A_2 \times A_2/A_2$ | $\frac{1}{2}A_1/A_2 + \frac{1}{2}A_2/A_2$ | $2vw$ |
| $A_2/A_2 \times A_2/A_2$ | $A_2/A_2$ | $w^2$ |

Suppose the initial proportions of the genotypes are $u$ for $A_1/A_1$, $v$ for $A_1/A_2$, and $w$ for $A_2/A_2$. Under the stated assumptions, the next generation will be composed as shown in Table 1.2. The entries in Table 1.2 yield for the three genotypes $A_1/A_1$, $A_1/A_2$, and $A_2/A_2$ the new frequencies

$$u^2 + uv + \frac{1}{4}v^2 = \left(u + \frac{1}{2}v\right)^2$$

$$uv + 2uw + \frac{1}{2}v^2 + vw = 2\left(u + \frac{1}{2}v\right)\left(\frac{1}{2}v + w\right)$$

$$\frac{1}{4}v^2 + vw + w^2 = \left(\frac{1}{2}v + w\right)^2,$$

respectively. If we define the frequencies of the two alleles $A_1$ and $A_2$ as $p_1 = u + \frac{v}{2}$ and $p_2 = \frac{v}{2} + w$, then $A_1/A_1$ occurs with frequency $p_1^2$, $A_1/A_2$ with frequency $2p_1p_2$, and $A_2/A_2$ with frequency $p_2^2$. After a second round of random mating, the frequencies of the genotypes $A_1/A_1$, $A_1/A_2$, and $A_2/A_2$ are

$$\left(p_1^2 + \frac{1}{2}2p_1p_2\right)^2 = \left[p_1(p_1 + p_2)\right]^2$$

$$= p_1^2$$

$$2\left(p_1^2 + \frac{1}{2}2p_1p_2\right)\left(\frac{1}{2}2p_1p_2 + p_2^2\right) = 2p_1(p_1 + p_2)p_2(p_1 + p_2)$$

$$= 2p_1 p_2$$

$$\left(\frac{1}{2} 2p_1 p_2 + p_2^2\right)^2 = \left[p_2(p_1 + p_2)\right]^2$$

$$= p_2^2.$$

Thus, after a single round of random mating, genotype frequencies stabilize at the Hardy-Weinberg proportions.

We may deduce the same result by considering the gamete population. $A_1$ gametes have frequency $p_1$ and $A_2$ gametes frequency $p_2$. Since random union of gametes is equivalent to random mating, $A_1/A_1$ is present in the next generation with frequency $p_1^2$, $A_1/A_2$ with frequency $2p_1 p_2$, and $A_2/A_2$ with frequency $p_2^2$. In the gamete pool from this new generation, $A_1$ again occurs with frequency $p_1^2 + p_1 p_2 = p_1(p_1 + p_2) = p_1$ and $A_2$ with frequency $p_2$. In other words, stability is attained in a single generation. This random union of gametes argument generalizes easily to more than two alleles.

Hardy-Weinberg equilibrium is a bit more subtle for X-linked loci. Consider a locus on the X chromosome and any allele at that locus. At generation $n$ let the frequency of the given allele in females be $q_n$ and in males be $r_n$. Under our stated assumptions for Hardy-Weinberg equilibrium, one can show that $q_n$ and $r_n$ converge quickly to the value $p = \frac{2}{3} q_0 + \frac{1}{3} r_0$. Twice as much weight is attached to the initial female frequency since females have two X chromosomes while males have only one.

Because a male always gets his X chromosome from his mother, and his mother precedes him by one generation,

$$r_n = q_{n-1}. \tag{1.1}$$

Likewise, the frequency in females is the average frequency for the two sexes from the preceding generation; in symbols,

$$q_n = \frac{1}{2} q_{n-1} + \frac{1}{2} r_{n-1}. \tag{1.2}$$

Equations (1.1) and (1.2) together imply

$$\frac{2}{3} q_n + \frac{1}{3} r_n = \frac{2}{3}\left(\frac{1}{2} q_{n-1} + \frac{1}{2} r_{n-1}\right) + \frac{1}{3} q_{n-1}$$

$$= \frac{2}{3} q_{n-1} + \frac{1}{3} r_{n-1}. \tag{1.3}$$

It follows that the weighted average $\frac{2}{3} q_n + \frac{1}{3} r_n = p$ for all $n$.

From equations (1.2) and (1.3), we deduce that

$$q_n - p = q_n - \frac{3}{2} p + \frac{1}{2} p$$

$$= \frac{1}{2} q_{n-1} + \frac{1}{2} r_{n-1} - \frac{3}{2}\left(\frac{2}{3} q_{n-1} + \frac{1}{3} r_{n-1}\right) + \frac{1}{2} p$$

$$= -\frac{1}{2}q_{n-1} + \frac{1}{2}p$$

$$= -\frac{1}{2}(q_{n-1} - p).$$

Continuing in this manner,

$$q_n - p = \left(-\frac{1}{2}\right)^n (q_0 - p).$$

Thus the difference between $q_n$ and $p$ diminishes by half each generation, and $q_n$ approaches $p$ in a zigzag manner. The male frequency $r_n$ displays the same behavior but lags behind $q_n$ by one generation. In contrast to the autosomal case, it takes more than one generation to achieve equilibrium. However, equilibrium is still approached relatively fast. In the extreme case that $q_0 = .75$ and $r_0 = .12$, Figure 1.2 plots $q_n$ for a few representative generations.

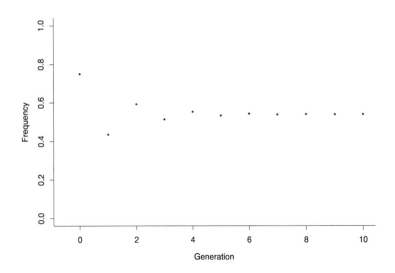

FIGURE 1.2. Approach to Equilibrium of $q_n$ as a Function of $n$

At equilibrium how do we calculate the frequencies of the various genotypes? Suppose we have two alleles $A_1$ and $A_2$ with equilibrium frequencies $p_1$ and $p_2$. Then the female genotypes $A_1/A_1$, $A_1/A_2$, and $A_2/A_2$ have frequencies $p_1^2$, $2p_1 p_2$, and $p_2^2$, respectively, just as in the autosomal case. In males the **hemizygous** genotypes $A_1$ and $A_2$ clearly have frequencies $p_1$ and $p_2$.

**Example 1.1** *Hardy-Weinberg Equilibrium for the $Xg(a)$ Locus*

The red cell antigen $Xg(a)$ is an X-linked dominant with a frequency in Caucasians of approximately $p = .65$. Thus, about .65 of all Caucasian males and about $p^2 + 2p(1 - p) = .88$ of all Caucasian females carry the antigen.    ∎

## 1.4   Linkage Equilibrium

Loci on nonhomologous chromosomes show independent segregation at meiosis. In contrast, genes at two physically close loci on the same chromosome tend to stick together during the formation of gametes. The recombination fraction $\theta$ between two loci is a monotone, nonlinear function of the physical distance separating them. In family studies in man or in breeding studies in other species, $\theta$ is the observable rather than physical distance. In Chapter 12 we show that $0 \le \theta \le \frac{1}{2}$. The upper bound of $\frac{1}{2}$ is attained by two loci on nonhomologous chromosomes.

The population genetics law of linkage equilibrium is of fundamental importance in theoretical calculations. Convergence to linkage equilibrium can be proved under the same assumptions used to prove Hardy-Weinberg equilibrium. Suppose that allele $A_i$ at locus $A$ has frequency $p_i$ and allele $B_j$ at locus $B$ has frequency $q_j$. Let $P_n(A_i B_j)$ be the frequency of chromosomes with alleles $A_i$ and $B_j$ among those gametes produced at generation $n$. Since recombination fractions almost invariably differ between the sexes, let $\theta_f$ and $\theta_m$ be the female and male recombination fractions, respectively, between the two loci. The average $\theta = (\theta_f + \theta_m)/2$ governs the rate of approach to linkage equilibrium.

We can express $P_n(A_i B_j)$ by conditioning on whether a gamete is an egg or a sperm and on whether nonrecombination or recombination occurs. If recombination occurs, then the gamete carries the two alleles $A_i$ and $B_j$ with equilibrium probability $p_i q_j$. Thus, the appropriate recurrence relation is

$$
\begin{aligned}
P_n(A_i B_j) &= \frac{1}{2}\left[(1 - \theta_f)P_{n-1}(A_i B_j) + \theta_f p_i q_j\right] \\
&\quad + \frac{1}{2}\left[(1 - \theta_m)P_{n-1}(A_i B_j) + \theta_m p_i q_j\right] \\
&= (1 - \theta)P_{n-1}(A_i B_j) + \theta p_i q_j.
\end{aligned}
$$

Note that this recurrence relation is valid when the two loci occur on nonhomologous chromosomes provided $\theta = \frac{1}{2}$ and we interpret $P_n(A_i B_j)$ as the probability that someone at generation $n$ receives a gamete bearing the two alleles $A_i$ and $B_j$. Subtracting $p_i q_j$ from both sides of the recurrence relation gives

$$
\begin{aligned}
P_n(A_i B_j) - p_i q_j &= (1 - \theta)[P_{n-1}(A_i B_j) - p_i q_j] \\
&\ \ \vdots \\
&= (1 - \theta)^n [P_0(A_i B_j) - p_i q_j].
\end{aligned}
$$

Thus, $P_n(A_i B_j)$ converges to $p_i q_j$ at the geometric rate $1 - \theta$. For two loci on different chromosomes, the deviation from linkage equilibrium is halved each generation. Equilibrium is approached much more slowly for closely spaced loci. Similar, but more cumbersome, proofs of convergence to linkage equilibrium can be given for three or more loci [1, 5, 9, 11]. Problem 7 explores the case of three loci.

## 1.5   Selection

The simplest model of evolution involves **selection** at an autosomal locus with two alleles $A$ and $a$. At generation $n$, let allele $A$ have population frequency $p_n$ and allele $a$ population frequency $q_n = 1 - p_n$. Under the usual assumptions of genetic equilibrium, we deduced the Hardy-Weinberg and linkage equilibrium laws. Now suppose that we relax the assumption of no selection by postulating different fitnesses $w_{A/A}$, $w_{A/a}$, and $w_{a/a}$ for the three genotypes. **Fitness** is a technical term dealing with the reproductive capacity rather than the longevity of people with a given genotype. Thus, $w_{A/A}/w_{A/a}$ is the ratio of the expected genetic contribution to the next generation of an $A/A$ individual to the expected genetic contribution of an $A/a$ individual. Since only fitness ratios are relevant, we can without loss of generality put $w_{A/a} = 1$, $w_{A/A} = 1 - r$, and $w_{a/a} = 1 - s$, provided of course that $r \le 1$ and $s \le 1$. Observe that $r$ and $s$ can be negative.

To explore the evolutionary dynamics of this model, we define the average fitness

$$
\begin{aligned}
\bar{w}_n &= (1 - r)p_n^2 + 2p_nq_n + (1 - s)q_n^2 \\
&= 1 - rp_n^2 - sq_n^2
\end{aligned}
$$

at generation $n$. Owing to our implicit assumption of random union of gametes, the Hardy-Weinberg proportions appear in the definition of $\bar{w}_n$ even though the allele frequency $p_n$ changes over time. Because $A/A$ individuals always contribute an $A$ allele whereas $A/a$ individuals do so only half of the time, the change in allele frequency $\Delta p_n = p_{n+1} - p_n$ can be expressed as

$$
\begin{aligned}
\Delta p_n &= \frac{(1 - r)p_n^2 + p_nq_n}{\bar{w}_n} - p_n \\
&= \frac{(1 - r)p_n^2 + p_nq_n - (1 - rp_n^2 - sq_n^2)p_n}{\bar{w}_n} \\
&= \frac{p_nq_n[s - (r + s)p_n]}{\bar{w}_n}.
\end{aligned}
\tag{1.4}
$$

At a **fixed point** $p_\infty \in [0, 1]$, we have $\Delta p_\infty = 0$. In view of equation (1.4), this can occur only when $p_\infty$ equals 0, 1, or possibly $\frac{s}{r+s}$. The third point is a legitimate fixed point if and only if $r$ and $s$ have the same sign. In the case $r > 0$ and $s \le 0$, the linear function $g(p) = s - (r + s)p$ satisfies $g(0) \le 0$ and $g(1) < 0$. It is therefore negative throughout the open interval $(0, 1)$, and equation (1.4) implies that $\Delta p_n < 0$ for all $p_n \in (0, 1)$. It follows that the decreasing sequence $p_n$ has a limit $p_\infty < 1$ when $p_0 < 1$. Equation (1.4) shows that $p_\infty > 0$ is inconsistent with $\Delta p_\infty = \lim_{n \to \infty} \Delta p_n = 0$. Hence, we arrive at the intuitively obvious conclusion that the $A$ allele is driven to extinction. In the opposite case $r \le 0$ and $s > 0$, the $a$ allele is driven to extinction.

When $r$ and $s$ have the same sign, it is helpful to consider the difference

$$p_{n+1} - \frac{s}{r+s} \;=\; \Delta p_n + p_n - \frac{s}{r+s}$$

$$=\; -\frac{(r+s)p_n q_n\left(p_n - \frac{s}{r+s}\right)}{1 - rp_n^2 - sq_n^2} + p_n - \frac{s}{r+s}$$

$$=\; \frac{1 - rp_n^2 - sq_n^2 - (r+s)p_n q_n}{1 - rp_n^2 - sq_n^2}\left(p_n - \frac{s}{r+s}\right)$$

$$=\; \frac{1 - rp_n - sq_n}{1 - rp_n^2 - sq_n^2}\left(p_n - \frac{s}{r+s}\right).$$

If both $r$ and $s$ are negative, then the factor

$$\lambda(p_n) \;=\; \frac{1 - rp_n - sq_n}{1 - rp_n^2 - sq_n^2}$$
$$>\; 1,$$

and $p_n - \frac{s}{r+s}$ has constant sign and grows in magnitude. Therefore, arguments similar to those given in the $r > 0$ and $s \le 0$ case imply that $\lim_{n\to\infty} p_n = 0$ for $p_0 < \frac{s}{r+s}$ and $\lim_{n\to\infty} p_n = 1$ for $p_0 > \frac{s}{r+s}$. The point $\frac{s}{r+s}$ is an **unstable equilibrium**.

If both $r$ and $s$ are positive, then $0 \le \lambda(p_n) < 1$, and $p_n - \frac{s}{r+s}$ has constant sign and declines in magnitude. In this case, $\lim_{n\to\infty} p_n = \frac{s}{r+s}$, and the point $\frac{s}{r+s}$ is a **stable equilibrium**. For $p_0 \approx \frac{s}{r+s}$,

$$\lambda(p_n) \;\approx\; \lambda\left(\frac{s}{r+s}\right)$$
$$=\; \frac{r+s-2rs}{r+s-rs},$$

and $p_n - \frac{s}{r+s} \approx \lambda(\frac{s}{r+s})^n (p_0 - \frac{s}{r+s})$. In other words, $p_n$ approaches its equilibrium value locally at the geometric rate $\lambda(\frac{s}{r+s})$.

The **rate of convergence** of $p_n$ to 0 or 1 depends on whether there is selection against the heterozygous genotype $A/a$. Consider the case $r > 0$ and $s < 0$ of selection against a dominant. Then $p_n \to 0$ and the approximation

$$p_{n+1} \;=\; \frac{(1-r)p_n^2 + p_n q_n}{1 - rp_n^2 - sq_n^2}$$
$$\approx\; \frac{p_n}{(1-s)}$$

for $p_n \approx 0$ makes it clear that $p_n$ approaches 0 locally at geometric rate $\frac{1}{1-s}$.

If $r > 0$ and $s = 0$, then $p_n \to 0$ still holds, but convergence no longer occurs at a geometric rate. Indeed, the equality

$$p_{n+1} \;=\; \frac{p_n(1 - rp_n)}{1 - rp_n^2}$$

entails

$$\frac{1}{p_{n+1}} - \frac{1}{p_n} = \frac{1}{p_n}\left(\frac{1 - rp_n^2}{1 - rp_n} - 1\right)$$

$$= \frac{r(1 - p_n)}{1 - rp_n}$$

$$\approx r.$$

It follows that for $p_0$ close to 0

$$\frac{1}{p_n} - \frac{1}{p_0} = \sum_{i=0}^{n-1}\left(\frac{1}{p_{i+1}} - \frac{1}{p_i}\right)$$

$$\approx nr.$$

This approximation implies the slow convergence

$$p_n \approx \frac{1}{nr + \frac{1}{p_0}}$$

for selection against a pure recessive.

**Heterozygote advantage** ($r$ and $s$ both positive) is the most interesting situation covered by this classic selection model. Geneticists have suggested that several recessive diseases are maintained at high frequencies by the mechanism of heterozygote advantage. The best evidence favoring this hypothesis exists for sickle cell anemia [2]. A single dose of the sickle cell gene appears to confer protection against malaria. The evidence is much weaker for a heterozygote advantage in Tay Sachs disease and cystic fibrosis. Geneticists have conjectured that these genes may protect carriers from tuberculosis and cholera, respectively [14].

## 1.6    Balance Between Mutation and Selection

**Mutations** furnish the raw material of evolutionary change. In practice, most mutations are either neutral or deleterious. We now briefly discuss the balance between deleterious mutations and selection. Consider first the case of a dominant disease. In the notation of the last section, let $a$ be the normal allele and $A$ the disease allele, and define the fitnesses of the three genotypes by $r > 0$ and $s < 0$. If the mutation rate from $a$ to $A$ is $\mu$, then equilibrium is achieved between the opposing forces of mutation and selection when

$$q_\infty = \frac{p_\infty q_\infty + (1 - s)q_\infty^2}{1 - rp_\infty^2 - sq_\infty^2}(1 - \mu).$$

If we multiply this equation by $1 - rp_\infty^2 - sq_\infty^2$ and divide it by $q_\infty$, we get

$$1 - rp_\infty^2 - sq_\infty^2 = (1 - sq_\infty)(1 - \mu).$$

Dropping the negligible term $rp_\infty^2$, we find that this quadratic has the approximate solution

$$q_\infty \approx \frac{1-\mu}{2}\left(1 + \sqrt{1 + \frac{4\mu}{s[1-\mu)]^2}}\right)$$

$$\approx \frac{1-\mu}{2}\left(1 + 1 + \frac{2\mu}{s[1-\mu]^2}\right)$$

$$\approx 1 + \frac{\mu(1-s)}{s},$$

which yields $p_\infty = 1 - q_\infty \approx \frac{\mu(1-s)}{-s}$. The corresponding equilibrium frequency of affecteds is $2p_\infty q_\infty \approx \frac{2\mu(1-s)}{-s}$.

For a recessive disease ($r > 0$ and $s = 0$), the balance equation becomes

$$q_\infty = \frac{p_\infty q_\infty + q_\infty^2}{\bar{w}_\infty}(1-\mu)$$

$$= \frac{q_\infty}{1 - rp_\infty^2}(1-\mu).$$

In other words, $1 - rp_\infty^2 = 1 - \mu$, which has solution $p_\infty = \sqrt{\mu/r}$. The frequency of affecteds at equilibrium is now $p_\infty^2 = \frac{\mu}{r}$. Thus given equal mutation rates, dominant and recessive diseases will afflict comparable numbers of people. In contrast, the underlying allele frequencies and rates of approach to equilibrium vary dramatically. Indeed, it is debatable whether any human population has existed long enough for the alleles at a recessive disease locus to achieve a balance between mutation and selection. Random sampling of gametes (**genetic drift**) and small initial population sizes (**founder effect**) play a much larger role in determining the frequency of recessive diseases in modern human populations.

## 1.7   Problems

1. In blood transfusions, compatibility at the ABO and Rh loci is important. These autosomal loci are unlinked. At the Rh locus the + allele codes for the presence of a red cell antigen and therefore is dominant to the − allele, which codes for the absence of the antigen. Suppose that the frequencies of the two Rh alleles are $q_+$ and $q_-$. Type $O-$ people are universal donors, and type $AB+$ people are universal recipients. Under genetic equilibrium, what are the population frequencies of these two types of people? (Reference [2] discusses these genetic systems and gives allele frequencies for some representative populations.)

2. Suppose in the Hardy-Weinberg model for an autosomal locus that the genotype frequencies for the two sexes differ. What is the ultimate frequency of a given allele? How long does it take genotype frequencies to stabilize at their Hardy-Weinberg values?

3. Consider an autosomal locus with $m$ alleles in Hardy-Weinberg equilibrium. If allele $A_i$ has frequency $p_i$, then show that a random non-inbred person is heterozygous with probability $1 - \sum_{i=1}^{m} p_i^2$. What is the maximum of this probability, and for what allele frequencies is this maximum attained?

4. In forensic applications of genetics, loci with high exclusion probabilities are typed. For a codominant locus with $n$ alleles, show that the probability of two random people having different genotypes is

$$e = \sum_{i=1}^{n-1} \sum_{j=i+1}^{n} 2p_i p_j (1 - 2p_i p_j) + \sum_{i=1}^{n} p_i^2 (1 - p_i^2)$$

under Hardy-Weinberg equilibrium [8]. Simplify this expression to

$$e = 1 - 2\left(\sum_{i=1}^{n} p_i^2\right)^2 + \sum_{i=1}^{n} p_i^4.$$

Prove that $e$ attains its maximum $e_{max} = 1 - \frac{2}{n^2} + \frac{1}{n^3}$ when $p_i = \frac{1}{n}$ for all $i$. For two independent loci with $\sqrt{n}$ alleles each, verify that the maximum exclusion probability based on exclusion at either locus is $1 - \frac{4}{n^2} + \frac{4}{n^{5/2}} - \frac{1}{n^3}$. How does this compare to the maximum exclusion probability for a single locus with $n$ equally frequent alleles when $n = 16$? What do you conclude about the information content of two loci versus one locus? (Hint: To prove the claim about $e_{max}$, note that without loss of generality one can assume $p_1 \leq p_2 \leq \cdots \leq p_n$. If $p_i < p_{i+1}$, then $e$ can be increased by replacing $p_i$ and $p_{i+1}$ by $p_i + x$ and $p_{i+1} - x$ for $x$ positive and sufficiently small.)

5. Moran [12] has posed a model for the approach of allele frequencies to Hardy-Weinberg equilibrium that permits generations to overlap. Let $u(t)$, $v(t)$, and $w(t)$ be the relative proportions of the genotypes $A_1/A_1$, $A_1/A_2$, and $A_2/A_2$ at time $t$. Assume that in the small time interval $(t, t + dt)$ a proportion $dt$ of the population dies and is replaced by the offspring of random matings from the residue of the population. In effect, members of the population have independent, exponentially distributed lifetimes of mean 1. The other assumptions for Hardy-Weinberg equilibrium remain in force.

(a) Show that for small $dt$

$$u(t + dt) = u(t)(1 - dt) + \left[u(t) + \frac{1}{2}v(t)\right]^2 dt + o(dt).$$

Hence,

$$u'(t) = -u(t) + \left[u(t) + \frac{1}{2}v(t)\right]^2.$$

(b) Similarly derive the differential equations

$$v'(t) = -v(t) + 2\left[u(t) + \frac{1}{2}v(t)\right]\left[\frac{1}{2}v(t) + w(t)\right]$$

$$w'(t) = -w(t) + \left[\frac{1}{2}v(t) + w(t)\right]^2.$$

(c) Let $p(t) = u(t) + \frac{1}{2}v(t)$ be the allele frequency of $A_1$. Verify that $p'(t) = 0$ and that $p(t) = p_0$ is constant.

(d) Show that

$$[u(t) - p_0^2]' = -[u(t) - p_0^2],$$

and so

$$u(t) - p_0^2 = [u(0) - p_0^2]e^{-t}.$$

(e) Similarly prove

$$v(t) - 2p_0(1 - p_0) = [v(0) - 2p_0(1 - p_0)]e^{-t}$$
$$w(t) - (1 - p_0)^2 = [w(0) - (1 - p_0)^2]e^{-t}.$$

(f) If time is measured in generations, then how many generations does it take for the departure from Hardy-Weinberg equilibrium to be halved?

6. Consider an X-linked version of the Moran model in the previous problem. Again let $u(t)$, $v(t)$, and $w(t)$ be the frequencies of the three female genotypes $A_1/A_1$, $A_1/A_2$, and $A_2/A_2$, respectively. Let $r(t)$ and $s(t)$ be the frequencies of the male genotypes $A_1$ and $A_2$.

(a) Verify the differential equations

$$r'(t) = -r(t) + u(t) + \frac{1}{2}v(t)$$

$$s'(t) = -s(t) + \frac{1}{2}v(t) + w(t)$$

$$u'(t) = -u(t) + r(t)\left[u(t) + \frac{1}{2}v(t)\right]$$

$$v'(t) = -v(t) + r(t)\left[\frac{1}{2}v(t) + w(t)\right] + s(t)\left[u(t) + \frac{1}{2}v(t)\right]$$

$$w'(t) = -w(t) + s(t)\left[\frac{1}{2}v(t) + w(t)\right].$$

(b) Show that the frequency $\frac{r(t)}{3} + \frac{2}{3}[u(t) + \frac{1}{2}v(t)]$ of the $A_1$ allele is constant.

(c) Let $p_0$ be the frequency of the $A_1$ allele. Demonstrate that

$$[r(t) - p_0]' = -\frac{3}{2}[r(t) - p_0],$$

and hence

$$r(t) - p_0 = [r(0) - p_0]e^{-\frac{3}{2}t}.$$

(d)  Use parts (a) and (c) to establish

$$\lim_{t \to \infty} \left[ u(t) + \frac{1}{2} v(t) \right] = p_0.$$

(e)  Show that

$$[(u(t) - p_0^2)e^t]'$$
$$= u'(t)e^t + u(t)e^t - p_0^2 e^t$$
$$= r(t) \left[ u(t) + \frac{1}{2} v(t) \right] e^t - p_0^2 e^t$$
$$= \left( p_0 + [r(0) - p_0]e^{-\frac{3}{2}t} \right)$$
$$\times \left( p_0 - \frac{1}{3} p_0 - \frac{1}{3}[r(0) - p_0]e^{-\frac{3}{2}t} \right) \frac{3}{2} e^t - p_0^2 e^t$$
$$= \left( p_0 + [r(0) - p_0]e^{-\frac{3}{2}t} \right)$$
$$\times \left( p_0 - \frac{1}{2}[r(0) - p_0]e^{-\frac{3}{2}t} \right) e^t - p_0^2 e^t$$
$$= \frac{p_0}{2}[r(0) - p_0]e^{-\frac{t}{2}} - \frac{1}{2}[r(0) - p_0]^2 e^{-2t}.$$

Thus,

$$u(t) - p_0^2 = [u(0) - p_0^2]e^{-t} + p_0[r(0) - p_0](e^{-t} - e^{-\frac{3}{2}t})$$
$$- \frac{1}{4}[r(0) - p_0]^2[e^{-t} - e^{-3t}].$$

It follows that $\lim_{t \to \infty} u(t) = p_0^2$.

(f)  Finally, show that

$$\lim_{t \to \infty} s(t) = 1 - p_0$$
$$\lim_{t \to \infty} v(t) = 2p_0(1 - p_0)$$
$$\lim_{t \to \infty} w(t) = (1 - p_0)^2.$$

7.  Consider three loci A—B—C along a chromosome. To model convergence to linkage equilibrium at these loci, select alleles $A_i$, $B_j$, and $C_k$ and denote their population frequencies by $p_i$, $q_j$, and $r_k$. Let $\theta_{AB}$ be the probability of recombination between loci $A$ and $B$ but not between $B$ and $C$. Define $\theta_{BC}$ similarly. Let $\theta_{AC}$ be the probability of simultaneous recombination between loci $A$ and $B$ and between loci $B$ and $C$. Finally, adopt the usual conditions for Hardy-Weinberg and linkage equilibrium.

(a)  Show that the gamete frequency $P_n(A_i B_j C_k)$ satisfies

$$P_n(A_iB_jC_k) \quad = \quad (1 - \theta_{AB} - \theta_{BC} - \theta_{AC})P_{n-1}(A_iB_jC_k)$$
$$+ \theta_{AB}\, p_i\, P_{n-1}(B_jC_k) + \theta_{BC}r_k\, P_{n-1}(A_iB_j)$$
$$+ \theta_{AC}q_j\, P_{n-1}(A_iC_k).$$

(b) Define the function

$$L_n(A_iB_jC_k) \quad = \quad P_n(A_iB_jC_k) - p_iq_jr_k - p_i[P_n(B_jC_k) - q_jr_k]$$
$$- r_k[P_n(A_iB_j) - p_iq_j] - q_j[P_n(A_iC_k) - p_ir_k].$$

Show that $L_n(A_iB_jC_k)$ satisfies

$$L_n(A_iB_jC_k) \quad = \quad (1 - \theta_{AB} - \theta_{BC} - \theta_{AC})L_{n-1}(A_iB_jC_k).$$

(Hint: Substitute for $P_n(B_jC_k) - q_jr_k$ and similar terms using the recurrence relation for two loci.)

(c) Argue that $\lim_{n\to\infty} L_n(A_iB_jC_k) = 0$. As a consequence, conclude that $\lim_{n\to\infty} P_n(A_iB_jC_k) = p_iq_jr_k$.

8. Consulting Problems 5 and 6, formulate a Moran model for approach to linkage equilibrium at two loci. In the context of this model, show that

$$P_t(A_iB_j) \quad = \quad e^{-\theta t}P_0(A_iB_j) + (1 - e^{-\theta t})p_iq_j,$$

where time $t$ is measured continuously.

9. To verify convergence to linkage equilibrium for a pair of X-linked loci $A$ and $B$, define $P_{nx}(A_iB_j)$ and $P_{ny}(A_iB_j)$ to be the frequencies of the $A_iB_j$ haplotype at generation $n$ in females and males, respectively. For the sake of simplicity, assume that both loci are in Hardy-Weinberg equilibrium and that the alleles $A_i$ and $B_j$ have frequencies $p_i$ and $q_j$. If $z_n$ denotes the column vector $[P_{nx}(A_iB_j), P_{ny}(A_iB_j)]^t$ and $\theta$ the female recombination fraction between the two loci, then demonstrate the recurrence relation

$$z_n \quad = \quad Mz_{n-1} + \theta p_iq_j \begin{pmatrix} \tfrac{1}{2} \\ 1 \end{pmatrix} \tag{1.5}$$

under the usual equilibrium conditions, where the matrix

$$M \quad = \quad \begin{pmatrix} \tfrac{1}{2}[1-\theta] & \tfrac{1}{2} \\ 1-\theta & 0 \end{pmatrix}.$$

Show that (1.5) can be recast as $w_n = Mw_{n-1}$ for $w_n = z_n - p_iq_j\mathbf{1}'$, where $\mathbf{1} = (1, 1)^t$. Solve this last recurrence and show that $\lim_{n\to\infty} w_n = 0$. (Hints: The matrix power $M^n$ can be simplified by diagonalizing $M$. Show that the eigenvalues $\omega_1$ and $\omega_2$ of $M$ are distinct and less than 1 in absolute value.)

10. Consider an autosomal dominant disease in a stationary population. If the fitness of normal $a/a$ people to the fitness of affected $A/a$ people is in the ratio $1 - s : 1$, then show that the average number of people ultimately affected by a new mutation is $\frac{1-s}{-s}$. (Hints: An $a/a$ person has on average 2 children while an $A/a$ person has on average $\frac{2}{1-s}$ children, half of whom are affected. Write and solve an equation counting the new mutant and the expected numbers of affecteds originating from each of his or her mutant children. Remember that $s < 0$.)

11. Consider a model for the mutation-selection balance at an X-linked locus. Let normal females and males have fitness 1, carrier females fitness $t_x$, and affected males fitness $t_y$. Also, let the mutation rate from the normal allele $a$ to the disease allele $A$ be $\mu$ in both sexes. It is possible to write and solve two equations for the equilibrium frequencies $p_{\infty x}$ and $p_{\infty y}$ of carrier females and affected males.

   (a) Derive the two approximate equations

   $$p_{\infty x} \approx 2\mu + p_{\infty x}\frac{1}{2}t_x + p_{\infty y}t_y$$

   $$p_{\infty y} \approx \mu + p_{\infty x}\frac{1}{2}t_x$$

   assuming the disease is rare.

   (b) Solve the two equations in (a).

   (c) When $t_x = 1$, show that the fraction of affected males representing new mutations is $\frac{1}{3}(1 - t_y)$. This fraction does not depend on the mutation rate.

   (d) If $t_x = 1$ and $t_y = 0$, then prove that $p_{\infty x} \approx 4\mu$ and $p_{\infty y} \approx 3\mu$.

# References

[1] Bennet JH (1954) On the theory of random mating. *Ann Eugen* 18:311–317

[2] Cavalli-Sforza LL, Bodmer WF (1971) *The Genetics of Human Populations*. Freeman, San Francisco

[3] Crow JF, Kimura M (1970) *An Introduction to Population Genetics Theory*. Harper and Row, New York

[4] Elandt-Johnson RC (1971) *Probability Models and Statistical Methods in Genetics*. Wiley, New York

[5] Geiringer H (1945) Further remarks on linkage theory in Mendelian heredity. *Ann Math Stat* 16:390–393

[6] Hartl DL, Clark AG (1989) *Principles of Population Genetics*, 2nd ed. Sinauer, Sunderland, MA

[7] Jacquard A (1974) *The Genetic Structure of Populations*. Springer-Verlag, New York

[8] Lange K (1991) Comment on "Inferences using DNA profiling in forensic identification and paternity cases" by DA Berry. *Stat Sci* 6:190–192

[9] Lange K (1993) A stochastic model for genetic linkage equilibrium. *Theor Pop Biol* 44:129–148

[10] Li CC (1976) *First Course in Population Genetics*. Boxwood Press, Pacific Grove, CA.

[11] Malécot G (1948) *Les Mathématiques de l'hérédité*. Masson et Cie, Paris

[12] Moran PAP (1962) *The Statistical Processes of Evolutionary Theory*. Clarendon Press, Oxford

[13] Nagylaki T (1992) *Introduction to Theoretical Population Genetics*. Springer-Verlag, Berlin

[14] Nesse RM (1995) When bad genes happen to good people. *Technology Review*, May/June: 32–40

# 2
# Counting Methods and the EM Algorithm

## 2.1  Introduction

In this chapter and the next, we undertake the study of estimation methods and their applications in genetics. Because of the complexity of genetic models, geneticists by and large rely on maximum likelihood estimators rather than on competing estimators derived from minimax, invariance, robustness, or Bayesian principles. A host of methods exists for numerically computing maximum likelihood estimates. Some of the most appealing involve simple counting arguments and the EM algorithm. Indeed, historically geneticists devised many special cases of the EM algorithm before it was generally formulated by Dempster et al. [5, 9]. Our initial example retraces some of the steps in the long march from concrete problems to an abstract algorithm applicable to an astonishing variety of statistical models.

## 2.2  Gene Counting

Suppose a geneticist takes a random sample from a population and observes the phenotype of each individual in the sample at some autosomal locus. How can the sample be used to estimate the frequency of an allele at the locus? If all alleles are codominant, the answer is obvious. Simply count the number of times the given allele appears in the sample, and divide by the total number of genes in the sample. Remember that there are twice as many genes as individuals.

TABLE 2.1. MN Blood Group Data

| Phenotype | Genotype | Number |
|-----------|----------|--------|
| $M$ | $M/M$ | 119 |
| $MN$ | $M/N$ | 76 |
| $N$ | $N/N$ | 13 |

**Example 2.1** *Gene Frequencies for the MN Blood Group*

The MN blood group has two codominant alleles $M$ and $N$. Crow [4] cites the data from Table 2.1 on 208 Bedouins of the Syrian desert. To estimate the frequency $p_M$ of the $M$ allele, we count two $M$ genes for each $M$ phenotype and one $M$ gene for each $MN$ phenotype. Thus, our estimate of $p_M$ is $\hat{p}_M = \frac{2 \times 119 + 76}{2 \times 208} = .755$. Similarly, $\hat{p}_N = \frac{2 \times 13 + 76}{2 \times 208} = .245$. Note that $\hat{p}_M + \hat{p}_N = 1$.    ■

In general, at a locus with $k$ codominant alleles, suppose we count $n_i$ alleles of type $i$ in a random sample of $n$ unrelated people. Then the ratio $\hat{p}_i = \frac{n_i}{2n}$ provides a desirable estimate of the frequency $p_i$ of allele $i$. Since the counts $(n_1, \ldots, n_k)$ follow a multinomial distribution, the expectation $E(\hat{p}_i) = \frac{2np_i}{2n} = p_i$. In other words, $\hat{p}_i$ is an unbiased estimator. By the law of large numbers, $\hat{p}_i$ is also a strongly consistent estimator. In passing, we also note the variance and covariance expressions

$$
\begin{aligned}
\text{Var}(\hat{p}_i) &= \frac{2np_i(1 - p_i)}{(2n)^2} \\
&= \frac{p_i(1 - p_i)}{2n} \\
\text{Cov}(\hat{p}_i, \hat{p}_j) &= -\frac{2np_i p_j}{(2n)^2} \\
&= -\frac{p_i p_j}{2n}.
\end{aligned}
$$

Finally, as observed in Problem 3, the $\hat{p}_i$ constitute the maximum likelihood estimates of the $p_i$.

This simple gene-counting argument encounters trouble if we consider a locus with recessive alleles because we can no longer infer genotypes from phenotypes. Consider the ABO locus, for instance. Suppose we observe $n_A$ people of type $A$, $n_B$ people of type $B$, $n_{AB}$ people of type $AB$, and $n_O$ people of type $O$. Let $n = n_A + n_B + n_{AB} + n_O$ be the total number of people in the random sample. If we want to estimate the frequency $p_A$ of the $A$ allele, we cannot say exactly how many of the $n_A$ people are homozygotes $A/A$ and how many are heterozygotes $A/O$. Thus, we are prevented from directly counting genes.

There is a way out of this dilemma that exploits Hardy-Weinberg equilibrium. If we knew the true allele frequencies $p_A$ and $p_O$, then we could correctly apportion the $n_A$ individuals of phenotype type $A$. Genotype $A/A$ has frequency $p_A^2$ in the population, while genotype $A/O$ has frequency $2p_A p_O$. Of the $n_A$ people of type $A$, we expect $n_{A/A} = n_A p_A^2/(p_A^2 + 2p_A p_O)$ people to have genotype $A/A$ and $n_{A/O} = n_A 2p_A p_O/(p_A^2 + 2p_A p_O)$ people to have genotype $A/O$. Employing circular reasoning, we now estimate $p_A$ by

$$\hat{p}_A = \frac{2n_{A/A} + n_{A/O} + n_{AB}}{2n}. \tag{2.1}$$

The trick now is to remove the circularity by iterating. Suppose we make an initial guess $p_{mA}$, $p_{mB}$, and $p_{mO}$ of the three allele frequencies at iteration 0. By analogy to the reasoning leading to (2.1), we attribute at iteration $m$

$$n_{m,A/A} = n_A \frac{p_{mA}^2}{p_{mA}^2 + 2p_{mA} p_{mO}}$$

people to genotype $A/A$ and

$$n_{m,A/O} = n_A \frac{2p_{mA} p_{mO}}{p_{mA}^2 + 2p_{mA} p_{mO}}$$

people to genotype $A/O$. We now update $p_{mA}$ by

$$p_{m+1,A} = \frac{2n_{m,A/A} + n_{m,A/O} + n_{AB}}{2n}. \tag{2.2}$$

The update for $p_{mB}$ is the same as (2.2) except for the interchange of the labels $A$ and $B$. The update for $p_{mO}$ is equally intuitive and preserves the counting requirement $p_{mA} + p_{mB} + p_{mO} = 1$. This iterative procedure continues until $p_{mA}$, $p_{mB}$, and $p_{mO}$ converge. Their converged values $p_{\infty A}$, $p_{\infty B}$, and $p_{\infty O}$ provide allele frequency estimates. This gene-counting algorithm [9] is a special case of the EM algorithm.

**Example 2.2** *Gene Frequencies for the ABO Blood Group*

As a practical example, let $n_A = 186$, $n_B = 38$, $n_{AB} = 13$, and $n_O = 284$. These are the types of 521 duodenal ulcer patients gathered by Clarke et al. [2]. As an initial guess, take $p_{0A} = .3$, $p_{0B} = .2$, and $p_{0O} = .5$. The gene-counting iterations can be done on a pocket calculator. It is evident from Table 2.2 that convergence occurs quickly. ∎

## 2.3   Description of the EM Algorithm

A sharp distinction is drawn in the EM algorithm between the observed, incomplete data $Y$ and the unobserved, complete data $X$ of a statistical experiment [5, 6, 10]. Some function $t(X) = Y$ collapses $X$ onto $Y$. For instance, if we represent $X$ as $(Y, Z)$, with $Z$ as the missing data, then $t$ is simply projection onto the $Y$-component of $X$. It should be stressed that the missing data can consist of more

TABLE 2.2. Iterations for ABO Duodenal Ulcer Data

| Iteration $m$ | $p_{mA}$ | $p_{mB}$ | $p_{mO}$ |
|:---:|:---:|:---:|:---:|
| 0 | .3000 | .2000 | .5000 |
| 1 | .2321 | .0550 | .7129 |
| 2 | .2160 | .0503 | .7337 |
| 3 | .2139 | .0502 | .7359 |
| 4 | .2136 | .0501 | .7363 |
| 5 | .2136 | .0501 | .7363 |

than just observations missing in the ordinary sense. In fact, the definition of $X$ is left up to the intuition and cleverness of the statistician. The general idea is to choose $X$ so that maximum likelihood becomes trivial for the complete data.

The complete data are assumed to have a probability density $f(X \mid \theta)$ that is a function of a parameter vector $\theta$ as well as of $X$. In the E step of the EM algorithm, we calculate the conditional expectation

$$Q(\theta \mid \theta_n) = \mathrm{E}[\ln f(X \mid \theta) \mid Y, \theta_n].$$

Here $\theta_n$ is the current estimated value of $\theta$. In the M step, we maximize $Q(\theta \mid \theta_n)$ with respect to $\theta$. This yields the new parameter estimate $\theta_{n+1}$, and we repeat this two-step process until convergence occurs. Note that $\theta$ and $\theta_n$ play fundamentally different roles in $Q(\theta \mid \theta_n)$.

The essence of the EM algorithm is that maximizing $Q(\theta \mid \theta_n)$ leads to an increase in the loglikelihood $\ln g(Y \mid \theta)$ of the observed data. This assertion is proved in the following theoretical section, which can be omitted by readers interested primarily in practical applications of the EM algorithm.

## 2.4   Ascent Property of the EM Algorithm

The entropy (or information) inequality at the heart of the EM algorithm is a consequence of Jensen's inequality, which relates convex functions to expectations. Recall that a twice-differentiable function $h(w)$ is convex on an interval $(a, b)$ if and only if $h''(w) \geq 0$ for all $w$ in $(a, b)$. If the defining inequality is strict, then $h(w)$ is said to be strictly convex.

**Proposition 2.1** *(Jensen's Inequality) Let $W$ be a random variable with values confined to the possibly infinite interval $(a, b)$. If $\mathrm{E}$ denotes expectation and $h(w)$ is convex on $(a, b)$, then $\mathrm{E}[h(W)] \geq h[\mathrm{E}(W)]$. For a strictly convex function, equality holds in Jensen's inequality if and only if $W = \mathrm{E}(W)$ almost surely.*

PROOF: Put $u = E(W)$. For $w$ in $(a, b)$, we have

$$h(w) \;=\; h(u) + h'(u)(w - u) + h''(v)\frac{(w - u)^2}{2}$$
$$\geq\; h(u) + h'(u)(w - u)$$

for some $v$ between $u$ and $w$. Note that $v$ is in $(a, b)$. Now substitute the random variable $W$ for the point $w$ and take expectations. It follows that

$$E[h(W)] \;\geq\; h(u) + h'(u)[E(W) - u]$$
$$=\; h(u).$$

If $h(w)$ is strictly convex, then the neglected term $h''(v)\frac{(w-u)^2}{2}$ is positive whenever $w \neq u$. ∎

**Proposition 2.2** *(Entropy Inequality) Let $f$ and $g$ be probability densities with respect to a measure $\mu$. Suppose $f > 0$ and $g > 0$ almost everywhere relative to $\mu$. If $E_f$ denotes expectation with respect to the probability measure $f d\mu$, then $E_f(\ln f) \geq E_f(\ln g)$, with equality only if $f = g$ almost everywhere relative to $\mu$.*

PROOF: Because $-\ln(w)$ is a strictly convex function on $(0, \infty)$, Jensen's inequality applied to the random variable $f/g$ implies

$$E_f(\ln f) - E_f(\ln g) \;=\; E_f\left(-\ln \frac{g}{f}\right)$$
$$\geq\; -\ln E_f\left(\frac{g}{f}\right)$$
$$=\; -\ln \int \frac{g}{f} f d\mu$$
$$=\; -\ln \int g d\mu$$
$$=\; 0.$$

Equality holds only if $\frac{g}{f} = E_f(\frac{g}{f})$ almost everywhere relative to $\mu$. However $E_f(\frac{g}{f}) = 1$. ∎

Reverting to the notation $Q(\theta \mid \theta_n) = E[\ln f(X \mid \theta) \mid Y = y, \theta_n]$ of the EM algorithm, let us next prove that

$$Q(\theta_n \mid \theta_n) - \ln g(y \mid \theta_n) \;\geq\; Q(\theta \mid \theta_n) - \ln g(y \mid \theta)$$

for all $\theta$ and $\theta_n$, where $g(y \mid \theta)$ is the likelihood of the observed data $Y = y$. To this end, note that both $\frac{f(x \mid \theta)}{g(y \mid \theta)}$ and $\frac{f(x \mid \theta_n)}{g(y \mid \theta_n)}$ are conditional densities of $X$ on $\{x : t(x) = y\}$ with respect to some measure $\mu_y$. The entropy inequality now indicates that

$$Q(\theta \mid \theta_n) - \ln g(y \mid \theta) \;=\; E\left(\ln\left[\frac{f(X \mid \theta)}{g(Y \mid \theta)}\right] \;\middle|\; Y = y, \theta_n\right)$$
$$\leq\; E\left(\ln\left[\frac{f(X \mid \theta_n)}{g(Y \mid \theta_n)}\right] \;\middle|\; Y = y, \theta_n\right)$$
$$=\; Q(\theta_n \mid \theta_n) - \ln g(y \mid \theta_n).$$

Thus, the difference $\ln g(y \mid \theta) - Q(\theta \mid \theta_n)$ attains its minimum when $\theta = \theta_n$. If we choose $\theta_{n+1}$ to maximize $Q(\theta \mid \theta_n)$, then it follows that

$$
\begin{aligned}
\ln g(y \mid \theta_{n+1}) &= Q(\theta_{n+1} \mid \theta_n) + [\ln g(y \mid \theta_{n+1}) - Q(\theta_{n+1} \mid \theta_n)] \\
&\geq Q(\theta_n \mid \theta_n) + [\ln g(y \mid \theta_n) - Q(\theta_n \mid \theta_n)] \\
&= \ln g(y \mid \theta_n),
\end{aligned}
$$

with strict inequality when $f(x \mid \theta_{n+1})/g(y \mid \theta_{n+1})$ and $f(x \mid \theta_n)/g(y \mid \theta_n)$ are different conditional densities or when $Q(\theta_{n+1} \mid \theta_n) > Q(\theta_n \mid \theta_n)$. This verifies the promised ascent property $\ln g(y \mid \theta_{n+1}) \geq \ln g(y \mid \theta_n)$ of the EM algorithm.

## 2.5    Allele Frequency Estimation by the EM Algorithm

Let us return to the ABO example and formalize gene counting as an EM algorithm. The observed numbers of people in each of the four phenotypic categories constitute the observed data $Y$, while the unknown numbers of people in each of the six genotypic categories constitute the complete data $X$. Let $n_{A/A}$ be the number of people of genotype $A/A$. Define $n_{A/O}$, $n_{B/B}$, and $n_{B/O}$ similarly and set $n = n_A + n_B + n_{AB} + n_O$. Note that the $n_{AB}$ people of phenotype $AB$ and the $n_O$ people of phenotype $O$ are already correctly assigned to their respective genotypes $A/B$ and $O/O$. With this notation the complete data loglikelihood becomes

$$
\begin{aligned}
\ln f(X \mid p) &= n_{A/A} \ln p_A^2 + n_{A/O} \ln(2p_A p_O) + n_{B/B} \ln p_B^2 \\
&\quad + n_{B/O} \ln(2p_B p_O) + n_{AB} \ln(2p_A p_B) + n_O \ln p_O^2 \quad (2.3) \\
&\quad + \ln \binom{n}{n_{A/A} \ n_{A/O} \ n_{B/B} \ n_{B/O} \ n_{AB} \ n_O}.
\end{aligned}
$$

In the E step of the EM algorithm, we take the expectation of $\ln f(X \mid p)$ conditional on the observed counts $n_A, n_B, n_{AB}$, and $n_O$ and the current parameter vector $p_m = (p_{mA}, p_{mB}, p_{mO})^t$. It is obvious that

$$
\begin{aligned}
n_{m,AB} &= E(n_{AB} \mid Y, p_m) \\
&= n_{AB} \\
n_{m,O} &= E(n_O \mid Y, p_m) \\
&= n_O.
\end{aligned}
$$

A moment's reflection also yields

$$
\begin{aligned}
n_{m,A/A} &= E(n_{A/A} \mid Y, p_m) \\
&= n_A \frac{p_{mA}^2}{p_{mA}^2 + 2p_{mA}p_{mO}} \\
n_{m,A/O} &= E(n_{AO} \mid Y, p_m) \\
&= n_A \frac{2p_{mA}p_{mO}}{p_{mA}^2 + 2p_{mA}p_{mO}}.
\end{aligned}
$$

The conditional expectations $n_{m,B/B}$ and $n_{m,B/O}$ are given by similar expressions.

The M step of the EM algorithm maximizes the $Q(p \mid p_m)$ function derived from (2.3) by replacing $n_{A/A}$ by $n_{m,A/A}$, and so forth. Maximization of $Q(p \mid p_m)$ can be accomplished by introducing a Lagrange multiplier and finding a stationary point of the unconstrained function

$$H(p, \lambda) \;=\; Q(p \mid p_m) + \lambda(p_A + p_B + p_O - 1).$$

Setting the partial derivatives

$$\frac{\partial}{\partial p_A} H(p, \lambda) \;=\; \frac{2n_{m,A/A}}{p_A} + \frac{n_{m,A/O}}{p_A} + \frac{n_{AB}}{p_A} + \lambda$$

$$\frac{\partial}{\partial p_B} H(p, \lambda) \;=\; \frac{2n_{m,B/B}}{p_B} + \frac{n_{m,B/O}}{p_B} + \frac{n_{AB}}{p_B} + \lambda$$

$$\frac{\partial}{\partial p_O} H(p, \lambda) \;=\; \frac{n_{m,A/O}}{p_O} + \frac{n_{m,B/O}}{p_O} + \frac{2n_O}{p_O} + \lambda$$

$$\frac{\partial}{\partial \lambda} H(p, \lambda) \;=\; p_A + p_B + p_O - 1$$

equal to 0 provides the unique stationary point of $H(p, \lambda)$. The solution of the resulting equations is

$$p_{m+1,A} \;=\; \frac{2n_{m,A/A} + n_{m,A/O} + n_{AB}}{2n}$$

$$p_{m+1,B} \;=\; \frac{2n_{m,B/B} + n_{m,B/O} + n_{AB}}{2n}$$

$$p_{m+1,O} \;=\; \frac{n_{m,A/O} + n_{m,B/O} + 2n_O}{2n}.$$

In other words, the EM update is identical to gene counting.

## 2.6   Classical Segregation Analysis by the EM Algorithm

Classical segregation analysis is used to test Mendelian segregation ratios in **nuclear** family data. A nuclear family consists of two parents and their common offspring. Usually the hypothesis of interest is that some rare disease shows an autosomal recessive or an autosomal dominant pattern of inheritance. Because the disease is rare, it is inefficient to collect families at random. Only families with at least one affected sibling enter a typical study. The families who come to the attention of an investigator are said to be **ascertained**. To test the Mendelian segregation ratio $p = \frac{1}{2}$ for an autosomal dominant disease or $p = \frac{1}{4}$ for an autosomal recessive disease, the investigator must correct for the ascertainment process. The simplest ascertainment model postulates that the number of ascertained siblings follows a binomial distribution with success probability $\pi$ and number of trials equal to the number of affected siblings. In effect, families are ascertained only through their affected siblings, and siblings come to the attention of the genetic investigator independently with common probability $\pi$ per sibling. The number of affecteds likewise follows a binomial distribution with success probability $p$ and number of trials equal to the number of siblings. The EM algorithm can be employed to estimate $p$ and $\pi$ jointly.

Suppose that the $k$th ascertained family has $s_k$ siblings, of whom $r_k$ are affected and $a_k$ are ascertained. The numbers $r_k$ and $a_k$ constitute the observed data $Y_k$ for the $k$th ascertained family. The missing data consist of the number of at-risk families that were missed in the ascertainment process and the corresponding statistics $r_k$ and $a_k = 0$ for each of these missing families. The likelihood of the observed data is

$$\prod_k \frac{\binom{s_k}{r_k} p^{r_k} (1-p)^{s_k-r_k} \binom{r_k}{a_k} \pi^{a_k} (1-\pi)^{r_k-a_k}}{1-(1-p\pi)^{s_k}},$$

where the product extends only over the ascertained families. The denominator $1-(1-p\pi)^{s_k}$ in this likelihood is the probability that a family with $s_k$ siblings is ascertained.

These denominators disappear in the complete data likelihood

$$\prod_k \binom{s_k}{r_k} p^{r_k} (1-p)^{s_k-r_k} \binom{r_k}{a_k} \pi^{a_k} (1-\pi)^{r_k-a_k}$$

because we no longer condition on the event of ascertainment for each family. This simplification is partially offset by the added complication that the product now extends over both the ascertained families and the at-risk unascertained families. If $\theta = (p, \pi)$, $r_{mk} = E(r_k \mid Y_k, \theta_m)$, and $a_{mk} = E(a_k \mid Y_k, \theta_m)$, then the E step of the EM algorithm amounts to forming

$$Q(\theta \mid \theta_m) = \sum_k [r_{mk} \ln p + (s_k - r_{mk}) \ln(1-p)$$
$$+ a_{mk} \ln \pi + (r_{mk} - a_{mk}) \ln(1-\pi)].$$

The M step requires solving the equations

$$\sum_k \left[ \frac{r_{mk}}{p} - \frac{s_k - r_{mk}}{1-p} \right] = 0$$

$$\sum_k \left[ \frac{a_{mk}}{\pi} - \frac{r_{mk} - a_{mk}}{1-\pi} \right] = 0.$$

The EM updates are therefore

$$p_{m+1} = \frac{\sum_k r_{mk}}{\sum_k s_k} \qquad (2.4)$$

$$\pi_{m+1} = \frac{\sum_k a_{mk}}{\sum_k r_{mk}}. \qquad (2.5)$$

We need to reduce the sums in the updates (2.4) and (2.5) to sums over the ascertained families alone. To achieve this goal, first note that the sum $\sum_k a_{mk} = \sum_k a_k$ automatically excludes contributions from the unascertained families. To simplify the other sums, consider the $k$th ascertained family. If we view ascertainment as a sampling process in which unascertained families of size $s_k$ are discarded one by one until the $k$th ascertained family is finally ascertained, then the number of unascertained families discarded before reaching the $k$th ascertained family follows a shifted geometric distribution with success probability $1 - (1 - p\pi)^{s_k}$. The sampling process discards on average

$$\frac{(1 - p\pi)^{s_k}}{1 - (1 - p\pi)^{s_k}}$$

unascertained families before reaching the $k$th ascertained family. Once this ascertained family is reached, the sampling process for the $(k + 1)$th ascertained family begins.

How many affected siblings are contained in the unascertained families corresponding to the $k$th ascertained family? The expected number of affected siblings in one such unascertained family is

$$e_k = \frac{\sum_{j=0}^{s_k} j \binom{s_k}{j} p^j (1 - p)^{s_k - j} (1 - \pi)^j}{(1 - p\pi)^{s_k}}.$$

A little calculus shows that

$$
\begin{aligned}
e_k &= \frac{d}{dt} \frac{[1 - p + p(1 - \pi)t]^{s_k}}{(1 - p\pi)^{s_k}} \Big|_{t=1} \\
&= \frac{s_k [1 - p + p(1 - \pi)t]^{s_k - 1} p(1 - \pi)}{(1 - p\pi)^{s_k}} \Big|_{t=1} \\
&= \frac{s_k p(1 - \pi)}{1 - p\pi}.
\end{aligned}
$$

The expected number of affected siblings in the unascertained families corresponding to the $k$th ascertained family is given by the product

$$\frac{s_k p(1 - \pi)}{1 - p\pi} \frac{(1 - p\pi)^{s_k}}{1 - (1 - p\pi)^{s_k}} = \frac{s_k p(1 - \pi)(1 - p\pi)^{s_k - 1}}{1 - (1 - p\pi)^{s_k}}$$

of the expected number of affecteds per unascertained family times the expected number of unascertained families.

These considerations lead us to rewrite the updates (2.4) and (2.5) as

$$p_{m+1} = \frac{\sum_k \left[ r_k + \frac{s_k p_m (1 - \pi_m)(1 - p_m \pi_m)^{s_k - 1}}{1 - (1 - p_m \pi_m)^{s_k}} \right]}{\sum_k s_k [1 + \frac{(1 - p_m \pi_m)^{s_k}}{1 - (1 - p_m \pi_m)^{s_k}}]}$$

$$\pi_{m+1} = \frac{\sum_k a_k}{\sum_k \left[ r_k + \frac{s_k p_m (1 - \pi_m)(1 - p_m \pi_m)^{s_k - 1}}{1 - (1 - p_m \pi_m)^{s_k}} \right]},$$

where all sums extend over the ascertained families alone.

**Example 2.3** *Segregation Analysis of Cystic Fibrosis*

The cystic fibrosis data of Crow [3] displayed in Table 2.3 offer an opportunity to apply the EM algorithm. In this table the column labeled "Families $n$" refers to the number of families showing a particular configuration of affected and ascertained siblings. For these data the maximum likelihood estimate $\hat{p} = .2679$ is consistent with the theoretical value of $p = 1/4$ for an autosomal recessive. Starting from $p = \pi = 1/2$, the EM algorithm takes about 20 iterations to converge to the maximum likelihood estimates $\hat{p} = .2679$ and $\hat{\pi} = .3594$.  ∎

TABLE 2.3. Cystic Fibrosis Data

| Siblings $s$ | Affecteds $r$ | Ascertaineds $a$ | Families $n$ |
|:---:|:---:|:---:|:---:|
| 10 | 3 | 1 | 1 |
| 9 | 3 | 1 | 1 |
| 8 | 4 | 1 | 1 |
| 7 | 3 | 2 | 1 |
| 7 | 3 | 1 | 1 |
| 7 | 2 | 1 | 1 |
| 7 | 1 | 1 | 1 |
| 6 | 2 | 1 | 1 |
| 6 | 1 | 1 | 1 |
| 5 | 3 | 3 | 1 |
| 5 | 3 | 2 | 1 |
| 5 | 2 | 1 | 5 |
| 5 | 1 | 1 | 2 |
| 4 | 3 | 2 | 1 |
| 4 | 3 | 1 | 2 |
| 4 | 2 | 1 | 4 |
| 4 | 1 | 1 | 6 |
| 3 | 2 | 2 | 3 |
| 3 | 2 | 1 | 3 |
| 3 | 1 | 1 | 10 |
| 2 | 2 | 2 | 2 |
| 2 | 2 | 1 | 4 |
| 2 | 1 | 1 | 18 |
| 1 | 1 | 1 | 9 |

## 2.7 Problems

1. At some autosomal locus with two alleles $R$ and $r$, let $R$ be dominant to $r$. Suppose a random sample of $n$ people contains $n_r$ people with the recessive

genotype $r/r$. Prove that $\sqrt{n_r/n}$ is the maximum likelihood estimate of the frequency of allele $r$ under Hardy-Weinberg equilibrium.

2. Color blindness is an X-linked recessive trait. Suppose that in a random sample there are $f_B$ normal females, $f_b$ color-blind females, $m_B$ normal males, and $m_b$ color-blind males. If $n = 2f_B + 2f_b + m_B + m_b$ is the number of genes in the sample, then show that under Hardy-Weinberg equilibrium the maximum likelihood estimate of the frequency of the color-blindness allele is

$$\hat{p}_b = \frac{-m_B + \sqrt{m_B^2 + 4n(m_b + 2f_b)}}{2n}.$$

Compute the estimate $\hat{p}_b = .0772$ for the data $f_B = 9032$, $f_b = 40$, $m_B = 8324$, and $m_b = 725$. These data represent an amalgamation of cases from two distinct forms of color blindness [4]. Protanopia, or red blindness, is determined by one X-linked locus, and deuteranopia, or green blindness, by a different X-linked locus.

3. Consider a codominant, autosomal locus with $k$ alleles. In a random sample of $n$ people, let $n_i$ be the number of genes of allele $i$. Show that the gene-counting estimates $\hat{p}_i = n_i/(2n)$ are maximum likelihood estimates.

4. In forensic applications of DNA fingerprinting, match probabilities $p_i^2$ for homozygotes and $2p_i p_j$ for heterozygotes are computed [1]. In practice, the frequencies $p_i$ can only be estimated. Assuming codominant alleles and the estimates $\hat{p}_i = n_i/(2n)$ given in the previous problem, show that the natural match probability estimates satisfy

$$\mathrm{E}(\hat{p}_i^2) = p_i^2 + \frac{p_i(1 - p_i)}{2n}$$

$$\mathrm{Var}(\hat{p}_i^2) = \frac{4p_i^3(1 - p_i)}{2n} + O\left(\frac{1}{n^2}\right)$$

$$\mathrm{E}(2\hat{p}_i \hat{p}_j) = 2p_i p_j - \frac{2p_i p_j}{2n}$$

$$\mathrm{Var}(2\hat{p}_i \hat{p}_j) = \frac{4p_i p_j}{2n}[p_i + p_j - 4p_i p_j] + O\left(\frac{1}{n^2}\right).$$

(Hint: The $n_i$ have joint moment-generating function $(\sum_i p_i e^{s_i})^{2n}$.)

5. Consider two loci in Hardy-Weinberg equilibrium, but possibly not in linkage equilibrium. Devise an EM algorithm for estimating the gamete frequencies $p_{AB}$, $p_{Ab}$, $p_{aB}$, and $p_{ab}$, where $A$ and $a$ are the two alleles at the first locus and $B$ and $b$ are the two alleles at the second locus [13]. In a random sample of $n$ individuals, let $n_{AABB}$ denote the observed number of individuals of genotype $A/A$ at the first locus and of genotype $B/B$ at the second locus. Denote the eight additional observed double-genotype frequencies

similarly. The only one of these observed numbers entailing any ambiguity is $n_{AaBb}$; for individuals of this genotype, phase cannot be discerned. Now show that the EM update for $p_{AB}$ is

$$p_{m+1,AB} = \frac{2n_{AABB} + n_{AABb} + n_{AaBB} + n_{mAB/ab}}{2n}$$

$$n_{mAB/ab} = n_{AaBb} \frac{2p_{mAB}p_{mab}}{2p_{mAB}p_{mab} + 2p_{mAb}p_{maB}}.$$

There are similar updates for $p_{Ab}$, $p_{aB}$, and $p_{ab}$, but these can be dispensed with if one notes that for all $m$

$$p_A = p_{mAB} + p_{mAb}$$
$$p_B = p_{mAB} + p_{maB}$$
$$1 = p_{mAB} + p_{mAb} + p_{maB} + p_{mab},$$

where $p_A$ and $p_B$ are the gene-counting estimates of the frequencies of alleles $A$ and $B$. Implement this EM algorithm on the mosquito data [13] given in Table 2.4. You should find that $\hat{p}_{AB} = .73$.

TABLE 2.4. Mosquito Data at the Idh1 and Mdh Loci

| $n_{AABB} = 19$ | $n_{AABb} = 5$ | $n_{AAbb} = 0$ |
|---|---|---|
| $n_{AaBB} = 8$ | $n_{AaBb} = 8$ | $n_{Aabb} = 0$ |
| $n_{aaBB} = 0$ | $n_{aaBb} = 0$ | $n_{aabb} = 0$ |

6. In a genetic linkage experiment, $AB/ab$ animals are crossed to measure the recombination fraction $\theta$ between two loci with alleles $A$ and $a$ at the first locus and alleles $B$ and $b$ at the second locus. In this design the dominant alleles $A$ and $B$ are in the coupling phase. Verify that the offspring of an $AB/ab \times AB/ab$ mating fall into the four categories $AB$, $Ab$, $aB$, and $ab$ with probabilities $\pi_1 = \frac{1}{2} + \frac{(1-\theta)^2}{4}$, $\pi_2 = \frac{1-(1-\theta)^2}{4}$, $\pi_3 = \frac{1-(1-\theta)^2}{4}$, and $\pi_4 = \frac{(1-\theta)^2}{4}$, respectively. Devise an EM algorithm to estimate $\theta$, and apply it to the counts

$$(y_1, y_2, y_3, y_4) = (125, 18, 20, 34)$$

observed on 197 offspring of such matings. You should find the maximum likelihood estimate $\hat{\theta} = .2082$ [8]. (Hints: Split the first category into two so that there are five categories for the complete data. Reparameterize by setting $\phi = (1-\theta)^2$.)

7. In an inbred population, the inbreeding coefficient $f$ is the probability that two genes of a random person at some locus are both copies of the same

ancestral gene. Assume that there are $k$ codominant alleles and that $p_i$ is the frequency of allele $A_i$. Show that $fp_i + (1 - f)p_i^2$ is the frequency of a homozygous genotype $A_i/A_i$ and $(1 - f)2p_ip_j$ is the frequency of a heterozygous genotype $A_i/A_j$. Suppose that we observe $n_{ij}$ people of genotype $A_i/A_j$ in a random sample. Formulate an EM algorithm for the estimation of the parameters $f, p_1, \ldots, p_k$ from the observed data.

8. Consider the data from the *London Times* [11] for the years 1910 to 1912 reproduced in Table 2.5. The two columns labeled "Deaths $i$" refer to the number of deaths to women 80 years and older reported by day. The columns labeled "Frequency $n_i$" refer to the number of days with $i$ deaths. A Poisson distribution gives a poor fit to these data, possibly because of different patterns of deaths in winter and summer. A mixture of two Poissons provides a much better fit. Under the Poisson admixture model, the likelihood of the observed data is

$$\prod_{i=0}^{9}\left[\alpha e^{-\mu_1}\frac{\mu_1^i}{i!} + (1 - \alpha)e^{-\mu_2}\frac{\mu_2^i}{i!}\right]^{n_i},$$

where $\alpha$ is the admixture parameter and $\mu_1$ and $\mu_2$ are the means of the two Poisson distributions.

TABLE 2.5. Death Notices from the *London Times*

| Deaths $i$ | Frequency $n_i$ | Deaths $i$ | Frequency $n_i$ |
|---|---|---|---|
| 0 | 162 | 5 | 61 |
| 1 | 267 | 6 | 27 |
| 2 | 271 | 7 | 8 |
| 3 | 185 | 8 | 3 |
| 4 | 111 | 9 | 1 |

Formulate an EM algorithm for this model. Let $\theta = (\alpha, \mu_1, \mu_2)^t$ and

$$z_i(\theta) = \frac{\alpha e^{-\mu_1}\mu_1^i}{\alpha e^{-\mu_1}\mu_1^i + (1 - \alpha)e^{-\mu_2}\mu_2^i}$$

be the posterior probability that a day with $i$ deaths belongs to Poisson population 1. Show that the EM algorithm is given by

$$\alpha_{m+1} = \frac{\sum_i n_i z_i(\theta_m)}{\sum_i n_i}$$

$$\mu_{m+1,1} = \frac{\sum_i n_i z_i(\theta_m)i}{\sum_i n_i z_i(\theta_m)}$$

$$\mu_{m+1,2} = \frac{\sum_i n_i[1 - z_i(\theta_m)]i}{\sum_i n_i[1 - z_i(\theta_m)]}.$$

From the initial estimates $\alpha_0 = .3$, $\mu_{01} = 1$. and $\mu_{02} = 2.5$, compute via the EM algorithm the maximum likelihood estimates $\hat{\alpha} = .3599$, $\hat{\mu}_1 = 1.2561$, and $\hat{\mu}_2 = 2.6634$. Note how slowly the EM algorithm converges in this example.

9. In the EM algorithm, demonstrate the identity

$$\frac{\partial}{\partial \theta_i} Q(\theta \mid \theta_n)|_{\theta = \theta_n} = \frac{\partial}{\partial \theta_i} L(\theta_n)$$

at any interior point $\theta_n$ of the parameter domain. Here $L(\theta)$ is the loglikelihood of the observed data $Y$.

10. Suppose that the complete data in the EM algorithm involve $N$ binomial trials with success probability $\theta$ per trial. Here $N$ can be random or fixed. If $M$ trials result in success, then the complete data likelihood can be written as $\theta^M (1 - \theta)^{N-M} c$, where $c$ is an irrelevant constant. The E step of the EM algorithm amounts to forming

$$Q(\theta \mid \theta_n) = E(M \mid Y, \theta_n) \ln \theta + E(N - M \mid Y, \theta_n) \ln(1 - \theta) + \ln c.$$

The binomial trials are hidden because only a function $Y$ of them is directly observed. Show in this setting that the EM update is given by either of the two equivalent expressions

$$\begin{aligned}
\theta_{n+1} &= \frac{E(M \mid Y, \theta_n)}{E(N \mid Y, \theta_n)} \\
&= \theta_n + \frac{\theta_n(1 - \theta_n)}{E(N \mid Y, \theta_n)} \frac{d}{d\theta} L(\theta_n),
\end{aligned}$$

where $L(\theta)$ is the loglikelihood of the observed data $Y$ [7, 12]. (Hint: Use Problem 9.)

11. As an example of the hidden binomial trials theory sketched in Problem 10, consider a random sample of twin pairs. Let $u$ of these pairs consist of male pairs, $v$ consist of female pairs, and $w$ consist of opposite sex pairs. A simple model to explain these data involves a random Bernoulli choice for each pair dictating whether it consists of identical or nonidentical twins. Suppose that identical twins occur with probability $p$ and nonidentical twins with probability $1 - p$. Once the decision is made as to whether the twins are identical or not, then sexes are assigned to the twins. If the twins are identical, one assignment of sex is made. If the twins are nonidentical, then two independent assignments of sex are made. Suppose boys are chosen with probability $q$ and girls with probability $1 - q$. Model these data as hidden binomial trials. Using the result of Problem 10, give the EM algorithm for estimating $p$ and $q$. What other problems from this chapter involve hidden binomial trials?

12. In the spirit of Problem 10, formulate a model for hidden Poisson or exponential trials [12]. If the number of trials is $N$ and the mean per trial is $\theta$, then show that the EM update in the Poisson case is

$$\theta_{n+1} \;=\; \theta_n + \frac{\theta_n}{\mathrm{E}(N \mid Y, \theta_n)}\frac{d}{d\theta}L(\theta_n)$$

and in the exponential case is

$$\theta_{n+1} \;=\; \theta_n + \frac{\theta_n^2}{\mathrm{E}(N \mid Y, \theta_n)}\frac{d}{d\theta}L(\theta_n),$$

where $L(\theta)$ is the loglikelihood of the observed data $Y$.

# References

[1] Chakraborty R, Srinivasan MR, Daiger SP (1993) Evaluation of standard error and confidence interval of estimated multilocus genotype probabilities, and their applications in DNA forensics. *Amer J Hum Genet* 52:60–70

[2] Clarke CA, Price-Evans DA, McConnell RB, Sheppard PM (1959) Secretion of blood group antigens and peptic ulcers. *Brit Med J* 1:603–607

[3] Crow JF (1965) Problems of ascertainment in the analysis of family data. *Epidemiology and Genetics of Chronic Disease*. Public Health Service Publication 1163, Neel JV, Shaw MW, Schull WJ, editors, Department of Health, Education, and Welfare, Washington, DC

[4] Crow JF (1986) *Basic Concepts in Population, Quantitative, and Ecological Genetics*. Freeman, San Francisco

[5] Dempster AP, Laird NM, Rubin DB (1977) Maximum likelihood from incomplete data via the EM algorithm (with discussion). *J Roy Stat Soc B* 39:1–38

[6] Little RJA, Rubin DB (1987) *Statistical Analysis with Missing Data*. Wiley, New York

[7] Ott J (1977) Counting methods (EM algorithm) in human pedigree analysis: linkage and segregation analysis. *Ann Hum Genet* 40:443–454

[8] Rao CR (1975) *Linear Statistical Inference and Its Applications*, 2nd ed. Wiley, New York

[9] Smith CAB (1957) Counting methods in genetical statistics. *Ann Hum Genet* 21:254–276

[10] Tanner MA (1993) *Tools for Statistical Inference: Methods for the Exploration of Posterior Distributions and Likelihood Functions*, 2nd ed. Springer-Verlag, New York

[11] Titterington DM, Smith AFM, Makov UE (1985) *Statistical Analysis of Finite Mixture Distributions*. Wiley, New York

[12] Weeks DE, Lange K (1989) Trials, tribulations, and triumphs of the EM algorithm in pedigree analysis. *IMA J Math Appl Med Biol* 6:209–232

[13] Weir BS (1990) *Genetic Data Analysis*. Sinauer, Sunderland, MA

# 3
# Newton's Method and Scoring

## 3.1 Introduction

This chapter explores some alternatives to maximum likelihood estimation by the EM algorithm. **Newton's method** and **scoring** usually converge faster than the EM algorithm. However, the trade-offs of programming ease, numerical stability, and speed of convergence are complex, and statistical geneticists should be fluent in a variety of numerical optimization techniques for finding maximum likelihood estimates. Outside the realm of maximum likelihood, Bayesian procedures have much to offer in small to moderate-sized problems. For those uncomfortable with pulling prior distributions out of thin air, **empirical Bayes** procedures can be an appealing compromise between classical and Bayesian methods. This chapter illustrates some of these well-known themes in the context of allele frequency estimation and linkage analysis.

## 3.2 Newton's Method

In iterating toward a maximum point $\hat{\theta}$, Newton's method and scoring rely on quadratic approximations to the loglikelihood $L(\theta)$ of a model. To motivate Newton's method, let us define the **score** $dL(\theta)$ to be the differential or row vector of first partial derivatives of $L(\theta)$ and the **observed information** $-d^2L(\theta)$ to be the Hessian matrix of second partial derivatives of $-L(\theta)$. A second-order Taylor's expansion around the current point $\theta_n$ gives

$$L(\theta) \quad \approx \quad L(\theta_n) + dL(\theta_n)(\theta - \theta_n)$$

$$+\frac{1}{2}(\theta - \theta_n)' d^2 L(\theta_n)(\theta - \theta_n). \tag{3.1}$$

In Newton's method one maximizes the quadratic approximation on the right of (3.1) by setting its gradient

$$dL(\theta_n)' + d^2 L(\theta_n)(\theta - \theta_n) \;\; = \;\; \mathbf{0}$$

and solving for the next iterate

$$\theta_{n+1} \;\; = \;\; \theta_n - d^2 L(\theta_n)^{-1} dL(\theta_n)'.$$

Obviously, any stationary point of $L(\theta)$ is a fixed point of Newton's algorithm.

There are two potential problems with Newton's method. First, it can be expensive computationally to evaluate the observed information. Second, far from $\hat{\theta}$, Newton's method is equally happy to head uphill or downhill. In other words, Newton's method is not an **ascent algorithm** in the sense that $L(\theta_{n+1}) > L(\theta_n)$. To generate an ascent algorithm, we can replace the observed information $-d^2 L(\theta_n)$ by a positive definite approximating matrix $A_n$. With this substitution, the proposed increment $\Delta\theta_n = A_n^{-1} dL(\theta_n)'$, if sufficiently contracted, forces an increase in $L(\theta)$. For a nonstationary point, this assertion follows from the first-order Taylor's expansion

$$L(\theta_n + \alpha\Delta\theta_n) - L(\theta_n) \;\; = \;\; dL(\theta_n)\alpha\Delta\theta_n + o(\alpha)$$
$$= \;\; \alpha dL(\theta_n) A_n^{-1} dL(\theta_n)' + o(\alpha),$$

where the error ratio $\frac{o(\alpha)}{\alpha} \to 0$ as the positive contraction constant $\alpha \to 0$. Thus, a positive definite modification of the observed information combined with some form of **backtracking** leads to an ascent algorithm. The simplest form of backtracking is **step-halving**. If the initial increment $\Delta\theta_n$ does not produce an increase in $L(\theta)$, then try $\frac{1}{2}\Delta\theta_n$. If $\frac{1}{2}\Delta\theta_n$ fails, then try $\frac{1}{4}\Delta\theta_n$, and so forth.

## 3.3   Scoring

A variety of ways of approximating the observed information exists. The method of **steepest ascent** replaces the observed information by the identity matrix $I$. The usually more efficient scoring algorithm replaces the observed information by the expected information $J(\theta) = \mathrm{E}[-d^2 L(\theta)]$. The alternative representation of $J(\theta)$ as the covariance matrix $\mathrm{Var}[dL|\theta]$ shows that it is nonnegative definite [17]. An extra dividend of scoring is that the inverse matrix $J(\hat{\theta})^{-1}$ immediately supplies the asymptotic variances and covariances of the maximum likelihood estimate $\hat{\theta}$ [17]. Scoring and Newton's method share this advantage since the observed information is asymptotically equivalent to the expected information under reasonably natural assumptions. The available evidence indicates that the observed information matrix is slightly superior to the expected information matrix in estimating parameter asymptotic standard errors [7].

It is possible to compute $J(\theta)$ explicitly for **exponential families** of densities [9]. Such densities take the form

$$f(x \mid \theta) \;=\; g(x)e^{\beta(\theta)+h(x)^t \gamma(\theta)} \tag{3.2}$$

relative to some measure $\nu$, which in practice is usually either Lebesgue measure or counting measure. Most of the distributional families commonly encountered in statistics are exponential families. The score and expected information can be expressed in terms of the mean vector $\mu(\theta) = E[h(X)]$ and covariance matrix $\Sigma(\theta) = \mathrm{Var}[h(X)]$ of the sufficient statistic $h(X)$. Several authors [2, 3, 9] have noted the representations

$$dL(\theta) \;=\; [h(x) - \mu(\theta)]^t \Sigma(\theta)^{-1} d\mu(\theta) \tag{3.3}$$
$$J(\theta) \;=\; d\mu(\theta)^t \Sigma(\theta)^{-1} d\mu(\theta), \tag{3.4}$$

where $d\mu(\theta)$ is the matrix of partial derivatives of $\mu(\theta)$. If the vector $\gamma(\theta)$ in definition (3.2) is linear in $\theta$, then

$$J(\theta) = -d^2 L(\theta) = -d^2 \beta(\theta),$$

and scoring coincides with Newton's method.

Although we will not stop to derive the general formulas (3.3) and (3.4), it is instructive to consider the special case of a multinomial distribution with $m$ trials and success probability $p_i$ for category $i$. If $X = (X_1, \ldots, X_l)^t$ denotes the random vector of counts and $\theta$ the model parameters, then the loglikelihood of the observed data $X = x$ is

$$L(\theta) \;=\; \sum_{i=1}^{l} x_i \ln p_i(\theta) + \ln \binom{m}{x_1 \ldots x_l},$$

and consequently the score vector $dL(\theta)$ has entries

$$\frac{\partial}{\partial \theta_j} L(\theta) \;=\; \sum_{i=1}^{l} \frac{x_i}{p_i(\theta)} \frac{\partial}{\partial \theta_j} p_i(\theta).$$

Because $\frac{\partial}{\partial \theta_j} \sum_{i=1}^{l} p_i(\theta) = \frac{\partial}{\partial \theta_j} 1 = 0$, the expected information matrix $J(\theta)$ has entries

$$J(\theta)_{jk} \;=\; E\left[-\frac{\partial^2}{\partial \theta_j \partial \theta_k} L(\theta)\right]$$
$$=\; \sum_{i=1}^{l} E(X_i) \frac{1}{p_i(\theta)^2} \frac{\partial}{\partial \theta_j} p_i(\theta) \frac{\partial}{\partial \theta_k} p_i(\theta)$$
$$-\; \sum_{i=1}^{l} E(X_i) \frac{1}{p_i(\theta)} \frac{\partial^2}{\partial \theta_j \partial \theta_k} p_i(\theta)$$

$$= m \sum_{i=1}^{l} \frac{1}{p_i(\theta)} \frac{\partial}{\partial \theta_j} p_i(\theta) \frac{\partial}{\partial \theta_k} p_i(\theta) \qquad (3.5)$$

$$- m \sum_{i=1}^{l} \frac{\partial^2}{\partial \theta_j \partial \theta_k} p_i(\theta)$$

$$= m \sum_{i=1}^{l} \frac{1}{p_i(\theta)} \frac{\partial}{\partial \theta_j} p_i(\theta) \frac{\partial}{\partial \theta_k} p_i(\theta).$$

These results for the multinomial distribution are summarized in Table 3.1, which displays the loglikelihood, score vector, and expected information matrix for some commonly applied exponential families. In the table, $X = x$ represents a single observation from the binomial, Poisson, and exponential families. The mean $E(X)$ is denoted by $\mu$ for the Poisson and exponential distributions. For the binomial family, we express the mean $E(X) = mp$ in terms of the number of trials $m$ and the success probability $p$ per trial. This is similar to the conventions adopted above for the multinomial family. Finally, the differentials $dp$, $dp_i$, and $d\mu$ appearing in the table are row vectors of partial derivatives with respect to the entries of $\theta$.

TABLE 3.1. Score and Information for Some Exponential Families

| Distribution | $L(\theta)$ | $dL(\theta)$ | $J(\theta)$ |
|---|---|---|---|
| Binomial | $x \ln p +$ $(m - x) \ln(1 - p)$ | $\frac{x - mp}{p(1-p)} dp$ | $\frac{m}{p(1-p)} dp^t dp$ |
| Multinomial | $\sum_i x_i \ln p_i$ | $\sum_i \frac{x_i}{p_i} dp_i$ | $\sum_i \frac{m}{p_i} dp_i^t dp_i$ |
| Poisson | $-\mu + x \ln \mu$ | $-d\mu + \frac{x}{\mu} d\mu$ | $\frac{1}{\mu} d\mu^t d\mu$ |
| Exponential | $-\ln \mu - \frac{x}{\mu}$ | $-\frac{1}{\mu} d\mu + \frac{x}{\mu^2} d\mu$ | $\frac{1}{\mu^2} d\mu^t d\mu$ |

**Example 3.1** *Inbreeding in Northeast Brazil*

Data cited by Yasuda [19] on haptoglobin genotypes from 1,948 people from northeast Brazil are recorded in column 2 of Table 3.2. The haptoglobin locus has three codominant alleles $G_1$, $G_2$, and $G_3$ and six corresponding genotypes. The slight excess of homozygotes in these data suggests inbreeding. Now the degree of inbreeding in a population is captured by the inbreeding coefficient $f$, which is formally defined as the probability that the two genes of a random person at a given locus are copies of the same ancestral gene. Column 3 of Table 3.2 gives theoretical haptoglobin genotype frequencies under the usual conditions necessary for genetic equilibrium except that inbreeding is now allowed. To illustrate how these frequencies are derived by conditioning, consider the homozygous genotype $G_1/G_1$. If the two genes of a random person are copies of the same ancestral gene, then the two genes are $G_1$ alleles with probability $p_1$, the population frequency of the $G_1$ allele. On the other hand, if the two genes are not copies of the same ancestral gene, then they are independently the $G_1$ allele with probability $p_1^2$. Thus, $G_1/G_1$ has frequency $fp_1 + (1 - f)p_1^2$. For a heterozygous genotype such as $G_1/G_2$, it is impossible for the genes to be copies of the same ancestral gene, and the appropriate genotype frequency is $(1 - f)2p_1p_2$.

TABLE 3.2. Brazilian Genotypes at the Haptoglobin Locus

| Genotype | Observed Number | Genotype Frequency |
|----------|-----------------|--------------------|
| $G_1/G_1$ | 108 | $fp_1 + (1 - f)p_1^2$ |
| $G_1/G_2$ | 196 | $(1 - f)2p_1p_2$ |
| $G_1/G_3$ | 429 | $(1 - f)2p_1p_3$ |
| $G_2/G_2$ | 143 | $fp_2 + (1 - f)p_2^2$ |
| $G_2/G_3$ | 513 | $(1 - f)2p_2p_3$ |
| $G_3/G_3$ | 559 | $fp_3 + (1 - f)p_3^2$ |

Because $p_3 = 1 - p_1 - p_2$, this model effectively has only the three parameters $(p_1, p_2, f)$. From the initial values, $(p_{01}, p_{02}, f_0) = (\frac{1}{3}, \frac{1}{3}, .02)$, scoring converges in five iterations to the maximum likelihood estimates

$$(\hat{p}_1, \hat{p}_2, \hat{f}) \;=\; (.2157, .2554, .0431).$$

If we invert the expected information matrix, then the asymptotic standard errors of $\hat{p}_1$, $\hat{p}_2$, and $\hat{f}$ are .0067, .0071, and .0166, respectively. If we invert the observed information matrix, the first two standard errors remain the same and the third changes to .0165. The asymptotic correlations of $\hat{f}$ with $\hat{p}_1$ and $\hat{p}_2$ are less than .02 in absolute value regardless of how they are computed.    ∎

## 3.4    Application to the Design of Linkage Experiments

In addition to being useful in the scoring algorithm, expected information provides a criterion for the rational design of genetic experiments. In animal breeding, it is possible to set up test matings for the detection of linkage and estimation of recombination fractions. Consider two linked, codominant loci $A$ and $B$ with alleles $A_1$ and $A_2$ and $B_1$ and $B_2$, respectively. The simplest experimental design is the phase-known, **double-backcross** mating $A_1B_1/A_2B_2 \times A_1B_1/A_1B_1$. This mating notation conveys, for example, that the left parent has one haplotype with alleles $A_1$ and $B_1$ and another haplotype with alleles $A_2$ and $B_2$. Offspring of this mating can be categorized as recombinant with probability $\theta$ and nonrecombinant with probability $1 - \theta$. In view of equation (3.5), the expected information per offspring is

$$
\begin{aligned}
J(\theta) &= \frac{1}{\theta} + \frac{1}{1 - \theta} \\
&= \frac{1}{\theta(1 - \theta)}.
\end{aligned} \tag{3.6}
$$

The efficiencies of mating designs can be compared based on their expected information numbers $J(\theta)$ [16]. The phase-known, **double-intercross** mating $A_1 B_1/A_2 B_2 \times A_1 B_1/A_2 B_2$ offers an alternative to the double-backcross mating. Table 3.3 shows nine phenotypic categories and their associated probabilities (column 3) for offspring of this mating. Since some of these probabilities are identical, the corresponding categories can be collapsed. Thus, categories 1 and 9 can be combined into a single category with probability $(1 - \theta)^2/2$; categories 2, 4, 6, and 8 can be combined into a single category with probability $2\theta(1 - \theta)$; and categories 5 and 7 can be combined into a single category with probability $\theta^2/2$. Category 3 has a unique probability. Based on these four redefined categories and formula (3.5), the expected information per offspring is

$$
J(\theta) = 4 + \frac{2(1 - 2\theta)^2}{\theta(1 - \theta)} + \frac{2(1 - 2\theta)^2}{\theta^2 + (1 - \theta)^2}. \tag{3.7}
$$

TABLE 3.3. Offspring Probabilities for a Double-Intercross Mating

| Category $i$ | Phenotype | $c \times c \ \ p_i$ | $c \times r \ \ p_i$ |
|:---:|:---:|:---:|:---:|
| 1 | $A_1/A_1, B_1/B_1$ | $\frac{(1-\theta)^2}{4}$ | $\frac{\theta(1-\theta)}{4}$ |
| 2 | $A_1/A_1, B_1/B_2$ | $\frac{\theta(1-\theta)}{2}$ | $\frac{\theta^2+(1-\theta)^2}{4}$ |
| 3 | $A_1/A_2, B_1/B_2$ | $\frac{\theta^2+(1-\theta)^2}{2}$ | $\theta(1 - \theta)$ |
| 4 | $A_1/A_2, B_1/B_1$ | $\frac{\theta(1-\theta)}{2}$ | $\frac{\theta^2+(1-\theta)^2}{4}$ |
| 5 | $A_1/A_1, B_2/B_2$ | $\frac{\theta^2}{4}$ | $\frac{\theta(1-\theta)}{4}$ |
| 6 | $A_1/A_2, B_2/B_2$ | $\frac{\theta(1-\theta)}{2}$ | $\frac{\theta^2+(1-\theta)^2}{4}$ |
| 7 | $A_2/A_2, B_1/B_1$ | $\frac{\theta^2}{4}$ | $\frac{\theta(1-\theta)}{4}$ |
| 8 | $A_2/A_2, B_1/B_2$ | $\frac{\theta(1-\theta)}{2}$ | $\frac{\theta^2+(1-\theta)^2}{4}$ |
| 9 | $A_2/A_2, B_2/B_2$ | $\frac{(1-\theta)^2}{4}$ | $\frac{\theta(1-\theta)}{4}$ |

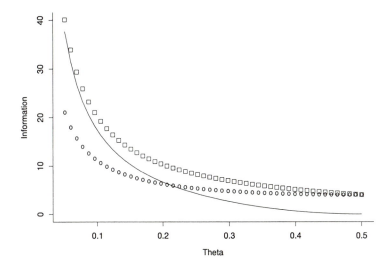

FIGURE 3.1. Graph of Linkage Information Numbers

Besides comparing the double-backcross mating to the **coupling × coupling**, double-intercross mating, we can compare both to the phase-known, **coupling×repulsion**, double-intercross mating $A_1B_1/A_2B_2 \times A_1B_2/A_2B_1$. Column 4 of Table 3.3 now provides the correct probabilities for the nine phenotypic categories. The odd-numbered categories of Table 3.3 collapse to a single category with probability $2\theta(1-\theta)$, and the even-numbered categories to a single category with probability $\theta^2 + (1-\theta)^2$. In this case, the expected information reduces to

$$J(\theta) = \frac{2(1-2\theta)^2}{\theta(1-\theta)} + \frac{4(1-2\theta)^2}{\theta^2 + (1-\theta)^2}. \tag{3.8}$$

In Figure 3.1 we plot the information numbers (3.6), (3.7), and (3.8) as functions of $\theta$ in circles, in boxes, and as a smooth curve, respectively. (See also Table 3.8 of [16].) Inspection of these curves shows that both intercross designs have nearly twice the information content as the backcross design for $\theta$ small. Beyond $\theta = .1$, the intercross designs begin to degrade relative to the backcross design. In the neighborhood of $\theta = .5$, the backcross design and the coupling×coupling, double-intercross design have about equivalent information while the coupling×repulsion, double-intercross design is of no practical value. In general, if one design has $\alpha$ times as much information per offspring as a second design, then it takes $\alpha$ times as many offspring for the second design to achieve the same precision in estimating $\theta$ as the first design.

## 3.5   Quasi-Newton Methods

**Quasi-Newton** methods of maximum likelihood update the current approximation $A_n$ to the observed information $-d^2 L(\theta_n)$ by a low-rank perturbation satisfying a **secant condition**. The secant condition originates from the first-order Taylor's approximation

$$dL(\theta_n)^t - dL(\theta_{n+1})^t \approx d^2 L(\theta_{n+1})(\theta_n - \theta_{n+1}).$$

If we set

$$\begin{aligned} g_n &= dL(\theta_n)^t - dL(\theta_{n+1})^t \\ s_n &= \theta_n - \theta_{n+1}, \end{aligned}$$

then the secant condition is $-A_{n+1} s_n = g_n$. The unique symmetric, rank-one update to $A_n$ satisfying the secant condition is furnished by Davidon's formula [5]

$$A_{n+1} = A_n - c_n v_n v_n^t, \tag{3.9}$$

with constant $c_n$ and vector $v_n$ specified by

$$\begin{aligned} c_n &= \frac{1}{(g_n + A_n s_n)^t s_n} \tag{3.10} \\ v_n &= g_n + A_n s_n. \end{aligned}$$

Until recently, symmetric rank-two updates such as those associated with Davidon, Fletcher, and Powell (DFP) or with Broyden, Fletcher, Goldfarb, and Shanno (BFGS) were considered superior to the more parsimonious update (3.9). However, numerical analysts [4, 10] are now beginning to appreciate the virtues of Davidon's formula. To put it into successful practice, monitoring $A_n$ for positive definiteness is necessary. An immediate concern is that the constant $c_n$ is undefined when the inner product $(g_n + A_n s_n)^t s_n = 0$. In such situations, or when $(g_n + A_n s_n)^t s_n$ is small compared to $v_n^t v_n$, one can ignore the secant requirement and simply take $A_{n+1} = A_n$.

If $A_n$ is positive definite and $c_n \leq 0$, then $A_{n+1}$ is certainly positive definite. If $c_n > 0$, then it may be necessary to shrink $c_n$ to maintain positive definiteness. In order for $A_{n+1}$ to be positive definite, it is necessary that

$$\begin{aligned} v_n^t A_n^{-1} [A_n - c_n v_n v_n^t] A_n^{-1} v_n &= v_n^t A_n^{-1} v_n [1 - c_n v_n^t A_n^{-1} v_n] \\ &> 0. \end{aligned}$$

In other words, $1 - c_n v_n^t A_n^{-1} v_n > 0$ must hold. Conversely, this condition is sufficient to insure positive definiteness of $A_{n+1}$. This fact can be most easily demonstrated by noting the Sherman-Morrison formula [14]

$$[A_n - c_n v_n v_n^t]^{-1} = A_n^{-1} + \frac{c_n}{1 - c_n v_n^t A_n^{-1} v_n} A_n^{-1} v_n [A_n^{-1} v_n]^t. \tag{3.11}$$

Formula (3.11) shows that $[A_n - c_n v_n v_n']^{-1}$ exists and is positive definite under the stated condition. Since the inverse of a positive definite matrix is positive definite, it follows that $A_n - c_n v_n v_n'$ is positive definite as well. This necessary and sufficient condition suggests that $c_n$ be replaced by $\min\{c_n, (1 - \epsilon)/(v_n' A_n^{-1} v_n)\}$ in updating $A_n$, where $\epsilon$ is some constant in $(0, 1)$.

## 3.6   The Dirichlet Distribution

In this section we briefly discuss the Dirichlet distribution. Application is made in the next section to an empirical Bayes procedure for estimating allele frequencies when genotype data are available from several different populations. As is often the case, the Bayes procedure provides an interesting and useful alternative to maximum likelihood estimation.

The **Dirichlet distribution** is a natural generalization of the beta distribution [11]. To generate a Dirichlet random vector $Y = (Y_1, \ldots, Y_k)'$, we take $k$ independent gamma random variables $X_1, \ldots, X_k$ of unit scale and form the ratios

$$Y_i = \frac{X_i}{\sum_{j=1}^k X_j}.$$

By "unit scale" we mean that $X_i$ has density $x_i^{\alpha_i - 1} e^{-x_i} / \Gamma(\alpha_i)$ on $(0, \infty)$ for some $\alpha_i > 0$. Clearly, each $Y_i \geq 0$ and $\sum_{i=1}^k Y_i = 1$.

We can find the joint density of $(Y_1, \ldots, Y_{k-1})$ by considering the larger random vector $Z = (Y_1, \ldots, Y_{k-1}, S)'$, where $S = \sum_{i=1}^k X_i$. The inverse transformation

$$x = T(z)$$
$$= \begin{pmatrix} y_1 s \\ \vdots \\ y_k s \end{pmatrix}$$

with $y_k = 1 - \sum_{i=1}^{k-1} y_i$ has Jacobian

$$\det(dT) = \det \begin{pmatrix} s & 0 & \cdots & 0 & y_1 \\ 0 & s & \cdots & 0 & y_2 \\ \vdots & \vdots & \ddots & \vdots & \vdots \\ 0 & 0 & \cdots & s & y_{k-1} \\ -s & -s & \cdots & -s & y_k \end{pmatrix}$$

$$= \det \begin{pmatrix} s & 0 & \cdots & 0 & y_1 \\ 0 & s & \cdots & 0 & y_2 \\ \vdots & \vdots & \ddots & \vdots & \vdots \\ 0 & 0 & \cdots & s & y_{k-1} \\ 0 & 0 & \cdots & 0 & 1 \end{pmatrix}$$

$$= s^{k-1}.$$

It follows that the density function of $Z$ is

$$\left[\prod_{i=1}^{k} \frac{y_i^{\alpha_i-1}}{\Gamma(\alpha_i)}\right] s^{\sum_{i=1}^{k} \alpha_i - 1} e^{-s}.$$

Integrating out the variable $s$, we find that $(Y_1, \ldots, Y_{k-1})^t$ has density

$$\frac{\Gamma(\alpha.)}{\prod_{i=1}^{k} \Gamma(\alpha_i)} \prod_{i=1}^{k} y_i^{\alpha_i-1}, \tag{3.12}$$

where $\alpha. = \sum_{i=1}^{k} \alpha_i$. It is more convenient to think of the density (3.12) as applying to the whole random vector $Y$. From this perspective, the density exists relative to the uniform measure on the unit simplex

$$\Delta_k = \left\{ (y_1, \ldots, y_k)^t : y_1 > 0, \ldots, y_k > 0, \sum_{i=1}^{k} y_i = 1 \right\}.$$

Once the density (3.12) is in hand, the elegant moment formula

$$
\begin{aligned}
\mathrm{E}\left(\prod_{i=1}^{k} Y_i^{m_i}\right) &= \frac{\Gamma(\alpha.)}{\prod_{i=1}^{k} \Gamma(\alpha_i)} \int_{\Delta_k} \prod_{i=1}^{k} y_i^{m_i+\alpha_i-1} dy \\
&= \frac{\Gamma(\alpha.)}{\Gamma(m.+\alpha.)} \prod_{i=1}^{k} \frac{\Gamma(m_i+\alpha_i)}{\Gamma(\alpha_i)}
\end{aligned}
\tag{3.13}
$$

follows immediately from the fact that the density has total mass 1. The moment formula (3.13) and the factorial property $\Gamma(t+1) = t\Gamma(t)$ of the gamma function together yield the mean $\mathrm{E}(Y_i) = \alpha_i/\alpha.$.

## 3.7    Empirical Bayes Estimation of Allele Frequencies

Consider a locus with $k$ codominant alleles. If in a sample of $n$ people allele $i$ appears $n_i$ times, then the maximum likelihood estimate of the $i$th allele frequency is $n_i/(2n)$. This classical estimate based on the multinomial distribution can be contrasted to a Bayes estimate using a Dirichlet prior for the allele frequencies $p_1, \ldots, p_k$ [12].

The Dirichlet prior is a **conjugate prior** for the multinomial distribution [13]. This means that if the allele frequency vector $p = (p_1, \ldots, p_k)^t$ has a Dirichlet prior with parameters $\alpha_1, \ldots, \alpha_k$, then taking into account the data, $p$ has a Dirichlet posterior with parameters $n_1 + \alpha_1, \ldots, n_k + \alpha_k$. We deduce this fact by applying the moment formula (3.13) in the conditional density computation

$$\frac{\frac{\Gamma(\alpha.)}{\prod_{i=1}^{k} \Gamma(\alpha_i)} \binom{2n}{n_1 \ldots n_k} \prod_{i=1}^{k} p_i^{n_i+\alpha_i-1}}{\frac{\Gamma(\alpha.)}{\prod_{i=1}^{k} \Gamma(\alpha_i)} \binom{2n}{n_1 \ldots n_k} \int_{\Delta_k} \prod_{i=1}^{k} q_i^{n_i+\alpha_i-1} dq}$$

$$= \frac{\Gamma(2n + \alpha_.)}{\prod_{i=1}^{k} \Gamma(n_i + \alpha_i)} \prod_{i=1}^{k} p_i^{n_i + \alpha_i - 1}.$$

The posterior mean $(n_i + \alpha_i)/(2n + \alpha_.)$ is a strongly consistent, asymptotically unbiased estimator of $p_i$.

The primary drawback of being Bayesian in this situation is that there is no obvious way of selecting a reasonable prior. However, if data from several distinct populations are available, then one can select an appropriate prior empirically. Consider the marginal distribution of the allele counts $(N_1, \ldots, N_k)^t$ in a sample of genes from a single population. Integrating out the prior on the allele frequency vector $p = (p_1, \ldots, p_k)^t$ yields the **predictive distribution** [15]

$$\Pr(N_1 = n_1, \ldots, N_k = n_k)$$

$$= \binom{2n}{n_1 \cdots n_k} \frac{\Gamma(\alpha_.)}{\Gamma(2n + \alpha_.)} \prod_{i=1}^{k} \frac{\Gamma(n_i + \alpha_i)}{\Gamma(\alpha_i)}. \tag{3.14}$$

This distribution is known as the **Dirichlet-multinomial distribution**. Its parameters are the $\alpha$'s rather than the $p$'s.

With independent data from several distinct populations, one can estimate the parameter vector $\alpha = (\alpha_1, \ldots, \alpha_k)^t$ of the Dirichlet-multinomial distribution by maximum likelihood. The estimated $\alpha$ can then be recycled to compute the posterior means of the allele frequencies for the separate populations. This interplay between frequentist and Bayesian techniques is typical of the empirical Bayes method.

To estimate the parameter vector $\alpha$ characterizing the prior, we again revert to Newton's method. We need the score $dL(\alpha)$ and the observed information $-d^2L(\alpha)$ for each population. Based on the likelihood (3.14), elementary calculus shows that the score has entries

$$\frac{\partial}{\partial \alpha_i} L(\alpha) = \psi(\alpha_.) - \psi(2n + \alpha_.) + \psi(n_i + \alpha_i) - \psi(\alpha_i), \tag{3.15}$$

where $\psi(s) = \frac{d}{ds} \ln \Gamma(s)$ is the digamma function [8]. The observed information has entries

$$-\frac{\partial^2}{\partial \alpha_i \partial \alpha_j} L(\alpha) = -\psi'(\alpha_.) + \psi'(2n + \alpha_.) \tag{3.16}$$

$$-1_{\{i=j\}}[\psi'(n_i + \alpha_i) - \psi'(\alpha_i)],$$

where $1_{\{i=j\}}$ is the indicator function of the event $\{i = j\}$, and where $\psi'(s)$ is the trigamma function $\frac{d^2}{ds^2} \ln \Gamma(s)$ [8]. The digamma and trigamma functions appearing in the expressions (3.15) and (3.16) should not be viewed as a barrier to computation since good software for evaluating these transcendental functions exists [1, 18].

Equation (3.16) for a single population can be summarized in matrix form by

$$-d^2L(\alpha) = D - c\mathbf{1}\mathbf{1}^t, \tag{3.17}$$

TABLE 3.4. Allele Counts in Four Subpopulations

| Allele | White | Black | Chicano | Asian |
|--------|-------|-------|---------|-------|
| 5 | 2 | 0 | 0 | 0 |
| 6 | 84 | 50 | 80 | 16 |
| 7 | 59 | 137 | 128 | 40 |
| 8 | 41 | 78 | 26 | 8 |
| 9 | 53 | 54 | 55 | 68 |
| 10 | 131 | 51 | 95 | 14 |
| 11 | 2 | 0 | 0 | 7 |
| 12 | 0 | 0 | 0 | 1 |
| Total $2n$ | 372 | 370 | 384 | 154 |

where $D$ is a diagonal matrix with $i$th diagonal entry

$$d_i = \psi'(\alpha_i) - \psi'(n_i + \alpha_i),$$

$c$ is the constant $\psi'(\alpha_.) - \psi'(2n + \alpha_.)$, and $\mathbf{1}$ is a column vector of all 1's. Because the trigamma function is decreasing [8], $d_i > 0$ when $n_i > 0$. For the same reason, $c > 0$. Since the representation (3.17) is preserved under finite sums, it holds, in fact, for the entire sample.

The observed information matrix (3.17) is the sum of a diagonal matrix, which is trivial to invert, plus a symmetric, rank-one perturbation. From our discussion of Davidon's symmetric, rank-one update, we know how to correct the observed information when it fails to be positive definite. A safeguarded Newton's method can be successfully implemented using the Sherman-Morrison formula to invert $-d^2 L(\alpha)$ or its substitute.

**Example 3.2** *Houston Data on the HUMTH01 Locus*

The data of Edwards et al. [6] on the eight alleles of the HUMTH01 locus on chromosome 11 are reproduced in Table 3.4. The allele names for this **tandem repeat** locus refer to numbers of repeat units. From the four separate Houston subpopulations of whites, blacks, Chicanos, and Asians, the eight $\alpha$'s are estimated by maximum likelihood to be .11, 4.63, 7.33, 2.97, 5.32, 5.26, .27, and .10. The large differences in the estimated $\alpha$'s suggest that arbitrarily invoking a **reference** prior with all $\alpha$'s equal would be a mistake in this problem.

Using the estimated $\alpha$'s, Table 3.5 compares the maximum likelihood estimates (first row) and posterior mean estimates (second row) of the allele frequencies within each subpopulation. It is noteworthy that all posterior means are within one standard error of the maximum likelihood estimates. (These standard errors are given in Table 2 of [6].) Nonetheless, the empirical Bayes procedure does tend to moderate the extremes in estimated allele frequencies seen in the different subpopulations. In particular, all posterior means are positive. The maximum likelihood estimates suggest that those alleles failing to appear in a sample are absent

TABLE 3.5. Classical and Bayesian Allele Frequency Estimates

| Allele | White | Black | Chicano | Asian |
|--------|-------|-------|---------|-------|
| 5 | .0054 | .0000 | .0000 | .0000 |
|   | .0053 | .0003 | .0003 | .0006 |
| 6 | .2258 | .1351 | .2083 | .1039 |
|   | .2227 | .1380 | .2064 | .1147 |
| 7 | .1586 | .3703 | .3333 | .2597 |
|   | .1667 | .3645 | .3301 | .2630 |
| 8 | .1102 | .2108 | .0677 | .0519 |
|   | .1105 | .2045 | .0707 | .0609 |
| 9 | .1425 | .1459 | .1432 | .4416 |
|   | .1465 | .1498 | .1471 | .4073 |
| 10 | .3522 | .1378 | .2474 | .0909 |
|    | .3424 | .1421 | .2445 | .1070 |
| 11 | .0054 | .0000 | .0000 | .0455 |
|    | .0057 | .0007 | .0007 | .0404 |
| 12 | .0000 | .0000 | .0000 | .0065 |
|    | .0002 | .0002 | .0002 | .0061 |
| Sample Size $2n$ | 372 | 370 | 384 | 154 |

in the corresponding subpopulation. The empirical Bayes estimates suggest more reasonably that such alleles are simply rare in the subpopulation. ∎

## 3.8   Problems

1. Let $f(x)$ be a real-valued function whose Hessian matrix $(\frac{\partial^2}{\partial x_i \partial x_j} f)$ is positive definite throughout some convex open set $U$ of $R^m$. For $u \neq 0$ and $x \in U$, show that the function $t \rightarrow f(x+tu)$ of the real variable $t$ is strictly convex on $\{t : x+tu \in U\}$. Use this fact to demonstrate that $f(x)$ can have at most one local minimum point on any convex subset of $U$.

2. Apply the result of Problem 1 to show that the loglikelihood of the observed data in the ABO example of Chapter 2 is strictly concave and therefore possesses a single global maximum. Why does the maximum occur on the interior of the feasible region?

3. Show that Newton's method converges in one iteration to the maximum of the quadratic function

$$L(\theta) \;=\; d + e^t\theta + \frac{1}{2}\theta^t F\theta$$

if the symmetric matrix $F$ is negative definite.

4. Verify the loglikelihood, score, and expected information entries in Table 3.1 for the binomial, Poisson, and exponential families.

5. A family of discrete density functions $p_n(\theta)$ defined on $\{0, 1, \ldots\}$ and indexed by a parameter $\theta > 0$ is said to be a power-series family if for all $n$

$$p_n(\theta) = \frac{c_n \theta^n}{g(\theta)}, \tag{3.18}$$

where $c_n \geq 0$ and where $g(\theta) = \sum_{k=0}^{\infty} c_k \theta^k$ is the appropriate normalizing constant. If $X_1, \ldots, X_m$ is a random sample from the discrete density (3.18) with observed values $x_1, \ldots, x_m$, then show that the maximum likelihood estimate of $\theta$ is a root of the equation

$$\frac{1}{m} \sum_{i=1}^{m} x_i = \frac{\theta g'(\theta)}{g(\theta)}.$$

Prove that the expected information in a single observation is

$$J(\theta) = \frac{\sigma^2(\theta)}{\theta^2},$$

where $\sigma^2(\theta)$ is the variance of the density (3.18).

6. Let the $m$ independent random variables $X_1, \ldots, X_m$ be normally distributed with means $\mu_i(\theta)$ and variances $\sigma^2/w_i$, where the $w_i$ are known constants. From observed values $X_1 = x_1, \ldots, X_m = x_m$, one can estimate the mean parameters $\theta$ and the variance parameter $\sigma^2$ simultaneously by the scoring algorithm. Prove that scoring updates $\theta$ by

$$\theta_{n+1} \tag{3.19}$$
$$= \theta_n + \left[ \sum_{i=1}^{m} w_i d\mu_i(\theta_n)^t d\mu_i(\theta_n) \right]^{-1} \sum_{i=1}^{m} w_i [x_i - \mu_i(\theta_n)] d\mu_i(\theta_n)^t$$

and $\sigma^2$ by

$$\sigma_{n+1}^2 = \frac{1}{m} \sum_{i=1}^{m} w_i [x_i - \mu_i(\theta_n)]^2.$$

In the least-squares literature, the scoring update of $\theta$ is better known as the Gauss-Newton algorithm.

7. In the Gauss-Newton algorithm (3.19), the matrix

$$\sum_{i=1}^{m} w_i d\mu_i(\theta_n)^t d\mu_i(\theta_n)$$

can be singular or nearly so. To cure this ill, Marquardt suggested substituting

$$A_n = \sum_{i=1}^{m} w_i d\mu_i(\theta_n)^t d\mu_i(\theta_n) + \lambda I$$

for it and iterating according to

$$\theta_{n+1} = \theta_n + A_n^{-1} \sum_{i=1}^{m} w_i [x_i - \mu_i(\theta_n)] d\mu_i(\theta_n)^t. \qquad (3.20)$$

Prove that the increment $\Delta\theta_n = \theta_{n+1} - \theta_n$ proposed in equation (3.20) minimizes the criterion

$$\frac{1}{2} \sum_{i=1}^{m} w_i [x_i - \mu_i(\theta_n) - d\mu_i(\theta_n)\Delta\theta_n]^2 + \frac{\lambda}{2} \|\Delta\theta_n\|_2^2.$$

8. Consider the quadratic function

$$L(\theta) = -(1, 1)\theta - \frac{1}{2}\theta^t \begin{pmatrix} 2 & 1 \\ 1 & 1 \end{pmatrix} \theta$$

defined on $R^2$. Compute the iterates of the quasi-Newton scheme

$$\theta_{n+1} = \theta_n + A_n^{-1} dL(\theta_n)^t$$

starting from $\theta_1 = (0, 0)^t$ and $A_1 = -\begin{pmatrix} 1 & 0 \\ 0 & 1 \end{pmatrix}$ and using Davidon's update (3.9).

9. Let the random vector $Y = (Y_1, \ldots, Y_k)^t$ follow a Dirichlet distribution with parameters $\alpha_1, \ldots, \alpha_k$. Compute $\mathrm{Var}(Y_i)$ and $\mathrm{Cov}(Y_i, Y_j)$ for $i \neq j$. Also show that $(Y_1 + Y_2, Y_3, \ldots, Y_k)^t$ has a Dirichlet distribution.

10. In the notation of Problem 9, find the score and expected information of a single observation from the Dirichlet distribution. (Hint: In calculating the expected information, take the expectation of the observed information rather than the covariance matrix of the score.)

11. Suppose $n$ unrelated people are sampled at a codominant locus with $k$ alleles. If $N_i = n_i$ genes of allele type $i$ are counted, and if a Dirichlet prior is assumed with parameters $\alpha_1, \ldots, \alpha_k$, then we have seen that the allele frequency vector $p = (p_1, \ldots, p_k)^t$ has a posterior Dirichlet distribution. Use formula (3.13) and show that

$$E(p_i^2 \mid N_1 = n_1, \ldots, N_k = n_k) = \frac{(n_i + \alpha_i)^{\overline{2}}}{(2n + \alpha_.)^{\overline{2}}}$$

$$\mathrm{Var}(p_i^2 \mid N_1 = n_1, \ldots, N_k = n_k) = \frac{(n_i + \alpha_i)^{\overline{4}}}{(2n + \alpha_{\cdot})^{\overline{4}}} - \left[\frac{(n_i + \alpha_i)^{\overline{2}}}{(2n + \alpha_{\cdot})^{\overline{2}}}\right]^2$$

$$\mathrm{E}(2p_i p_j \mid N_1 = n_1, \ldots, N_k = n_k) = 2\frac{(n_i + \alpha_i)(n_j + \alpha_j)}{(2n + \alpha_{\cdot})^{\overline{2}}}$$

$$\mathrm{Var}(2p_i p_j \mid N_1 = n_1, \ldots, N_k = n_k) = \frac{4(n_i + \alpha_i)^{\overline{2}}(n_j + \alpha_j)^{\overline{2}}}{(2n + \alpha_{\cdot})^{\overline{4}}}$$
$$- \left[\frac{2(n_i + \alpha_i)(n_j + \alpha_j)}{(2n + \alpha_{\cdot})^{\overline{2}}}\right]^2,$$

where $x^{\overline{r}} = x(x + 1) \cdots (x + r - 1)$ denotes a rising factorial power. It is interesting that the above mean expressions entail

$$\mathrm{E}(p_i^2 \mid N_1 = n_1, \ldots, N_k = n_k) > \tilde{p}_i^2$$
$$\mathrm{E}(2p_i p_j \mid N_1 = n_1, \ldots, N_k = n_k) < 2\tilde{p}_i \tilde{p}_j,$$

where $\tilde{p}_i$ and $\tilde{p}_j$ are the posterior means of $p_i$ and $p_j$.

# References

[1] Bernardo JM (1976) Algorithm AS 103: psi (digamma) function. *Appl Statist* 25:315–317

[2] Bradley EL (1973) The equivalence of maximum likelihood and weighted least squares estimates in the exponential family. *J Amer Stat Assoc* 68: 199–200

[3] Charnes A, Frome EL, Yu PL (1976) The equivalence of generalized least squares and maximum likelihood in the exponential family. *J Amer Stat Assoc* 71:169–171

[4] Conn AR, Gould NIM, Toint PL (1991) Convergence of quasi-Newton matrices generated by the symmetric rank one update. *Math Prog* 50:177–195

[5] Davidon WC (1959) Variable metric methods for minimization. *AEC Research and Development Report ANL–5990*, Argonne National Laboratory

[6] Edwards A, Hammond HA, Jin L, Caskey CT, Chakraborty R (1992) Genetic variation at five trimeric and tetrameric tandem repeat loci in four human population groups. *Genomics* 12:241–253

[7] Efron B, Hinkley DV (1978) Assessing the accuracy of the maximum likelihood estimator: Observed versus expected Fisher information. *Biometrika* 65:457–487

[8] Hille E (1959) *Analytic Function Theory Vol 1*. Blaisdell Ginn, New York

[9] Jennrich RI, Moore RH (1975) Maximum likelihood estimation by means of nonlinear least squares. *Proceedings of the Statistical Computing Section: American Statistical Association* 57–65

[10] Khalfan HF, Byrd RH, Schnabel RB (1993) A theoretical and experimental study of the symmetric rank-one update. *SIAM J Optimization* 3:1–24

[11] Kingman JFC (1993) *Poisson Processes*. Oxford University Press, Oxford

[12] Lange K (1995) Applications of the Dirichlet distribution to forensic match probabilities. *Genetica* 96:107–117

[13] Lee PM (1989) *Bayesian Statistics: An Introduction*. Edward Arnold, London.

[14] Miller KS (1987) *Some Eclectic Matrix Theory*. Robert E Krieger Publishing, Malabar, FL

[15] Mosimann JE (1962) On the compound multinomial distribution, the multivariate $\beta$-distribution, and correlations among proportions. *Biometrika* 49:65–82

[16] Ott J (1985) *Analysis of Human Genetic Linkage*. Johns Hopkins University Press, Baltimore

[17] Rao CR (1973) *Linear Statistical Inference and its Applications*, 2nd ed. Wiley, New York

[18] Schneider BE (1978) Algorithm AS 121: trigamma function. *Appl Statist* 27:97–99

[19] Yasuda N (1968) Estimation of the inbreeding coefficient from phenotype frequencies by a method of maximum likelihood scoring. *Biometrics* 24:915–934

# 4

# Hypothesis Testing and Categorical Data

## 4.1 Introduction

Most statistical geneticists are frequentists, and fairly traditional ones at that. In testing statistical hypotheses, they prefer pure significance tests or likelihood ratio tests based on large sample theory. Although one could easily dismiss this conservatism as undue reverence for Karl Pearson and R. A. Fisher, it is grounded in the humble reality of geneticists' inability to describe precise alternative hypotheses and to impose convincing priors. In the first part of this chapter, we will review by way of example the large sample methods summarized so admirably by Cavalli-Sforza and Bodmer [4] and Elandt-Johnson [8]. Then we will move on to modern elaborations of frequentist tests for contingency tables. The novelty here lies not in geneticists' inference philosophy, but in designing tests sensitive to certain types of departures from randomness and in computing $p$-values. Good algorithms permit exact or nearly exact computation of $p$-values and consequently relieve our anxieties about large sample approximations.

## 4.2 Hypotheses About Genotype Frequencies

An obvious question of interest to a geneticist is whether a trait satisfies Hardy-Weinberg equilibrium in a particular population. If the trait is not in Hardy-Weinberg equilibrium, then two explanations are possible. First, the genetic model for the trait may be incorrect. For instance, a one-locus model is inappropriate for a two-locus trait. If the model is basically correct, then the further population assumptions necessary for Hardy-Weinberg equilibrium may not be met. Thus, forces such as selection and migration may be distorting the Hardy-Weinberg proportions.

Our aim in this section is to discuss simple likelihood methods for testing Hardy-Weinberg proportions. We emphasize likelihood ratio tests rather than the usual chi-square tests. The two types of tests are similar, but likelihood ratio tests extend more naturally to other statistical settings. Our exposition assumes familiarity with basic notions of large sample theory.

**Example 4.1** *ABO Ulcer Data*

Consider the ABO duodenal ulcer data presented earlier and repeated in column 2 of Table 4.1. If we do not assume Hardy-Weinberg equilibrium, then each of the four phenotypes $A$, $B$, $AB$, and $O$ is assigned a corresponding frequency $q_A$, $q_B$, $q_{AB}$, and $q_O$, with no implied functional relationship among them except for $q_A + q_B + q_{AB} + q_O = 1$. The maximum likelihood estimates of these frequencies are the sample proportions $\hat{q}_A = \frac{n_A}{n} = \frac{186}{521}$, $\hat{q}_B = \frac{n_B}{n} = \frac{38}{521}$, $\hat{q}_{AB} = \frac{n_{AB}}{n} = \frac{13}{521}$, and $\hat{q}_O = \frac{n_o}{n} = \frac{284}{521}$. Under Hardy-Weinberg equilibrium, gene counting provides the maximum likelihood estimates $\hat{p}_A = .2136$, $\hat{p}_B = .0501$, and $\hat{p}_O = .7363$. Denote the vector of maximum likelihood estimates for the two hypotheses by $\hat{q}$ and $\hat{p}$, respectively, and the corresponding maximum likelihoods by $L(\hat{q})$ and $L(\hat{p})$. The likelihood ratio test involves the statistic

$$
\begin{aligned}
2\ln\frac{L(\hat{q})}{L(\hat{p})} &= 2\ln\frac{\hat{q}_A^{n_A}\hat{q}_B^{n_B}\hat{q}_{AB}^{n_{AB}}\hat{q}_O^{n_O}}{(\hat{p}_A^2 + 2\hat{p}_A\hat{p}_O)^{n_A}(\hat{p}_B^2 + 2\hat{p}_B\hat{p}_O)^{n_B}(2\hat{p}_A\hat{p}_B)^{n_{AB}}(\hat{p}_O^2)^{n_O}} \\
&= 2n_A\ln\frac{\hat{q}_A}{\hat{p}_A^2 + 2\hat{p}_A\hat{p}_O} + 2n_B\ln\frac{\hat{q}_B}{\hat{p}_B^2 + 2\hat{p}_B\hat{p}_O} \\
&\quad + 2n_{AB}\ln\frac{\hat{q}_{AB}}{2\hat{p}_A\hat{p}_B} + 2n_O\ln\frac{\hat{q}_O}{\hat{p}_O^2} \\
&= 2\,(1.578 - 1.625 - 1.740 + 1.983) \\
&= .393.
\end{aligned}
$$

This statistic is approximately distributed as a $\chi^2$ distribution with degrees of freedom equaling the difference in the number of independent parameters between the full hypothesis and the Hardy-Weinberg subhypothesis. In this case the degrees of freedom are $3 - 2 = 1$. The likelihood ratio is not significant at the .05 level based on comparison with a $\chi_1^2$ distribution. Thus, we provisionally accept Hardy-Weinberg equilibrium in this population of ulcer patients.

The ABO ulcer data come from a study that also includes data on normal controls [5]. Table 4.1 provides the more comprehensive data. It appears that there may be too many $O$-type individuals among the ulcer patients. We can test this conjecture by testing whether allele frequencies differ between ulcer patients and normal controls. Let $p$, $q$, and $r$ denote the vector of allele frequencies among patients, controls, and the combined sample, respectively. To test the hypothesis $p = q$,

TABLE 4.1. ABO Data on Ulcer Patients and Controls

| Phenotype | Ulcer Patients | Normal Controls |
|-----------|----------------|-----------------|
| A         | 186            | 279             |
| B         | 38             | 69              |
| AB        | 13             | 17              |
| O         | 284            | 315             |

we compute separate maximum likelihoods $L_u(\hat{p})$, $L_n(\hat{q})$, and $L_c(\hat{r})$ for the ulcer patients, normal controls, and combined sample under the assumption of Hardy-Weinberg equilibrium. The appropriate likelihood ratio statistic is

$$\chi_2^2 = 2 \ln \frac{L_u(\hat{p})L_n(\hat{q})}{L_c(\hat{r})}$$
$$= 2 \ln L_u(\hat{p}) + 2 \ln L_n(\hat{q}) - 2 \ln L_c(\hat{r}).$$

The degrees of freedom of the $\chi^2$ are the difference $4 - 2 = 2$ between the number of independent parameters for the two populations treated separately versus them combined.

Gene counting for the normal controls yields the maximum likelihood estimates $\hat{q}_A = .2492$, $\hat{q}_B = .0655$, and $\hat{q}_O = .6853$ and for the combined sample $\hat{r}_A = .2335$, $\hat{r}_B = .0588$, and $\hat{r}_O = .7077$. Straightforward computations yield

$$\ln L_u(\hat{p}) = -173.903 - 189.955 - 97.750 - 49.963$$
$$= -511.571$$
$$\ln L_n(\hat{q}) = -238.134 - 253.114 - 163.050 - 58.161$$
$$= -712.459$$
$$\ln L_c(\hat{r}) = -414.198 - 443.848 - 261.644 - 107.846$$
$$= -1227.536.$$

Hence, the homogeneity $\chi_2^2 = 2 (-511.571 - 712.459 + 1227.536) = 7.012$. This statistic is significant at the .05 level but not at the .01 level. Subsequent studies have substantiated the association between duodenal ulcer and blood type $O$.  ∎

**Example 4.2** *Color Blindness*

The data for this color-blindness example were mentioned in Problem 2 of Chapter 2. If Hardy-Weinberg equilibrium does not hold, then we postulate a probability $q_B$ for normal females, $q_b$ for color-blind females, $r_B$ for normal males, and $r_b$ for color-blind males. The only functional relationship tying these frequencies together are the constraints $q_B + q_b = 1$ and $r_B + r_b = 1$. If in a random sample there are $f_B$ normal females, $f_b$ color-blind females, $m_B$ normal males, and $m_b$ color-blind males, then the likelihood of the sample is

$$\binom{f_B + f_b}{f_B} q_B^{f_B} q_b^{f_b} \binom{m_B + m_b}{m_B} r_B^{m_B} r_b^{m_b}.$$

Maximizing this likelihood leads to the four estimates $\hat{q}_B = \frac{f_B}{f_B+f_b}$, $\hat{q}_b = \frac{f_b}{f_B+f_b}$, $\hat{r}_B = \frac{m_B}{m_B+m_b}$, and $\hat{r}_b = \frac{m_b}{m_B+m_b}$.

Under the Hardy-Weinberg restrictions, Problem 2 of Chapter 2 shows how to compute the maximum likelihood estimates of the allele frequency $p_b$. Testing Hardy-Weinberg equilibrium with the data $f_B = 9032$, $f_b = 40$, $m_B = 8324$, and $m_b = 725$ requires computing the approximate chi-square statistic

$$
\begin{aligned}
\chi_1^2 &= 2\ln \frac{\hat{q}_B^{f_B}\,\hat{q}_b^{f_b}\,\hat{r}_B^{m_B}\,\hat{r}_b^{m_b}}{(1-\hat{p}_b^2)^{f_B}(\hat{p}_b^2)^{f_b}(1-\hat{p}_b)^{m_B}(\hat{p}_b)^{m_b}} \\
&= 2f_B \ln \frac{\hat{q}_B}{1-\hat{p}_b^2} + 2f_b \ln \frac{\hat{q}_b}{\hat{p}_b^2} + 2m_B \ln \frac{\hat{r}_B}{1-\hat{p}_b} + 2m_b \ln \frac{\hat{r}_b}{\hat{p}_b} \\
&= 2\,(14.115 - 12.081 - 26.144 + 26.669) \\
&= 5.118.
\end{aligned}
$$

This chi-square statistic has $2 - 1 = 1$ degree of freedom and is significant at the .025 level. In fact, there are two different common forms of color blindness in humans. A two-locus X-linked model does provide an adequate fit to these data. ∎

## 4.3   Other Multinomial Problems in Genetics

Historically, chi-square tests have been the preferred method of testing hypotheses about multinomial data with known probabilities per category. Chi-square tests are appropriate when no clear alternative suggests itself. However, in many genetics problems the most reasonable alternative is some type of clustering of observations in one or a few categories. In such situations, tests for detecting excess counts in a few categories should be conducted. Ewens et al. [9] highlight the $Z_{\max}$ test in an application to **in situ hybridization**, a form of physical mapping of genes to particular chromosome regions. This application is characterized by fairly large observed counts in most categories and an excess count in a single category. Other applications, such as measuring the nonrandomness of chromosome breakpoints in cancer [6], involve lower counts per category and excess counts in several categories.

For relatively sparse multinomial data with known but unequal probabilities per category, other statistics besides $Z_{\max}$ are useful. For instance, the number of categories $W_d$ with $d$ or more observations can be a sensitive indicator of clustering. Problems in detecting nonrandomness in mutations in different proteins or in amino acids along a single protein afford interesting opportunities for applying the $W_d$ statistic [12, 24]. When the variance and mean of $W_d$ are approximately equal, then $W_d$ is approximately Poisson [3, 14]. In practice, this asymptotic approximation should be checked by applying an exact numerical algorithm for computing $p$-values.

## 4.4   The $Z_{\max}$ Test

Consider a multinomial experiment with $n$ trials and $m$ categories. Denote the probability of category $i$ by $p_i$ and the random number of outcomes in category $i$ by $N_i$. The $Z_{\max}$ statistic [9, 10] is defined by

$$
Z_{\max} \quad = \quad \max_{1 \leq i \leq m} \frac{N_i - np_i}{\sqrt{np_i(1 - p_i)}}.
$$

This statistic is designed to detect departures from the multinomial assumptions caused by the clustering of the observations in one or a few categories. Consequently, a one-sided test is appropriate, and the multinomial model is rejected when $Z_{\max}$ is too large. The specific form of the $Z_{\max}$ statistic is suggested by the fact that the category specific statistics

$$
Z_i \quad = \quad \frac{N_i - np_i}{\sqrt{np_i(1 - p_i)}}
$$

are standardized to have mean 0 and variance 1. Furthermore, when $n$ is large, each $Z_i$ is approximately normally distributed. The usual rule of thumb $np_i \geq 3$ for normality is helpful, particularly if a continuity correction is added to $Z_i$.

To compute $p$-values for $Z_{\max}$, let $z_{\max}$ be the observed value of the statistic, and define the events $A_i = \{Z_i \geq z_{\max}\}$. Then

$$
\Pr(Z_{\max} \geq z_{\max}) \quad = \quad \Pr\left( \bigcup_{i=1}^{m} A_i \right)
$$

$$
\leq \quad \sum_{i=1}^{m} \Pr(A_i) \tag{4.1}
$$

$$
\approx \quad m[1 - \Phi(z_{\max})],
$$

where $\Phi$ is the standard normal distribution function. Alternatively, each $\Pr(A_i)$ can be computed exactly as a right-tail probability of a binomial distribution with $n$ trials and success probability $p_i$.

The upper bound (4.1) can be supplemented by the lower bound

$$
\Pr\left( \bigcup_{i=1}^{m} A_i \right) \quad \geq \quad \sum_{i=1}^{m} \Pr(A_i) - \sum_{i<j} \Pr(A_i \cap A_j)
$$

$$
\geq \quad \sum_{i=1}^{m} \Pr(A_i) - \sum_{i<j} \Pr(A_i) \Pr(A_j) \tag{4.2}
$$

$$
= \quad \sum_{i=1}^{m} \Pr(A_i) + \frac{1}{2} \sum_{i=1}^{m} \Pr(A_i)^2 - \frac{1}{2} \left[ \sum_{i=1}^{m} \Pr(A_i) \right]^2
$$

$$
\approx \quad m[1 - \Phi(z_{\max})] - \frac{m(m-1)}{2}[1 - \Phi(z_{\max})]^2.
$$

If $m[1 - \Phi(z_{max})]$ is small, then the bound (4.1) will be an excellent approximation to the $p$-value.

The first inequality in (4.2) is an example of an inclusion-exclusion bound. To prove it, take expectations in the inequality

$$1_{\cup_{i=1}^m A_i} \geq \sum_{i=1}^m 1_{A_i} - \sum_{i<j} 1_{A_i} 1_{A_j} \qquad (4.3)$$

involving indicator functions. To establish the inequality (4.3), suppose that a sample point belongs to exactly $k$ of the events $A_i$. If $k = 0$, then inequality (4.3) is trivial. If $k > 0$, then inequality (4.3) becomes $1 \geq k - \binom{k}{2}$, which is logically equivalent to $k^2 - 3k + 2 = (k - 2)(k - 1) \geq 0$. The replacement $\Pr(A_i \cap A_j) \leq \Pr(A_i)\Pr(A_j)$ in (4.2) can be rigorously justified [13, 17] as sketched in Problem 3. Note that this inequality reflects the negative correlation of the multinomial components $N_i$.

Ewens et al. [9] suggest that if the $Z_{max}$ test is highly significant, then the category $i$ with largest component $Z_i$ should be removed and the $Z_{max}$ statistic recalculated. This entails replacing $n$ by $n - N_i$ and each $p_j$ by $p_j/(1 - p_i)$ for $j \neq i$ and computing a new $Z_{max}$ for the reduced data. This procedure is repeated until all outlying categories have been identified and $Z_{max}$ is no longer significant.

**Example 4.3** *Application to In Situ Hybridization*

TABLE 4.2. $Z_{max}$ Test for the ZYF Probe in *Macropus eugenii*

| Segment | Proportion $p_i$ | Grains $n_i$ | Statistic $z_i$ |
|---------|------------------|--------------|-----------------|
| 1p | 0.042 | 24 | 3.666 |
| 1q | 0.189 | 37 | -2.406 |
| 2p | 0.019 | 4 | -0.571 |
| 2q | 0.136 | 25 | -2.261 |
| 3/4p | 0.104 | 35 | 1.174 |
| 3/4q | 0.178 | 44 | -0.886 |
| 5p | 0.031 | 29 | 7.030 |
| 5q | 0.097 | 28 | 0.190 |
| 6p | 0.048 | 11 | -0.670 |
| 6q | 0.062 | 11 | -1.564 |
| 7 | 0.053 | 19 | 1.126 |
| Xp | 0.011 | 4 | 0.534 |
| Xq | 0.018 | 3 | -0.911 |
| Y | 0.012 | 5 | 0.908 |

In situ hybridization is a technique for mapping unique sequence DNA probes to particular chromosomal regions [9]. In **metaphase spreads**, chromosomes are highly contracted and can be distinguished on the basis of size, position of their **centromeres**, and characteristic banding patterns. To map a probe, the DNA trapped within a metaphase spread on a microscope slide is denatured in situ and hybridized with a tritium-labeled or fluorescent-labeled probe. A photographic emulsion immediately above the spread registers the presence of the probe on one or more chromosomes.

When a human probe is hybridized to chromosomes of another mammalian species, the probe and corresponding conserved sequence on the mammalian chromosome may be sufficiently different that the hybridization signal is weak. In such cases the probe can appear to hybridize preferentially to several different chromosomal regions. To pick out the real peaks of hybridization from purely random peaks, Ewens et al. [9] apply the $Z_{max}$ test. Table 4.2 reproduces their data on the hybridization of the human ZYF probe, a zinc finger protein probe on the Y chromosome, to homologous regions of the chromosomes of the Australian marsupial *Macropus eugenii*. Fourteen chromosomal segments and 279 hybridization events appear in the table. The observed $z_{max}$ statistic of 7.030 is significant at the .001 level and confirms the presence of a ZYF homologue on the p arm of chromosome 5 of the marsupial. Recalculation of the $Z_{max}$ statistic with segment 5p omitted shows a second significant site on region 1p. Further analysis identifies no other significant regions.                                                                                    ■

## 4.5    The $W_d$ Statistic

Another useful statistic is the number of categories $W_d$ having $d$ or more observations, where $d$ is some fixed positive integer. This statistic has mean $\lambda = \sum_{i=1}^{m} \mu_i$, where

$$\mu_i = \sum_{k=d}^{n} \binom{n}{k} p_i^k (1 - p_i)^{n-k}$$

is the probability that the count of category $i$ satisfies $N_i \geq d$. If the variance of $W_d$ is close to $\lambda$, then as discussed in Problem 4, $W_d$ follows an approximate Poisson distribution with mean $\lambda$ [3].

As a supplement to this approximation, it is possible to compute the distribution function $\Pr(W_d \leq j)$ recursively by adapting a technique of Sandell [19]. Once this is done, the $p$-value of an experimental result $w_d$ can be recovered via $\Pr(W_d \geq w_d) = 1 - \Pr(W_d \leq w_d - 1)$. The recursive scheme can be organized by defining $t_{j,k,l}$ to be the probability that $W_d \leq j$ given $k$ trials and $l$ categories. The indices $j$, $k$, and $l$ are confined to the ranges $0 \leq j \leq w_d - 1$, $0 \leq k \leq n$, and $1 \leq l \leq m$. The $l$ categories implicit in $t_{j,k,l}$ refer to the first $l$ of the overall $m$ categories; the $i$th of these $l$ categories is assigned the conditional probability $p_i/(p_1 + \cdots + p_l)$.

With these definitions in mind, note first the obvious initial values (a) $t_{0,k,1} = 1$ for $k < d$, (b) $t_{0,k,1} = 0$ for $k \geq d$, and (c) $t_{j,k,1} = 1$ for $j > 0$. Now beginning with $l = 1$, compute $t_{j,k,l}$ recursively by conditioning on how many observations fall in category $l$. Since at most $d - 1$ observations can fall in category $l$ without increasing $W_d$ by 1, the recurrence relation for $j = 0$ is

$$
t_{0,k,l}
$$

$$
= \sum_{i=0}^{\min\{d-1,k\}} \binom{k}{i} \left( \frac{p_l}{p_1 + \cdots + p_l} \right)^i \left( 1 - \frac{p_l}{p_1 + \cdots + p_l} \right)^{k-i} t_{0,k-i,l-1},
$$

and the recurrence relation for $j > 0$ is

$$
t_{j,k,l}
$$

$$
= \sum_{i=0}^{\min\{d-1,k\}} \binom{k}{i} \left( \frac{p_l}{p_1 + \cdots + p_l} \right)^i \left( 1 - \frac{p_l}{p_1 + \cdots + p_l} \right)^{k-i} t_{j,k-i,l-1}
$$

$$
+ \sum_{i=d}^{k} \binom{k}{i} \left( \frac{p_l}{p_1 + \cdots + p_l} \right)^i \left( 1 - \frac{p_l}{p_1 + \cdots + p_l} \right)^{k-i} t_{j-1,k-i,l-1}.
$$

These recurrence relations jointly permit replacing the matrix $(t_{j,k,l-1})$ by the matrix $(t_{j,k,l})$. At the end of this recursive scheme on $l = 2, \ldots, m$, we extract the desired probability $t_{w_d-1,n,m}$.

This algorithm for computing the distribution function of $W_d$ relies on evaluation of binomial probabilities $b_{i,k} = \binom{k}{i} r^i (1 - r)^{k-i}$. The naive way of computing the $b_{i,k}$ is to evaluate the binomial coefficient separately and then multiply it by the two appropriate powers. The recurrence relations $b_{i,k} = r b_{i-1,k-1} + (1-r) b_{i,k-1}$ for $0 < i < k$ and the boundary recurrences $b_{0,k} = (1 - r) b_{0,k-1}$ and $b_{k,k} = r b_{k-1,k-1}$ offer a faster and more stable method. To start the recurrence, use the initial conditions $b_{0,1} = 1 - r$ and $b_{1,1} = r$. It is noteworthy that the binomial recurrence increments the number of trials $k$ whereas the recurrence for the distribution function of $W_d$ increments the number of categories $l$.

**Example 4.4** *Mutations in Hemoglobin $\alpha$*

Mutations in the human hemoglobin molecule have been observed in many populations. Vogel and Motulsky [24] tabulate 66 mutations in the 141 amino acids of the hemoglobin $\alpha$ chain. Of these 141 amino acids, 16 show two or more mutations. With all $p_i = \frac{1}{141}$, the mean of $W_2$ is $\lambda = 11.3$. Under the Poisson approximation for $W_2$, the associated p-value is .11. In this example the Poisson approximation is poor, and the exact algorithm yields the more impressive p-value of .028. Thus, the data suggest nonrandomness. It may be that some amino acids are so essential for hemoglobin function that mutations in these amino acids are immediately eliminated by evolution. ∎

## 4.6    Exact Tests of Independence

The problem of testing linkage equilibrium is equivalent to a more general statistical problem of testing for independence in contingency tables. To translate into the usual statistical terminology, one need only equate "locus" to "factor," "allele" to "level," and "linkage equilibrium" to "independence." In exact inference, one conditions on the marginal counts of a contingency table. In the linkage equilibrium setting, this means conditioning on the allele counts at each locus. Suppose we sample $n$ independent haplotypes defined on $m$ loci. Recall that a haplotype $\mathbf{i} = (i_1, \ldots, i_m)$ is just an $m$-tuple of allele choices at the participating loci. If the frequency of allele $k$ at locus $j$ is $p_{jk}$, then under linkage equilibrium the haplotype $\mathbf{i} = (i_1, \ldots, i_m)$ has probability

$$p_{\mathbf{i}} = \prod_{j=1}^{m} p_{ji_j},$$

and the haplotype counts $\{n_{\mathbf{i}}\}$ from the sample follow a multinomial distribution with parameters $(n, \{p_{\mathbf{i}}\})$. The marginal allele counts $\{n_{jk}\}$ at any locus $j$ likewise follow a multinomial distribution with parameters $(n, \{p_{jk}\})$. Since under the null hypothesis of linkage equilibrium, marginal counts are independent from locus to locus, the conditional distribution of the haplotype counts is

$$
\begin{aligned}
\Pr(\{n_{\mathbf{i}}\} \mid \{n_{jk}\}) &= \frac{\binom{n}{\{n_{\mathbf{i}}\}} \prod_{\mathbf{i}} p_{\mathbf{i}}^{n_{\mathbf{i}}}}{\prod_{j=1}^{m} \binom{n}{\{n_{jk}\}} \prod_k (p_{jk})^{n_{jk}}} \\
&= \frac{\binom{n}{\{n_{\mathbf{i}}\}}}{\prod_{j=1}^{m} \binom{n}{\{n_{jk}\}}}.
\end{aligned}
\tag{4.4}
$$

One of the pleasant facts of exact inference is that the multivariate **Fisher-Yates distribution** (4.4) does not depend on the unknown allele frequencies. Problem 8 indicates how to compute its moments [15].

We can also derive the Fisher-Yates distribution by a counting argument involving a sample space distinct from the space of haplotype counts. Consider an $m \times n$ matrix whose rows correspond to loci and whose columns correspond to haplotypes. At locus $j$ there are $n$ genes with $n_{jk}$ genes representing allele $k$. If we uniquely label each of these $n$ genes, then there are $n!$ distinguishable permutations of the genes in row $j$. The uniform sample space consists of the $(n!)^m$ matrices derived from the $n!$ permutations of each of the $m$ rows. Each such matrix is assigned probability $1/(n!)^m$. For instance, if we distinguish duplicate alleles by a superscript $*$, then the $3 \times 4$ matrix

$$
\begin{pmatrix}
a_1 & a_2 & a_1^* & a_2^* \\
b_3 & b_1 & b_1^* & b_2 \\
c_2 & c_1 & c_3 & c_2^*
\end{pmatrix}
\tag{4.5}
$$

for $m = 3$ loci and $n = 4$ haplotypes represents one out of $(4!)^3$ equally likely matrices and yields the nonzero haplotype counts

$$n_{a_1 b_3 c_2} = 1$$
$$n_{a_2 b_1 c_1} = 1$$
$$n_{a_1 b_1 c_3} = 1$$
$$n_{a_2 b_2 c_2} = 1.$$

To count the number of matrices consistent with a haplotype count vector $\{n_i\}$, note that the haplotypes can be assigned to the columns of a typical matrix from the uniform space in $\binom{n}{\{n_i\}}$ ways. Within each such assignment, there are $\prod_{j=1}^{m} \prod_k n_{jk}!$ permutations of the genes of the various allele types among the available positions for each allele type. It follows that the haplotype count vector $\{n_i\}$ has probability

$$\Pr(\{n_i\}) = \frac{\binom{n}{\{n_i\}} \prod_{j=1}^{m} \prod_k n_{jk}!}{(n!)^m}$$

$$= \frac{\binom{n}{\{n_i\}}}{\prod_{j=1}^{m} \binom{n}{\{n_{jk}\}}}.$$

In other words, we recover the Fisher-Yates distribution.

This alternative representation yields a device for random sampling from the Fisher-Yates distribution [16]. If we arrange our observed haplotypes in an $m \times n$ matrix as described above and randomly permute the entries within each row, then we get a new matrix whose haplotype counts are drawn from the Fisher-Yates distribution. For example, appropriate permutations within each row of the matrix (4.5) produce the matrix

$$\begin{pmatrix} a_1 & a_1^* & a_2 & a_2^* \\ b_1 & b_1^* & b_2 & b_3 \\ c_2 & c_2^* & c_1 & c_3 \end{pmatrix}$$

with nonzero haplotype counts

$$n_{a_1 b_1 c_2} = 2$$
$$n_{a_2 b_2 c_1} = 1$$
$$n_{a_2 b_3 c_3} = 1.$$

Iterating this permutation procedure $r$ times generates an independent, random sample $Z_1, \ldots, Z_r$ from the Fisher-Yates distribution. In practice, it suffices to permute all rows except the bottom row $m$ because haplotype counts do not depend on the order of the haplotypes in a haplotype matrix such as (4.5). Given the observed value $T_{\text{obs}}$ of a test statistic $T$ for linkage equilibrium, we estimate the corresponding $p$-value by the sample average $\frac{1}{r} \sum_{l=1}^{r} 1_{\{T(Z_l) \geq T_{\text{obs}}\}}$.

In Fisher's exact test, the statistic $T$ is the negative of the Fisher-Yates probability (4.4). Thus, the null hypothesis of linkage equilibrium (independence) is rejected if the observed Fisher-Yates probability is too low. The chi-square statistic

$\sum_i \frac{[n_i - E(n_i)]^2}{E(n_i)}$ is also reasonable for testing independence, provided we estimate its $p$-value by random sampling and do not foolishly rely on the standard chi-square approximation. As noted in Problem 8, the expectation $E(n_i) = n \prod_{j=1}^{m} (n_{ji_j}/n)$.

**Example 4.5** *Chromosome-11 Haplotype Data*

Weir and Brooks [25] construct 184 haplotypes on 8 chromosome-11 markers from phenotype data on 24 Utah pedigrees. Omitting the two markers BEGl-Hind3 and ADJ-BCl and the two individuals 1353-8600 and 1355-8516 due to incomplete typing, we wind up with 180 full haplotypes on 6 pertinent markers. These markers possess 2, 2, 10, 5, 3, and 2 alleles, respectively. The data can be summarized in a six-dimensional contingency table by giving the counts $n_i$ for each possible haplotype $\mathbf{i} = (i_1, \ldots, i_6)$. Since there are $2 \times 2 \times 10 \times 5 \times 3 \times 2 = 1,200$ haplotypes in all, the table is very sparse, and large sample methods of testing linkage equilibrium are suspect. The chi-square statistic $\chi^2 = \sum_i \frac{[n_i - E(n_i)]^2}{E(n_i)}$ has an observed value of 1,517 for these data. This corresponds to a large sample $p$-value of essentially 0. On the other hand, the empirical $p$-value calculated from 3,999 independent samples of the $\chi^2$ statistic is $.1332 \pm .0057$ [16]. Although the grossly misleading large sample result is hardly surprising in this extreme case, it does remind us of the limitations of large sample approximations and the remedies offered by modern computing. ∎

Readers should be aware that there are other methods for calculating $p$-values associated with exact tests on contingency tables. Agresti [1] surveys the deterministic algorithms useful on small to intermediate-sized tables. For the large, sparse tables encountered in testing Hardy-Weinberg and linkage equilibrium, Markov chain Monte Carlo methods can be even faster than the random permutation method described above [11, 16].

## 4.7   The Transmission/Disequilibrium Test

Example 4.1 on the association between the ABO system and duodenal ulcer depended on detecting a difference in allele frequencies between patients and normal controls. In a racially homogeneous population like that of Britain at the time of the Clarke et al. study [5], this is a reasonable procedure. However, in racially mixed societies like that of the United States, associations can result from population stratification rather than direct causation of alleles at a candidate locus or linkage disequilibrium between the alleles at a marker locus and deleterious alleles at a nearby disease-predisposing locus. Thus, if a disease is concentrated in one racial or ethnic group, then that group's allele frequencies at a marker will predominate in the affecteds regardless of whether or not the marker is linked to a disease locus. If normal controls are not matched by ethnicity to affecteds, then transmission association can be easily confused with ethnic association.

The transmission/disequilibrium test [21, 22] neatly circumvents these misleading ethnic associations by exploiting the internal controls provided by parents. If marker data are collected on the parents of an affected as well as on the affected himself, then one can determine for a codominant marker which of the maternal and paternal alleles are passed to the affected and which are not. The only ambiguity arises when both parents and the child share the same heterozygous genotype. Even in this case one can still count the number of alleles of each type passed to the affected. In the transmission/disequilibrium test, the marker alleles potentially contributed by heterozygous parents to sampled affecteds are arranged in a $2 \times m$ contingency table, with one row counting parental alleles passed to affecteds and the other row counting parental alleles not passed to affecteds. The $m$ columns correspond to the $m$ different alleles seen among the parents. It seems reasonable in this scheme to exclude contributions from homozygous parents because these tell us nothing about transmission distortion.

In analyzing contingency table data of this sort, we should explicitly condition on the parental genotypes. This eliminates ethnic association. Once we have done this, there is no harm in counting alleles transmitted to affected siblings or to related, affected individuals scattered throughout an extended pedigree. The two inviolable rules to observe are that both parents of an affected must be typed and that marker typing should done in one part of a family without regard to the outcomes of marker typing in another part of the family.

The transmission/disequilibrium test for two alleles permits exact calculation of $p$-values [21]. In generalizing the test to multiple alleles, this convenience is sacrificed, but one can approach the problem of calculating approximate $p$-values by standard permutation techniques. The question of an appropriate test statistic also becomes murky unless we consider rather simple, and probably unrealistic, alternative hypotheses. We will suggest two statistics that are intuitively reasonable. Both are based on computing a standardized residual for each cell of the $2 \times m$ table. Let $N_{ij}$ be the count appearing in row $i$ and column $j$ of the table. If $h_j$ heterozygous parents carry allele $j$, then under the null hypothesis of Mendelian transmission, $N_{ij}$ is binomially distributed with $h_j$ trials and success probability $\frac{1}{2}$. The standardized residual corresponding to $N_{ij}$ is therefore

$$Z_{ij} = \frac{N_{ij} - \frac{h_j}{2}}{\sqrt{\frac{h_j}{4}}}.$$

The chi-square statistic $\chi^2 = \sum_{i=1}^{2} \sum_{j=1}^{m} Z_{ij}^2$ furnishes an omnibus test for departure from the null hypothesis of Mendelian segregation to affecteds. The maximum standardized residual $Z_{\max} = \max_{i,j} Z_{ij}$ should be sensitive to preferential transmission of a single allele to affecteds just as in the in situ hybridization problem. Because $Z_{2j} = -Z_{1j}$, the statistic $Z_{\max}$ coincides with $\max_{i,j} |Z_{ij}|$.

The conditional probability space involved in testing the null hypothesis is complicated. Verbally we can describe its sample points as those tables that could have been generated by transmission from the parents with their given genotypes to their affected offspring. The constraints imposed by conditioning on parental genotypes not only fix the margins on the table but also couple the fate of alleles shared by parents. Under the null hypothesis, each relevant transmission event is independent and equally likely to involve either gene of the transmitting parent. Except for biallelic markers [21], it is difficult to compute the exact distribution of either proposed test statistic in this setting. However, we can easily sample from the underlying probability space by randomly selecting for each affected what maternal and paternal genes are transmitted to him. Once these random segregation choices are made, then a new table is constructed by counting the number of alleles of each type transmitted to affecteds. If we let $T_i$ be the value of the statistic $T$ for the $i$th randomly generated table from a sample of $n$ such independent tables, then the $p$-value of the observed statistic $T_{\text{obs}}$ can be approximated by the sample proportion

$$\widehat{\Pr}(T \geq T_{\text{obs}}) = \frac{1}{n} \sum_{i=1}^{n} 1_{\{T_i \geq T_{\text{obs}}\}}.$$

TABLE 4.3. Transmission/Disequilibrium Test for Costa Rican AT Families

| Transmission | Allele | | | | | | | | | |
| Pattern | 1 | 3 | 4 | 5 | 7 | 8 | 10 | 11 | 20 | 21 |
|---|---|---|---|---|---|---|---|---|---|---|
| Transmitted | 3 | 0 | 22 | 0 | 1 | 0 | 0 | 0 | 0 | 2 |
| Not Transmitted | 0 | 4 | 0 | 4 | 3 | 4 | 1 | 1 | 2 | 9 |

**Example 4.6** *Ataxia-telangiectasia (AT) in Costa Rica*

Table 4.3 summarizes marker data on 16 Costa Rican children afflicted with the recessive disease ataxia-telangiectasia (AT). At the chromosome-11 marker D11S1817, 28 of their 32 fully typed parents are heterozygous. Inspection of Table 4.3 strongly suggests that at the very least allele 4 of this marker is preferentially transmitted to affecteds. This suspicion is confirmed by the two permutation tests. Out of $10^6$ independent trials, none of the simulated statistics was as large as the corresponding observed statistics $\chi^2 = 92.91$ and $Z_{\text{max}} = 4.69$. In fact, there are just a handful of different AT mutations segregating in this population isolate. Each mutation is defined by a unique haplotype signature involving marker D11S1817 and several other markers closely linked to the AT locus [23].    ∎

## 4.8    Problems

1. Test for Hardy-Weinberg equilibrium in the MN Syrian data presented in Chapter 2.

2. Table 4.4 lists frequencies of coat colors among cats in Singapore [20]. Assuming an X-linked locus with two alleles, estimate the two allele fre-

quencies by gene counting. Test for Hardy-Weinberg equilibrium using a
likelihood ratio test.

TABLE 4.4. Coat Colors among Singapore Cats

| Females | | | Males | |
|---|---|---|---|---|
| Dark t/t | Calico t/y | Yellow y/y | Dark t | Yellow y |
| 63 | 55 | 12 | 74 | 38 |

3. Let $(N_1, \ldots, N_m)$ be the outcome vector for a multinomial experiment with
$n$ trials and $m$ categories. Prove that

$$\Pr(N_1 \leq t_1, \ldots, N_m \leq t_m) \leq \prod_{i=1}^{m} \Pr(N_i \leq t_i) \qquad (4.6)$$

$$\Pr(N_1 \geq t_1, \ldots, N_m \geq t_m) \leq \prod_{i=1}^{m} \Pr(N_i \geq t_i) \qquad (4.7)$$

for all integers $t_1, \ldots, t_m$. If all $t_k = 0$ in (4.7) except for $t_i$ and $t_j$, conclude
that

$$\Pr(N_i \geq t_i, N_j \geq t_j) \leq \Pr(N_i \geq t_i) \Pr(N_j \geq t_j)$$

as stated in the text. (Hints: It suffices to show that inequality (4.6) holds for
$n = 1$ and that the set of random vectors satisfying (4.6) is closed under the
formation of sums of independent random vectors. For (4.7) consider the
vectors $-N_1, \ldots, -N_m$.)

4. Using the Chen-Stein method and probabilistic coupling, Barbour et al. [3]
show that the statistic $W_d$ satisfies the inequality

$$\sup_{A \subset \mathcal{N}} |\Pr(W_d \in A) - \Pr(Z \in A)| \leq \frac{1 - e^{-\lambda}}{\lambda} [\lambda - \mathrm{Var}(W_d)], \qquad (4.8)$$

where $Z$ is a Poisson random variable having exactly the same expectation
$\lambda = \sum_{i=1}^{m} \mu_i$ as $W_d$, and where $\mathcal{N}$ denotes the set $\{0, 1, \ldots\}$ of nonnegative
integers. Prove that

$$\lambda - \mathrm{Var}(W_d) = \sum_i \mu_i^2 - \sum_i \sum_{j \neq i} \mathrm{Cov}(1_{\{N_i \geq d\}}, 1_{\{N_j \geq d\}}).$$

In view of Problem 3, the random variables $1_{\{N_i \geq d\}}$ and $1_{\{N_j \geq d\}}$ are negatively
correlated. It follows that the bound (4.8) is only useful when the number
$\lambda^{-1}(1 - e^{-\lambda}) \sum_i \mu_i^2$ is small. What is the value of $\lambda^{-1}(1 - e^{-\lambda}) \sum_i \mu_i^2$
for the hemoglobin data when $d = 2$? Careful estimates of the difference
$\lambda - \mathrm{Var}(W_d)$ are provided in [3].

5. Consider a multinomial model with $m$ categories, $n$ trials, and probability $p_i$ attached to category $i$. Express the distribution function of the maximum number of counts $\max_i N_i$ observed in any category in terms of the distribution functions of the $W_d$. How can the algorithm for computing the distribution function of $W_d$ be simplified to give an algorithm for computing a $p$-value of $\max_i N_i$?

6. Continuing Problem 5, define the statistic $U_d$ to be the number of categories $i$ with $N_i < d$. Express the right-tail probability $\Pr(U_d \geq j)$ in terms of the distribution function of $W_d$. This gives a method for computing $p$-values of the statistic $U_d$. In some circumstances $U_d$ has an approximate Poisson distribution. What do you conjecture about these circumstances?

7. The nonparametric linkage test of de Vries et al. [7] uses affected sibling data. Consider a nuclear family with $s$ affected sibs and a heterozygous parent with genotype $a/b$ at some marker locus. Let $n_a$ and $n_b$ count the number of affected sibs receiving the $a$ and $b$ alleles, respectively, from the parent. If the other parent is typed, then this determination is always possible unless both parents and the child are simultaneously of genotype $a/b$. de Vries et al. [7] suggest the statistic $T = |n_a - n_b|$. Under the null hypothesis of independent transmission of the disease and marker genes, Badner et al. [2] show that $T$ has mean and variance

$$E(T) \;=\; \begin{cases} s(\tfrac{1}{2})^s \binom{s}{\frac{s}{2}} & s \text{ even} \\ s(\tfrac{1}{2})^{s-1}\binom{s-1}{\frac{s-1}{2}} & s \text{ odd} \end{cases}$$

$$\operatorname{Var}(T) \;=\; s - E(T)^2.$$

Prove these formulas. If there are $n$ such parents (usually two per family), and the $i$th parent has statistic $T_i$, then the overall statistic

$$\frac{\sum_{i=1}^n [T_i - E(T_i)]}{\sqrt{\sum_{i=1}^n \operatorname{Var}(T_i)}}$$

should be approximately standard normal. A one-sided test is appropriate because the $T_i$ tend to increase in the presence of linkage between the marker locus and a disease predisposing locus. (Hint: The identities

$$\sum_{i=0}^{\frac{s}{2}-1} \binom{s}{i} \;=\; 2^{s-1} - \frac{\binom{s}{\frac{s}{2}}}{2}$$

$$\sum_{i=0}^{\frac{s}{2}-1} i\binom{s}{i} \;=\; s\left[2^{s-2} - \binom{s-1}{\frac{s}{2}}\right]$$

for $s$ even and similar identities for $s$ odd are helpful.)

8. To compute moments under the Fisher-Yates distribution (4.4), let

$$u^{\underline{r}} = u(u-1)\cdots(u-r+1)$$

be a falling factorial power, and let $\{l_i\}$ be a collection of nonnegative integers indexed by the haplotypes $\mathbf{i} = (i_1\ldots,i_m)$. With the notation $l = \sum_{\mathbf{i}} l_{\mathbf{i}}$ and $l_{jk} = \sum_{\mathbf{i}} 1_{\{i_j=k\}} l_{\mathbf{i}}$, show that

$$\mathrm{E}\Big(\prod_{\mathbf{i}} n_{\mathbf{i}}^{l_{\mathbf{i}}}\Big) = \frac{\prod_{j=1}^{m} \prod_k (n_{jk})^{\underline{l_{jk}}}}{(n^{\underline{l}})^{m-1}}.$$

In particular, verify that $\mathrm{E}(n_{\mathbf{i}}) = n \prod_{j=1}^{m} \frac{n_{ji_j}}{n}$.

9. A geneticist phenotypes $n$ unrelated people at each of $m$ loci with codominant alleles and records a vector $\mathbf{i} = (i_1/i_1^*, \ldots, i_m/i_m^*)$ of genotypes for each person. Because phase is unknown, $\mathbf{i}$ cannot be resolved into two haplotypes. The data gathered can be summarized by the number of people $n_{\mathbf{i}}$ counted for each genotype vector $\mathbf{i}$. Let $n_{jk}$ be the number of alleles of type $k$ at locus $j$ observed in the sample, and let $n_h$ be the total number of heterozygotes observed over all loci. Assuming genetic equilibrium, prove that the distribution of the counts $\{n_{\mathbf{i}}\}$ conditional on the allele totals $\{n_{jk}\}$ is

$$\Pr(\{n_{\mathbf{i}}\} \mid \{n_{jk}\}) = \frac{\binom{n}{\{n_{\mathbf{i}}\}} 2^{n_h}}{\prod_{j=1}^{m} \binom{2n}{\{n_{jk}\}}}. \tag{4.9}$$

The moments of the distribution (4.9) are computed in [16]; just as with haplotype count data, all allele frequencies cancel.

10. Describe and program an efficient algorithm for generating random permutations of the set $\{1, \ldots, n\}$. How many calls of a random number generator are involved? How many interchanges of two numbers? You might wish to compare your results to the algorithm in [18].

# References

[1] Agresti A (1992) A survey of exact inference for contingency tables. *Stat Sci* 7:131–177.

[2] Badner JA, Chakravarti A, Wagener DK (1984) A test of nonrandom segregation. *Genetic Epidemiology* 1:329–340

[3] Barbour AD, Holst L, Janson S (1992) *Poisson Approximation*. Oxford University Press, Oxford

[4] Cavalli-Sforza LL, Bodmer WF (1971) *The Genetics of Human Populations*. Freeman, San Francisco

[5] Clarke CA, Price Evans DA, McConnell RB, Sheppard PM (1959) Secretion of blood group antigens and peptic ulcers. *Brit Med J* 1:603–607

[6] De Braekeleer M, Smith B (1988) Two methods for measuring the non-randomness of chromosome abnormalities. *Ann Hum Genet* 52:63–67

[7] de Vries RRP, Lai A, Fat RFM, Nijenhuis LE, van Rood JJ (1976) HLA-linked genetic control of host response to Mycobacterium leprae. *Lancet* 2:1328–1330

[8] Elandt-Johnson RC (1971) *Probability Models and Statistical Methods in Genetics*. Wiley, New York

[9] Ewens WJ, Griffiths RC, Ethier SN, Wilcox SA, Graves JAM (1992) Statistical analysis of in situ hybridization data: Derivation and use of the $z_{max}$ test. *Genomics* 12:675–682

[10] Fuchs C, Kenett R (1980) A test for detecting outlying cells in the multinomial distribution and two-way contingency tables. *J Amer Stat Assoc* 75:395–398

[11] Guo S-W, Thompson E (1992) Performing the exact test of Hardy-Weinberg proportion for multiple alleles. *Biometrics* 48:361–372.

[12] Hanash SM, Boehnke M, Chu EHY, Neel JV, Kuick RD (1988) Nonrandom distribution of structural mutants in ethylnitrosourea-treated cultured human lymphoblastoid cells. *Proc Natl Acad Sci USA* 85:165–169

[13] Joag-Dev K, Proschan F (1983) Negative association of random variables with applications. *Ann Stat* 11:286–295

[14] Kolchin VF, Sevast'yanov BA, Chistyakov VP (1978) *Random Allocations*. Winston, Washington DC

[15] Lange, K (1993) A stochastic model for genetic linkage equilibrium. *Theor Pop Biol* 44:129–148

[16] Lazzeroni LC, Lange K (1997) Markov chains for Monte Carlo tests of genetic equilibrium in multidimensional contingency tables. *Ann Stat* (in press)

[17] Mallows CL (1968). An inequality involving multinomial probabilities. *Biometrika* 55:422–424

[18] Nijenhuis A, Wilf HS (1978) *Combinatorial Algorithms for Computers and Calculators*, 2nd ed. Academic Press, New York

[19] Sandell D (1991) Computing probabilities in a generalized birthday problem. *Math Scientist* 16:78–82

[20] Searle AG (1959) A study of variation in Singapore cats. *J Genet* 56:111–127

[21] Spielman RS, McGinnis RE, Ewens WJ (1993) Transmission test for linkage disequilibrium: The insulin gene region and Insulin-Dependent Diabetes Mellitus (IDDM). *Amer J Hum Genet* 52:506–516

[22] Terwilliger JD, Ott J (1992) A haplotype-based 'haplotype relative risk' approach to detecting allelic associations. *Hum Hered* 42:337-346

[23] Uhrhammer N, Lange E, Porras E, Naiem A, Chen X, Sheikhavandi S, Chiplunkar S, Yang L, Dandekar S, Liang T, Patel N, Teraoka S, Udar N, Calvo N, Concannon P, Lange K, Gatti RA (1995) Sublocalization of an ataxia-telangiectasia gene distal to D11S384 by ancestral haplotyping of Costa Rican families. *Amer J Hum Genet* 57:103-111

[24] Vogel F, Motulsky AG (1986) *Human Genetics: Problems and Approaches*, 2nd ed. Springer-Verlag, Berlin

[25] Weir BS, Brooks LD (1986) Disequilibrium on human chromosome 11p. *Genet Epidemiology Suppl* 1:177–183.

# 5
# Genetic Identity Coefficients

## 5.1 Introduction

Genetic identity coefficients are powerful theoretical tools for genetic analysis. Geneticists have devised these indices to measure the degree of inbreeding of a single individual and the degree of relatedness of a pair of relatives. Since the degree of inbreeding of a single individual can be summarized by the relationship between his or her parents, we will focus on identity coefficients for relative pairs. These coefficients pertain to a generic autosomal locus and depend only on the relevant pedigree connecting two relatives and not on any phenotypes observed in the pedigree. In Chapter 6 we will investigate the applications of identity coefficients. Readers desiring motivation for the combinatorial problems attacked here may want to glance at Chapter 6 first.

## 5.2 Kinship and Inbreeding Coefficients

Two genes $G_1$ and $G_2$ are **identical by descent** (i.b.d.) if one is a physical copy of the other or if they are both physical copies of the same ancestral gene. Two genes are **identical by state** if they represent the same allele. Identity by descent implies identity by state, but not conversely. The simplest measure of relationship between two relatives $i$ and $j$ is their **kinship coefficient** $\Phi_{ij}$. Malécot [12] defined this index to be the probability that a gene selected randomly from $i$ and a gene selected randomly from the same autosomal locus of $j$ are i.b.d. The kinship coefficient takes into account the common ancestry of $i$ and $j$ but not their observed phenotypes at any particular locus. When $i$ and $j$ are the same person, the same gene can be drawn twice because kinship sampling is done with replacement. The **inbreeding coefficient** $f_i$ of an individual $i$ is the probability that his or her two genes at any autosomal locus are i.b.d.; inbreeding sampling is done without replacement. Since $\Phi_{ii} = \frac{1}{2}(1 + f_i)$ and $f_i = \Phi_{kl}$, where $k$ and $l$ are the parents of $i$, an inbreeding coefficient entails no new information. Note that $f_i = 0$ unless $i$'s parents $k$ and $l$ are related. If $f_i > 0$, then $i$ is said to be **inbred**.

TABLE 5.1. Condensed Coefficients of Identity

| Relationship | $\Delta_7$ | $\Delta_8$ | $\Delta_9$ | $\Phi$ |
|---|---|---|---|---|
| Parent–Offspring | 0 | 1 | 0 | $\frac{1}{4}$ |
| Half Siblings | 0 | $\frac{1}{2}$ | $\frac{1}{2}$ | $\frac{1}{8}$ |
| Full Siblings | $\frac{1}{4}$ | $\frac{1}{2}$ | $\frac{1}{4}$ | $\frac{1}{4}$ |
| First Cousins | 0 | $\frac{1}{4}$ | $\frac{3}{4}$ | $\frac{1}{16}$ |
| Double First Cousins | $\frac{1}{16}$ | $\frac{6}{16}$ | $\frac{9}{16}$ | $\frac{1}{8}$ |
| Second Cousins | 0 | $\frac{1}{16}$ | $\frac{15}{16}$ | $\frac{1}{64}$ |
| Uncle–Nephew | 0 | $\frac{1}{2}$ | $\frac{1}{2}$ | $\frac{1}{8}$ |

The last column of Table 5.1 lists kinship coefficients for several common types of relative pairs. The table also contains probabilities for other identity coefficients. Before defining these additional indices of relationship, let us focus on a simple algorithm for computing kinship coefficients. This algorithm produces the kinship coefficient for every possible pair in a pedigree. These coefficients can be arranged in a symmetric matrix $\Phi$ with $\Phi_{ij}$ as the entry in row $i$ and column $j$. To compute $\Phi$ we first number the people in the pedigree in such a way that every parent precedes his or her children. Any person should have either both or neither of his or her parents present in the pedigree. To avoid ambiguity, it is convenient to assume that all pedigree founders are non-inbred and unrelated.

The matrix $\Phi$ is constructed starting with the $1 \times 1$ submatrix in its upper left corner. This submatrix is iteratively expanded by adding a partial row and column as each successive pedigree member is encountered. To make this precise, consider the numbered individuals in sequence. If the current individual $i$ is a founder, then set $\Phi_{ii} = \frac{1}{2}$, reflecting the assumption that founders are not inbred. For each previously considered person $j$, also set $\Phi_{ij} = \Phi_{ji} = 0$, reflecting the fact that $j$ can never be a descendant of $i$ due to our numbering convention. If $i$ is not a founder, then let $i$ have parents $k$ and $l$. It is clear that $\Phi_{ii} = \frac{1}{2} + \frac{1}{2}\Phi_{kl}$ because in sampling the genes of $i$ we are equally likely to choose either the

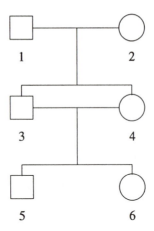

FIGURE 5.1. A Brother–Sister Mating

same gene twice or both maternally and paternally derived genes once. Likewise, $\Phi_{ij} = \Phi_{ji} = \frac{1}{2}\Phi_{jk} + \frac{1}{2}\Phi_{jl}$ because we are equally likely to compare either the maternal gene of $i$ or the paternal gene of $i$ to a randomly drawn gene from $j$. These rules increase the extent of $\Phi$ by an additional diagonal entry and the corresponding partial row and column up to the diagonal entry. This recursive process is repeated until the matrix $\Phi$ is fully defined.

To see the algorithm in action, consider Figure 5.1. The pedigree depicted there involves a brother–sister mating. Its kinship matrix

$$
\Phi = \begin{pmatrix}
\frac{1}{2} & 0 & \frac{1}{4} & \frac{1}{4} & \frac{1}{4} & \frac{1}{4} \\
0 & \frac{1}{2} & \frac{1}{4} & \frac{1}{4} & \frac{1}{4} & \frac{1}{4} \\
\frac{1}{4} & \frac{1}{4} & \frac{1}{2} & \frac{1}{4} & \frac{3}{8} & \frac{3}{8} \\
\frac{1}{4} & \frac{1}{4} & \frac{1}{4} & \frac{1}{2} & \frac{3}{8} & \frac{3}{8} \\
\frac{1}{4} & \frac{1}{4} & \frac{3}{8} & \frac{3}{8} & \frac{5}{8} & \frac{3}{8} \\
\frac{1}{4} & \frac{1}{4} & \frac{3}{8} & \frac{3}{8} & \frac{3}{8} & \frac{5}{8}
\end{pmatrix}
$$

is constructed by creating successively larger submatrices in the upper left corner of the final matrix.

Before proceeding further, let us pause to consider a counterexample illustrating a subtle point about the kinship algorithm. In the pedigree displayed in Figure 5.1, we have $\Phi_{35} \neq \frac{1}{2}\Phi_{15} + \frac{1}{2}\Phi_{25}$ in spite of the fact that 3 has parents 1 and 2. This paradox shows that the substitution rule for computing kinship coefficients should always operate on the higher-numbered person. The problem in this counterexample is that while the paternal (or maternal) gene passed to 3 is randomly chosen, once this choice is made, it limits what can pass to 5. The two random experiments of choosing a gene from 1 to pass to 3 and choosing a gene from 1 for kinship comparison with 5 are not one and the same.

While useful in many applications, the kinship coefficient $\Phi_{ij}$ does not completely summarize the genetic relation between two individuals $i$ and $j$. For instance, siblings and parent–offspring pairs share a common kinship coefficient of $\frac{1}{4}$. Recognizing the deficiencies of kinship coefficients, Gillois [2], Harris [3], and Jacquard [5] capitalized on earlier work of Cotterman [1] and introduced further genetic identity coefficients. Collectively, these new identity coefficients better discriminate between different types of relative pairs. Unfortunately, the traditional graph-tracing algorithms for computation of these identity coefficients are cumbersome compared to the simple algorithm just given for the computation of kinship coefficients [13]. We will explore more recent algorithms that approach the problem of computing identity coefficients obliquely by first defining generalized kinship coefficients and then relating these generalized kinship coefficients to the pairwise identity coefficients [9].

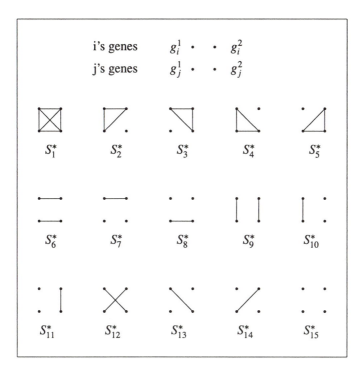

FIGURE 5.2. The Fifteen Detailed Identity States

## 5.3  Condensed Identity Coefficients

Consider the ordered genotypes $g_i^1/g_i^2$ and $g_j^1/g_j^2$ of two people $i$ and $j$ at some autosomal locus. The relation of identity by descent partitions these four genes into equivalence classes or blocks of i.b.d. genes. How many different partitions or identity states exist? Exhaustive enumeration gives a total of 15 partitions or **detailed identity states**. These are depicted in Figure 5.2, which is adapted from [6]. In Figure 5.2, dots correspond to genes and lines connect genes that are i.b.d. The detailed identity states range from the partition $S_1^*$ with only one block, where all four genes are i.b.d., to the partition $S_{15}^*$ with four blocks, where no genes are i.b.d. Several of the partitions are equivalent if maternally derived genes and paternally derived genes are interchanged in one or both of the two people $i$ and $j$. If the maternal and paternal origins of the two pairs of genes are ignored, then the 15 detailed identity states collapse to 9 **condensed identity states** [6]. Figure 5.3 depicts these nine states $S_1, \ldots, S_9$. Note that

$$
\begin{aligned}
S_3 &= S_2^* \cup S_3^* \\
S_5 &= S_4^* \cup S_5^* \\
S_7 &= S_9^* \cup S_{12}^* \\
S_8 &= S_{10}^* \cup S_{11}^* \cup S_{13}^* \cup S_{14}^*.
\end{aligned}
$$

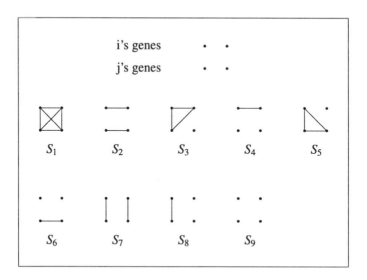

FIGURE 5.3. The Nine Condensed Identity States

Suppose $\Delta_k$ denotes the probability of condensed state $S_k$. Although it is not immediately obvious how to compute the **condensed identity coefficient** $\Delta_k$, some general patterns can easily be discerned. For example, $\Delta_1$, $\Delta_2$, $\Delta_3$, and $\Delta_4$ are all

0 when $i$ is not inbred. Likewise, $\Delta_1$, $\Delta_2$, $\Delta_5$, and $\Delta_6$ are 0 when $j$ is not inbred. The relation

$$\Phi_{ij} = \Delta_1 + \frac{1}{2}(\Delta_3 + \Delta_5 + \Delta_7) + \frac{1}{4}\Delta_8$$

is also easy to verify and provides an alternative method of computing the kinship coefficient $\Phi_{ij}$.

By ad-hoc reasoning, one can compute the $\Delta_k$ in simple cases. For example, Table 5.1 gives the nonzero $\Delta_k$ for some common pairs of relatives. A more complex example is afforded by the two offspring of the brother–sister mating in Figure 5.1. Straightforward but tedious calculations show that $\Delta_1 = \frac{1}{16}$, $\Delta_2 = \frac{1}{32}$, $\Delta_3 = \frac{1}{8}$, $\Delta_4 = \frac{1}{32}$, $\Delta_5 = \frac{1}{8}$, $\Delta_6 = \frac{1}{32}$, $\Delta_7 = \frac{7}{32}$, $\Delta_8 = \frac{5}{16}$, and $\Delta_9 = \frac{1}{16}$ in this case. Because inbreeding is rare, the three coefficients $\Delta_7$, $\Delta_8$, and $\Delta_9$ originally introduced by Cotterman [1] suffice for most practical purposes.

## 5.4  Generalized Kinship Coefficients

We generalize classical kinship coefficients by randomly sampling one gene from each person on an ordered list of $n$ people rather than one gene from each of two people [7, 8, 9, 17]. If a person is repeated in the list, then sampling is done with replacement. The ordered sequence of $n$ sampled genes $G_1, \ldots, G_n$ can be partitioned into nonoverlapping blocks whose constituent genes are i.b.d. A generalized kinship coefficient gives the probability that a particular partition occurs.

This simple verbal description necessarily involves complex notation. For instance, with four individuals $i$, $j$, $k$, and $l$, there are 15 partitions of the four sampled genes $G_i$, $G_j$, $G_k$, and $G_l$. Again these range from the single-block partition $\{G_i, G_j, G_k, G_l\}$, where all of the genes are i.b.d., to the four-block partition $\{G_i\}, \{G_j\}, \{G_k\}, \{G_l\}$, where no genes are i.b.d. We will denote the probability of any such partition by enclosing its blocks within parentheses preceded by $\Phi$, bearing in mind that this probability depends neither on the order of the blocks nor on the order of the sampled genes within a block. Thus, the two partitions mentioned above have probabilities $\Phi(\{G_i, G_j, G_k, G_l\})$ and $\Phi(\{G_i\}, \{G_j\}, \{G_k\}, \{G_l\})$, respectively. The classical kinship coefficient between $i$ and $j$ becomes $\Phi(\{G_i, G_j\})$ in this notation, and its complementary probability becomes $\Phi(\{G_i\}, \{G_j\})$.

## 5.5  From Kinship to Identity Coefficients

What is the relationship between generalized kinship coefficients and condensed identity coefficients? Suppose we randomly sample two genes $G_i^1$ and $G_i^2$ from $i$ and two genes $G_j^1$ and $G_j^2$ from $j$. Now consider one of the detailed identity states $S_k^*$ of Figure 5.2. If we imagine $G_i^1$ and $G_i^2$ occupying the upper two gene positions and $G_j^1$ and $G_j^2$ occupying the lower two gene positions, then $S_k^*$ defines a detailed identity state of the four randomly sampled genes. Corresponding to the 15 random detailed identity states are 9 random condensed identity states as shown in Figure 5.3. Denote the probability of a random condensed identity state $S_k$ by $\Psi_k$. The probability $\Psi_k$ is an integer multiple of a generalized kinship coefficient. For example,

$$\Psi_1 \quad = \quad \Phi(\{G_i^1, G_i^2, G_j^1, G_j^2\}),$$

and by symmetry

$$
\begin{aligned}
\Psi_8 \quad &= \quad \Phi(\{G_i^1, G_j^1\}\{G_i^2\}\{G_j^2\}) + \Phi(\{G_i^1, G_j^2\}\{G_i^2\}\{G_j^1\}) \\
&\quad + \Phi(\{G_i^2, G_j^1\}\{G_i^1\}\{G_j^2\}) + \Phi(\{G_i^2, G_j^2\}\{G_i^1\}\{G_j^1\}) \\
&= \quad 4\Phi(\{G_i^1, G_j^1\}\{G_i^2\}\{G_j^2\})
\end{aligned}
$$

since $S_8 = S_{10}^* \cup S_{11}^* \cup S_{13}^* \cup S_{14}^*$.

It is straightforward to express the $\Psi$'s in terms of the $\Delta$'s by conditioning on which condensed identity state the four original genes of $i$ and $j$ occupy. For instance,

$$\Psi_1 \quad = \quad \Delta_1 + \frac{1}{4}\Delta_3 + \frac{1}{4}\Delta_5 + \frac{1}{8}\Delta_7 + \frac{1}{16}\Delta_8. \tag{5.1}$$

To verify equation (5.1), suppose the four genes of $i$ and $j$ occur in condensed identity state $S_1$. Then the four randomly sampled genes fall in $S_1$ with probability 1. This accounts for the first term on the right of equation (5.1). The second term $\frac{1}{4}\Delta_3$ arises because if the four genes of $i$ and $j$ are in $S_3$, both $G_i^1$ and $G_j^2$ must be drawn from the lower left gene of $S_3$ to achieve state $S_1$ for the randomly sampled genes. Given condensed identity state $S_3$, $G_i^1$ and $G_j^2$ are so chosen with probability $\frac{1}{4}$. The term $\frac{1}{4}\Delta_5$ is accounted for similarly. The term $\frac{1}{8}\Delta_7$ arises because if the four genes of $i$ and $j$ are in $S_7$, the four randomly sampled genes must all be drawn from either the left-hand side of $S_7$ or the right-hand side of $S_7$. Finally, the term $\frac{1}{16}\Delta_8$ arises because if the four genes of $i$ and $j$ are in $S_8$, the four randomly sampled genes can only be drawn from the left-hand side of $S_8$. The remaining condensed identity states are incompatible with the random condensed identity state $S_1$. For example, there is no term involving $\Delta_2$ in equation (5.1) since $S_2$ does not permit identity by descent between any gene of $i$ and any gene of $j$.

Similar reasoning leads to the complete system of equations

$$
\begin{aligned}
\Psi_1 \quad &= \quad \Delta_1 + \frac{1}{4}\Delta_3 + \frac{1}{4}\Delta_5 + \frac{1}{8}\Delta_7 + \frac{1}{16}\Delta_8 \\
\Psi_2 \quad &= \quad \Delta_2 + \frac{1}{4}\Delta_3 + \frac{1}{2}\Delta_4 + \frac{1}{4}\Delta_5 + \frac{1}{2}\Delta_6 + \frac{1}{8}\Delta_7 + \frac{3}{16}\Delta_8 + \frac{1}{4}\Delta_9 \\
\Psi_3 \quad &= \quad \frac{1}{2}\Delta_3 + \frac{1}{4}\Delta_7 + \frac{1}{8}\Delta_8 \\
\Psi_4 \quad &= \quad \frac{1}{2}\Delta_4 + \frac{1}{8}\Delta_8 + \frac{1}{4}\Delta_9
\end{aligned}
$$

$$\Psi_5 = \frac{1}{2}\Delta_5 + \frac{1}{4}\Delta_7 + \frac{1}{8}\Delta_8 \tag{5.2}$$

$$\Psi_6 = \frac{1}{2}\Delta_6 + \frac{1}{8}\Delta_8 + \frac{1}{4}\Delta_9$$

$$\Psi_7 = \frac{1}{4}\Delta_7$$

$$\Psi_8 = \frac{1}{4}\Delta_8$$

$$\Psi_9 = \frac{1}{4}\Delta_9.$$

It is obvious that the matrix of coefficients appearing on the right of (5.2) is upper triangular. This allows us to backsolve for the $\Delta$'s in terms of the $\Psi$'s, beginning with $\Delta_9$ and working upward toward $\Delta_1$. The result is

$$\Delta_1 = \Psi_1 - \frac{1}{2}\Psi_3 - \frac{1}{2}\Psi_5 + \frac{1}{2}\Psi_7 + \frac{1}{4}\Psi_8$$

$$\Delta_2 = \Psi_2 - \frac{1}{2}\Psi_3 - \Psi_4 - \frac{1}{2}\Psi_5 - \Psi_6 + \frac{1}{2}\Psi_7 + \frac{3}{4}\Psi_8 + \Psi_9$$

$$\Delta_3 = 2\Psi_3 - 2\Psi_7 - \Psi_8$$

$$\Delta_4 = 2\Psi_4 - \Psi_8 - 2\Psi_9$$

$$\Delta_5 = 2\Psi_5 - 2\Psi_7 - \Psi_8 \tag{5.3}$$

$$\Delta_6 = 2\Psi_6 - \Psi_8 - 2\Psi_9$$

$$\Delta_7 = 4\Psi_7$$

$$\Delta_8 = 4\Psi_8$$

$$\Delta_9 = 4\Psi_9.$$

It follows that one can compute all of the condensed identity coefficients $\Delta_1, \ldots, \Delta_9$ by computing the coefficients $\Psi_1, \ldots, \Psi_9$. The algorithm developed in the next section for calculating generalized kinship coefficients immediately specializes to calculation of the $\Psi$'s.

## 5.6   Calculation of Generalized Kinship Coefficients

Generalized kinship coefficients (kinship coefficients for short) can be computed recursively by a straightforward algorithm having two phases [7, 9, 14, 17]. In the recursive phase of the algorithm, a currently required kinship coefficient is replaced by a linear combination of subsequently required kinship coefficients. This replacement is effected by moving upward through a pedigree and substituting randomly sampled parental genes for randomly sampled offspring genes. In the static phase of the algorithm, boundary kinship coefficients involving only randomly sampled genes from founders are evaluated. Not surprisingly, the algorithm is reminiscent of our earlier algorithm for computing ordinary kinship coefficients.

We again assume the members of a pedigree are numbered so that parents precede their offspring.

## Boundary Conditions

**Boundary Condition 1** If a founder, or indeed any person, is involved in three or more blocks, then $\Phi = 0$. This condition is obvious because a person has exactly two genes at a given autosomal locus.

**Boundary Condition 2** If two founders occur in the same block, then again put $\Phi = 0$. This condition follows because founders are by definition unrelated.

**Boundary Condition 3** If only founders contribute sampled genes and neither boundary condition 1 nor boundary condition 2 pertains, then $\Phi = (\frac{1}{2})^{m_1-m_2}$, where $m_1$ is the total number of sampled founder genes over all blocks, and $m_2$ is the total number of founders sampled. To verify this condition, imagine choosing one initial gene for each founder involved in $\Phi$. Since a founder cannot be inbred, a subsequent gene chosen for him or her must coincide with the initial choice if the two genes contribute to the same block. If they contribute to two different blocks, the subsequent gene must differ from the initial gene. In either case, the correct choice is made with probability $\frac{1}{2}$ independently of other choices.

## Recurrence Rules

Suppose $i$ is a nonfounder involved in a kinship coefficient $\Phi$. The three recurrence rules operate by substituting genes sampled from $i$'s parents $j$ and $k$ for genes sampled from $i$. It is required that no person involved in $\Phi$ be a descendant of $i$. According to our numbering convention, this requirement can be met by taking $i$ to be the highest-numbered person in $\Phi$. The form of the recurrence rules depends on whether $i$ belongs to one or two blocks. In the former case, suppose without loss of generality that $i$ occupies the first part of the first block. In the latter case, suppose that $i$ occupies the first parts of the first two blocks. It is noteworthy that all three recurrence rules preserve or diminish the number of sampled genes involved in the replacement kinship coefficients relative to the number of sampled genes involved in the current kinship coefficient. In stating the rules, we let $G_i$, $G_j$, and $G_k$ denote randomly sampled genes from $i$, $j$, and $k$, respectively.

**Recurrence Rule 1** Assume that only one gene $G_i$ is sampled from $i$. Then

$$\Phi[\{G_i, \ldots\}\{\} \ldots \{\}] = \frac{1}{2}\Phi[\{G_j, \ldots\}\{\} \ldots \{\}]$$
$$+ \frac{1}{2}\Phi[\{G_k, \ldots\}\{\} \ldots \{\}].$$

This rule follows because the gene drawn at random from $i$ is equally likely to be a gene drawn at random from either $j$ or $k$.

**Recurrence Rule 2** Assume that the genes $G_i^1, \ldots, G_i^s$ are sampled from $i$ for $s > 1$. If these genes occur in one block, then

$$\Phi[\{G_i^1, \ldots, G_i^s, \ldots\}\{\} \ldots \{\}]$$
$$= [1 - 2\left(\frac{1}{2}\right)^s]\Phi[\{G_j, G_k, \ldots\}\{\} \ldots \{\}]$$
$$+ \left(\frac{1}{2}\right)^s \Phi[\{G_j, \ldots\}\{\} \ldots \{\}]$$
$$+ \left(\frac{1}{2}\right)^s \Phi[\{G_k, \ldots\}\{\} \ldots \{\}].$$

In this rule the genes $G_i^1, \ldots, G_i^s$ are replaced, respectively, by single genes from both $j$ and $k$, by a single gene from $j$ only, or by a single gene from $k$ only. The three corresponding coefficients $1 - 2(\frac{1}{2})^s$, $(\frac{1}{2})^s$, and $(\frac{1}{2})^s$ are determined by binomial sampling with $s$ trials and success probability $\frac{1}{2}$.

**Recurrence Rule 3** Assume that the genes $G_i^1, \ldots, G_i^s, G_i^{s+1}, \ldots, G_i^{s+t}$ are sampled from $i$. If the first $s$ genes occur in one block and the remaining $t$ genes occur in another block, then

$$\Phi[\{G_i^1, \ldots, G_i^s, \ldots\}\{G_i^{s+1}, \ldots, G_i^{s+t}, \ldots\}\{\} \ldots \{\}]$$
$$= \left(\frac{1}{2}\right)^{s+t} \Phi[\{G_j, \ldots\}\{G_k, \ldots\}\{\} \ldots \{\}]$$
$$+ \left(\frac{1}{2}\right)^{s+t} \Phi[\{G_k, \ldots\}\{G_j, \ldots\}\{\} \ldots \{\}].$$

This rule follows because neither the maternal gene nor the paternal gene of $i$ can be present in both blocks. Again binomial sampling determines the coefficients $(\frac{1}{2})^{s+t}$.

**Example 5.1** *Sample Calculations for an Inbred Pedigree*

Consider the inbred siblings 5 and 6 in Figure 5.1. Let us compute the kinship coefficient $\frac{1}{4}\Psi_8 = \Phi(\{G_5^1, G_6^1\}\{G_5^2\}\{G_6^2\})$, where $G_k^l$ denotes the $l$th sampled gene of person $k$. Recurrence rule 3 and symmetry imply

$$\frac{1}{4}\Psi_8 = \frac{1}{4}\Phi(\{G_3^1, G_6^1\}\{G_4^1\}\{G_6^2\}) + \frac{1}{4}\Phi(\{G_4^1, G_6^1\}\{G_3^1\}\{G_6^2\})$$
$$= \frac{1}{2}\Phi(\{G_3^1, G_6^1\}\{G_4^1\}\{G_6^2\})$$
$$= \frac{1}{8}\Phi(\{G_3^1, G_3^2\}\{G_4^1\}\{G_4^2\}) + \frac{1}{8}\Phi(\{G_3^1, G_4^2\}\{G_4^1\}\{G_3^2\}).$$

Recurrence rules 2 and 3, boundary conditions 2 and 3, and symmetry permit us to express $A = \Phi(\{G_3^1, G_3^2\}\{G_4^1\}\{G_4^2\})$ as

$$A = \frac{1}{2}\Phi(\{G_1^1, G_2^1\}\{G_4^1\}\{G_4^2\}) + \frac{1}{4}\Phi(\{G_1^1\}\{G_4^1\}\{G_4^2\})$$

$$+ \frac{1}{4} \Phi(\{G_2^1\}\{G_4^1\}\{G_4^2\})$$

$$= 0 + \frac{1}{2} \Phi(\{G_1^1\}\{G_4^1\}\{G_4^2\})$$

$$= \frac{1}{8} \Phi(\{G_1^1\}\{G_1^2\}\{G_2^1\}) + \frac{1}{8} \Phi(\{G_1^1\}\{G_2^1\}\{G_1^2\})$$

$$= \frac{1}{4} \Phi(\{G_1^1\}\{G_1^2\}\{G_2^1\})$$

$$= \frac{1}{8}.$$

Although here we replace 3 by the grandparents 1 and 2 before we replace 4 by the same pair, this is permitted because 4 is not a descendant of 3. In like manner, $B = \Phi(\{G_3^1, G_4^2\}\{G_4^1\}\{G_3^2\})$ can be reduced to

$$B = \frac{1}{4} \Phi(\{G_1^1, G_4^2\}\{G_4^1\}\{G_2^1\}) + \frac{1}{4} \Phi(\{G_2^1, G_4^2\}\{G_4^1\}\{G_1^1\})$$

$$= \frac{1}{2} \Phi(\{G_1^1, G_4^2\}\{G_4^1\}\{G_2^1\})$$

$$= \frac{1}{8} \Phi(\{G_1^1, G_1^2\}\{G_2^1\}\{G_2^1\}) + \frac{1}{8} \Phi(\{G_1^1, G_2^2\}\{G_1^1\}\{G_2^1\})$$

$$= \frac{1}{8} \times \frac{1}{4} + \frac{1}{8} \times 0$$

$$= \frac{1}{32}.$$

These reductions yield $\frac{1}{4} \Psi_8 = \frac{1}{8} A + \frac{1}{8} B = \frac{5}{256}$ for this particular sibling pair. The condensed identity coefficient $\Delta_8 = 4\Psi_8 = \frac{5}{16}$ for the pair follows directly from equation (5.3). ∎

## 5.7  Problems

1. Consider two non-inbred relatives $i$ and $j$ with parents $k$ and $l$ and $m$ and $n$, respectively. Show that

$$\Delta_7 = \Phi_{km} \Phi_{ln} + \Phi_{kn} \Phi_{lm}$$

$$\Delta_8 = 4\Phi_{ij} - 2\Delta_7$$

$$\Delta_9 = 1 - \Delta_7 - \Delta_8,$$

where the condensed identity coefficients all pertain to the pair $i$ and $j$. Thus, in the absence of inbreeding, all nonzero condensed identity coefficients can be expressed in terms of ordinary kinship coefficients.

2. Given the assumptions and notation of Problem 1, demonstrate the inequality $4\Delta_7 \Delta_9 \leq \Delta_8^2$ [15]. This inequality puts an additional constraint on $\Delta_7$,

$\Delta_8$, and $\Delta_9$ besides the obvious nonnegativity requirements and the sum requirement $\Delta_7 + \Delta_8 + \Delta_9 = 1$. (Hints: Note first that

$$\Phi_{ij} = \frac{1}{2}\Delta_7 + \frac{1}{4}\Delta_8$$

$$= \frac{1}{4}\Phi_{km} + \frac{1}{4}\Phi_{kn} + \frac{1}{4}\Phi_{lm} + \frac{1}{4}\Phi_{ln}.$$

Next apply the inequality $(a + b)^2 \geq 4ab$ to prove $4\Delta_7 \leq (4\Phi_{ij})^2$; finally, rearrange.)

3. Calculate all nine condensed identity coefficients for the two inbred siblings 5 and 6 of Figure 5.1.

4. The Cholesky decomposition of a positive definite matrix $\Omega$ is the unique lower triangular matrix $L = (l_{ij})$ satisfying $\Omega = LL'$ and $l_{ii} > 0$ for all $i$. Let $\Phi$ be the kinship matrix of a pedigree with $n$ people numbered so that parents precede their children. The Cholesky decomposition $L$ of $\Phi$ can be defined inductively one row at a time starting with row 1. Given that rows $1, \ldots, i - 1$ have been defined and that $i$ has parents $r$ and $s$, define [4, 10]

$$l_{ij} = \begin{cases} 0 & j > i \\ \frac{1}{2}l_{rj} + \frac{1}{2}l_{sj} & j < i \\ (\Phi_{ii} - \sum_{k=1}^{i-1} l_{ik}^2)^{\frac{1}{2}} & j = i. \end{cases}$$

Prove by induction that $L$ is the Cholesky decomposition of $\Phi$. Why is $l_{ii}$ positive? (Hints: $\Phi_{ii} > \frac{1}{2}\Phi_{ri} + \frac{1}{2}\Phi_{si}$ and $\Phi_{ij} = \frac{1}{2}\Phi_{rj} + \frac{1}{2}\Phi_{sj}$ for $j < i$.)

5. Explicit diagonalization of the kinship matrix $\Phi$ of a pedigree is an unsolved problem in general. In this problem we consider the special case of a nuclear family with $n$ siblings. For convenience, number the parents 1 and 2 and the siblings $3, \ldots, n + 2$. Let $\mathbf{e}_i$ be the vector with 1 in position $i$ and 0 elsewhere. Show that the kinship matrix $\Phi$ for the nuclear family has one eigenvector $\mathbf{e}_1 - \mathbf{e}_2$ with eigenvalue $\frac{1}{2}$; exactly $n - 1$ orthogonal eigenvectors $\frac{1}{m-3}\sum_{j=3}^{m-1} \mathbf{e}_j - \mathbf{e}_m$, $4 \leq m \leq n+2$, with eigenvalue $\frac{1}{4}$; and one eigenvector

$$\mathbf{e}_1 + \mathbf{e}_2 + \frac{4\lambda - 2}{n}(\mathbf{e}_3 + \cdots + \mathbf{e}_{n+2})$$

with eigenvalue $\lambda$ for each of the two solutions of the quadratic equation

$$\lambda^2 - (\frac{1}{2} + \frac{n + 1}{4})\lambda + \frac{1}{8} = 0.$$

This accounts for $n + 2$ orthogonal eigenvectors and therefore diagonalizes $\Phi$.

6. Continuing Problem 5, we can extract some of the eigenvectors and eigenvalues of a kinship matrix of a general pedigree [16]. Consider a set of

individuals in the pedigree possessing the same inbreeding coefficient and
the same kinship coefficients with other pedigree members. Typical cases
are a set of siblings with no children and a married pair of pedigree founders
with shared offspring but no unshared offspring. Without loss of generality, number the members of the set $1, \ldots, m$ and the remaining pedigree
members $m + 1, \ldots, n$. Show that

(a) The kinship matrix $\Phi$ can be written as the partitioned matrix

$$
\Phi \;=\; \begin{pmatrix} a_1 I_m + a_2 \mathbf{1}\mathbf{1}' & \mathbf{1}b' \\ b\mathbf{1}' & C \end{pmatrix},
$$

where $\mathbf{1}$ is a column vector consisting of $m$ $1$'s, $I_m$ is the $m \times m$ identity
matrix, $a_1$ and $a_2$ are real constants, $b$ is a column vector with $n - m$
entries, and $C$ is the $(n - m) \times (n - m)$ kinship matrix of the $n - m$
pedigree members not in the designated set.

(b) The matrix $a_1 I_m + a_2 \mathbf{1}\mathbf{1}'$ has $\mathbf{1}$ as eigenvector with eigenvalue $a_1 + m a_2$
and $m - 1$ orthogonal eigenvectors

$$
u_i \;=\; \frac{1}{i-1} \sum_{j=1}^{i-1} \mathbf{e}_j - \mathbf{e}_i,
$$

$i = 2, \ldots, m$, with eigenvalue $a_1$. Note that each $u_i$ is perpendicular
to $\mathbf{1}$.

(c) The $m - 1$ partitioned vectors $\begin{pmatrix} u_i \\ \mathbf{0} \end{pmatrix}$ are orthogonal eigenvectors of
$\Phi$ with eigenvalue $a_1$.

7. We define the X-linked kinship coefficient $\Phi_{ij}$ between two relatives $i$ and
$j$ as the probability that a gene drawn randomly from an X-linked locus of $i$
is i.b.d. to a gene drawn randomly from the same X-linked locus of $j$. When
$i = j$, sampling is done with replacement. When either $i$ or $j$ is male, one
necessarily selects the maternal gene. Show how the algorithm of Section
5.2 can be modified to compute the X-linked kinship matrix $\Phi$ of a pedigree
[11].

8. **Selfing** is a mating system used extensively in plant breeding. As its name
implies, a plant is mated to itself, then one of its offspring is mated to
itself, and so forth. Let $f_n$ be the inbreeding coefficient of the relevant plant
after $n$ rounds of selfing. Show that $f_{n+1} = \frac{1}{2}(1 + f_n)$ and therefore that
$f_n = (2^n - 1)/2^n$.

9. Geneticists employ repeated sib mating to produce inbred lines of laboratory
animals such as mice. At generation 0, two unrelated animals are mated to
produce generation 1. A brother and sister of generation 1 are then mated
to produce generation 2, and so forth. Let $\phi_n$ be the kinship coefficient of

the brother–sister combination at generation $n$, and let $f_n$ be their common inbreeding coefficient. Demonstrate that $f_{n+1} = \phi_n$ and that

$$\phi_{n+1} = \frac{1}{2}\phi_n + \frac{1}{4}f_n + \frac{1}{4}$$
$$= \frac{1}{2}\phi_n + \frac{1}{4}\phi_{n-1} + \frac{1}{4}.$$

From this second-order difference equation, deduce that

$$\phi_n = 1 - \left(\frac{1}{2} + \frac{1}{\sqrt{5}}\right)\left(\frac{1+\sqrt{5}}{4}\right)^n - \left(\frac{1}{2} - \frac{1}{\sqrt{5}}\right)\left(\frac{1-\sqrt{5}}{4}\right)^n.$$

Thus, $\lim_{n\to\infty} \phi_n = \lim_{n\to\infty} f_n = 1$, and one random allele is fixed at each locus.

10. Wright proposed a path formula for computing inbreeding coefficients that can be generalized to computing kinship coefficients [15]. The pedigree formula is

$$\Phi_{ij} = \sum_{p_{ij}} \left(\frac{1}{2}\right)^{n(p_{ij})} [1 + f_{a(p_{ij})}],$$

where the sum extends over all pairs $p_{ij}$ of nonintersecting paths descending from a common ancestor $a(p_{ij})$ of $i$ and $j$ to $i$ and $j$, respectively, and where $n(p_{ij})$ is the number of people counted along the two paths. The common ancestor is counted only once. If $i = j$, there is only the degenerate pair of paths that start and end at $i$ but possess no arcs connecting a parent to a child. In this case, the formula reduces to the fact $\Phi_{ii} = \frac{1}{2}(1 + f_i)$. In general, a path is composed of arcs connecting parents to their children. Two paths intersect when they share a common arc. To get a feel for Wright's formula, verify it for the case of siblings of unrelated parents. Next prove it in general by induction. Note that although founders are allowed to be inbred, no two of them can be related. (Hint: Consider first the founders of a pedigree and then recursively each child of parents already taken into account.)

11. The definition of a generalized X-linked kinship coefficient exactly parallels the definition of a generalized kinship coefficient except that genes are sampled from a generic X-linked locus rather than a generic autosomal locus. Adapt the algorithm of Section 5.6 to the X-linked case by showing how to revise the boundary conditions and recurrence relations [18].

# References

[1] Cotterman CW (1940) A Calculus for Statistico-Genetics. Ph.D. thesis, Ohio State University. Published in *Genetics and Social Structure*. (1974) Ballonoff PA, editor, Academic Press, New York

[2] Gillois M (1964) La relation d'identité en génétique. *Ann Inst Henri Poincaré B* 2:1–94

[3] Harris DL (1964) Genotypic covariances between inbred relatives. *Genetics* 50:1319–1348

[4] Henderson CR (1976) A simple method for computing the inverse of the numerator relationship matrix used in prediction of breeding values. *Biometrics* 32:69–83

[5] Jacquard A (1966) Logique du calcul des coefficients d'identité entre deux individus. *Population (Paris)* 21:751–776

[6] Jacquard A (1974) *The Genetic Structure of Populations*. Springer-Verlag, New York

[7] Karigl G (1981) A recursive algorithm for the calculation of identity coefficients. *Ann Hum Genet* 45:299–305

[8] Karigl G (1982) Multiple genetic relationships: joint and conditional genotype probabilities. *Ann Hum Genet* 46:83–92

[9] Lange K, Sinsheimer JS (1992) Calculation of genetic identity coefficients. *Ann Hum Genet* 56:339–346

[10] Lange K, Westlake J, Spence MA (1976) Extensions to pedigree analysis. II. Recurrence risk calculation under the polygenic threshold model. *Hum Hered* 26:337–348

[11] Lange K, Westlake J, Spence MA (1976) Extensions to pedigree analysis. III. Variance components by the scoring method. *Ann Hum Genet* 39:485–491

[12] Malécot G (1948) *Les Mathématiques de l'Hérédité*. Masson et Cie, Paris

[13] Nadot R, Vaysseix G (1973) Apparentement et identité. Algorithme du calcul des coefficients d'identité, *Biometrics* 29:347–359

[14] Thompson EA (1983) Gene extinction and allelic origins in complex genealogies. *Proc R Soc London B* 219:241–251

[15] Thompson EA (1986) *Pedigree Analysis in Human Genetics*. Johns Hopkins University Press, Baltimore

[16] Thompson EA, Shaw RG (1990) Pedigree analysis for quantitative traits: Variance components without matrix inversion. *Biometrics* 46:399–413

[17] Weeks DE, Lange K (1988) The affected-pedigree-member method of linkage analysis. *Amer J Hum Genet* 42:315–326

[18] Weeks DE, Valappil TI, Schroeder M, Brown DL (1995) An X-linked version of the affected pedigree member method of linkage analysis. *Hum Hered* 45:25–33

# 6
# Applications of Identity Coefficients

## 6.1 Introduction

The current chapter discusses some applications of kinship and condensed identity coefficients. We commence with the simplest problem of genetic risk prediction involving just two relatives. This setting is artificial because practical genetic counseling usually takes into account information on a whole pedigree rather than information on just a single relative. We will revisit the question of genetic counseling when we explore algorithms for computing pedigree likelihoods.

Our applications of identity coefficients to the correlations between relatives, to risk ratios for qualitative diseases, and to robust linkage analysis are more relevant. Calculation of correlations between relatives forms the foundation of classical **biometrical** analyses of quantitative traits such as height, weight, and cholesterol level [3]. Due to the advent of molecular genetics and positional cloning strategies and to the controversies surrounding race and IQ, biometrical genetics has fallen out of fashion. Nonetheless, it is still a useful tool for exploratory analysis of quantitative traits. If one is mindful of its untestable assumptions and treats its results with caution, then biometrical genetics can offer remarkable insights into the nature and strength of genetic influences on quantitative traits.

Calculation of genetic risk ratios brings genetics into the mainstream of epidemiological thinking on qualitative diseases. Although the models employed to interpret risk ratios are simplistic, it is helpful to have simple models for benchmarks. If these models are ruled out for a disease, then geneticists should adopt robust methods for mapping genes predisposing people to the disease. This chapter ends by explaining one such robust technique for linkage analysis.

## 6.2  Genotype Prediction

One application of condensed identity coefficients involves predicting the genotype of person $j$ based on the observed genotype of person $i$. At an autosomal locus in Hardy-Weinberg equilibrium, suppose allele $a_k$ has population frequency $p_k$. To obtain the genotypic distribution of $j$ at this locus given $j$'s relationship to $i$ and $i$'s genotype, we condition on the various condensed identity states that $i$ and $j$ can jointly occupy. (Figure 5.3 of Chapter 5 depicts the nine possible states.) This conditioning yields

$$\Pr(j = a_m/a_n \mid i = a_k/a_l) = \sum_{r=1}^{9} \Pr(j = a_m/a_n \mid S_r, i = a_k/a_l)$$
$$\times \Pr(S_r \mid i = a_k/a_l).$$

If $i$ has heterozygous genotype $a_k/a_l$ and inbreeding coefficient $f_i$ [7], then states $S_1, \ldots, S_4$ are impossible, and

$$\Pr(S_r \mid i = a_k/a_l) = \frac{\Pr(S_r, i = a_k/a_l)}{\Pr(i = a_k/a_l)}$$
$$= \begin{cases} 0 & \text{for } r \leq 4 \\ \frac{\Delta_r 2 p_k p_l}{(1-f_i) 2 p_k p_l} & \text{for } r > 4 \end{cases}$$
$$= \begin{cases} 0 & \text{for } r \leq 4 \\ \frac{\Delta_r}{1-f_i} & \text{for } r > 4. \end{cases}$$

When $i$ is a homozygote $a_k/a_k$, states $S_1, \ldots, S_4$ come into play. In this case [2],

$$\Pr(S_r \mid i = a_k/a_k) = \frac{\Pr(S_r, i = a_k/a_k)}{\Pr(i = a_k/a_k)}$$
$$= \begin{cases} \frac{\Delta_r p_k}{f_i p_k + (1-f_i) p_k^2} & \text{for } r \leq 4 \\ \frac{\Delta_r p_k^2}{f_i p_k + (1-f_i) p_k^2} & \text{for } r > 4 \end{cases}$$
$$= \begin{cases} \frac{\Delta_r}{f_i + (1-f_i) p_k} & \text{for } r \leq 4 \\ \frac{\Delta_r p_k}{f_i + (1-f_i) p_k} & \text{for } r > 4. \end{cases}$$

Note that $\Pr(S_r \mid i = a_k/a_l) = \Pr(S_r \mid i = a_k/a_k) = \Delta_r$ when $f_i = 0$.

The conditional probabilities $\Pr(j = a_m/a_n \mid S_r, i = a_k/a_l)$ can be computed as follows [7]: In states $S_1$ and $S_7$, $j$ has the same genotype as $i$. In states $S_2$, $S_4$, $S_6$, and $S_9$, $j$'s genotype is independent of $i$'s genotype. In states $S_2$ and $S_6$, $j$ is also an obligate homozygote and has the homozygous genotype $a_m/a_m$ with probability $p_m$. In states $S_4$ and $S_9$, $j$'s genotype follows the Hardy-Weinberg law. In states $S_3$ and $S_8$, $j$ shares one gene in common with $i$; the shared gene is equally likely to be either of $i$'s two genes. The other gene of $j$ is drawn at random from the surrounding population. Thus, if $i$ is a heterozygote $a_k/a_l$ in state $S_8$, then $j$ has genotypes $a_k/a_r$ $(a_r \neq a_l)$, $a_l/a_r$ $(a_r \neq a_k)$, and $a_k/a_l$ with probabilities $p_r/2$,

$p_r/2$, and $p_k/2 + p_l/2$, respectively. If $i$ is a homozygote $a_k/a_k$ in states $S_3$ or $S_8$, then $j$ has genotype $a_k/a_r$ with probability $p_r$. Finally in state $S_5$, $j$ is again an obligate homozygote. If $i$ is $a_k/a_l$, then $j$ is equally likely to be either $a_k/a_k$ or $a_l/a_l$.

**Example 6.1** *Siblings at the ABO Locus*

Consider two non-inbred siblings $i$ and $j$ and their ABO phenotypes. Because $\Pr(S_r \mid i = A/B) = \Delta_r$, we have, for example,

$$\Pr(j = A/B \mid i = A/B)$$
$$= \frac{1}{4}\Pr(j = A/B \mid S_7, \ i = A/B) + \frac{1}{2}\Pr(j = A/B \mid S_8, \ i = A/B)$$
$$+ \frac{1}{4}\Pr(j = A/B \mid S_9, \ i = A/B)$$
$$= \frac{1}{4} \times 1 + \frac{1}{2}\left(\frac{1}{2}p_A + \frac{1}{2}p_B\right) + \frac{1}{4}2p_Ap_B.$$

Similarly,

$$\Pr(j = A \mid i = O/O)$$
$$= \Pr(j = A/A \mid i = O/O) + \Pr(j = A/O \mid i = O/O)$$
$$= \frac{1}{4} \times 0 + \frac{1}{2} \times 0 + \frac{1}{4}p_A^2 + \frac{1}{4} \times 0 + \frac{1}{2}p_A + \frac{1}{4}2p_Ap_O.$$

If we assign $i$ either phenotype $A$ or phenotype $B$, then we must decompose $i$'s phenotype into its constituent genotypes. Problem 2 addresses complications of this sort. ∎

## 6.3    Covariances for a Quantitative Trait

Consider a quantitative trait controlled by a single locus in Hardy-Weinberg equilibrium. Let the $k$th allele $a_k$ at the determining locus have population frequency $p_k$. In the absence of environmental effects, a non-inbred person with ordered genotype $a_k/a_l$ has constant trait value $\mu_{kl} = \mu_{lk}$. No generality is lost if we standardize all trait values so that the random value $X$ of a non-inbred person has mean $E(X) = \sum_k \sum_l \mu_{kl} p_k p_l = 0$. In quantitative genetics, an additive decomposition $\mu_{kl} = \alpha_k + \alpha_l$ is sought. Because such a decomposition may not be possible, the allelic contributions $\alpha_k$ are chosen to minimize the deviations $\delta_{kl} = \mu_{kl} - \alpha_k - \alpha_l$. The classical way of doing this is to minimize the sum of squares

$$\sum_k \sum_l \delta_{kl}^2 p_k p_l = \sum_k \sum_l (\mu_{kl} - \alpha_k - \alpha_l)^2 p_k p_l. \tag{6.1}$$

Setting the partial derivative of (6.1) with respect to $\alpha_k$ equal to 0 gives

$$0 = -4\sum_l (\mu_{kl} - \alpha_k - \alpha_l)p_k p_l$$

$$= -4p_k \sum_l \delta_{kl} p_l.$$

It follows that the optimal deviations satisfy $\sum_l \delta_{kl} p_l = 0$ for all $k$. Because $E(X) = 0$ and $\sum_k p_k = 1$, we also find that

$$
\begin{aligned}
0 &= \sum_k p_k \sum_l \delta_{kl} p_l \\
&= \sum_k \sum_l \delta_{kl} p_k p_l \\
&= \sum_k \sum_l \mu_{kl} p_k p_l - \sum_k \sum_l \alpha_k p_k p_l - \sum_k \sum_l \alpha_l p_k p_l \\
&= -2 \sum_k \alpha_k p_k.
\end{aligned}
$$

Using the fact $\sum_k \alpha_k p_k = 0$ just established, we now conclude that

$$
\begin{aligned}
0 &= \sum_l \delta_{kl} p_l \\
&= \sum_l \mu_{kl} p_l - \sum_l \alpha_k p_l - \sum_l \alpha_l p_l \\
&= \sum_l \mu_{kl} p_l - \alpha_k.
\end{aligned}
$$

In other words, $\alpha_k = \sum_l \mu_{kl} p_l$.

The above calculations can be carried out in a more abstract setting. Suppose $Z_1$ and $Z_2$ are independent random variables. Given a random variable $X$ with mean $E(X) = 0$, how can one choose functions $h_1$ and $h_2$ so that the mean squared error $E([X - h_1(Z_1) - h_2(Z_2)]^2)$ is minimized? This problem is easy to solve if one observes that $E[h_1(Z_1)] = E[h_2(Z_2)] = 0$ should hold and that

$$
\begin{aligned}
\text{Var}[X - h_1(Z_1) - h_2(Z_2)] &= \text{Var}(X) + \text{Var}[h_1(Z_1)] + \text{Var}[h_2(Z_2)] \\
&\quad - 2\,\text{Cov}[X, h_1(Z_1)] - 2\,\text{Cov}[X, h_2(Z_2)] \\
&= \text{Var}[X - h_1(Z_1)] + \text{Var}[X - h_2(Z_2)] \\
&\quad - \text{Var}(X).
\end{aligned}
$$

Now it is well known that $\text{Var}[X - h_i(Z_i)]$ is minimized by taking $h_i(Z_i)$ to be the conditional expectation $E(X \mid Z_i)$ of $X$ given $Z_i$ [4]. In the present case, $X$ is the trait value, $Z_1$ is the maternal allele, and $Z_2$ is the paternal allele. The solution $E(X \mid Z_i = a_k) = \sum_l \mu_{kl} p_l$ coincides with $\alpha_k$ given above. It is natural to introduce the additive genetic variance $\sigma_a^2 = 2\,\text{Var}[E(X \mid Z_i)]$ and the dominance genetic variance

$$\sigma_d^2 = \text{Var}[X - E(X \mid Z_1) - E(X \mid Z_2)].$$

Next suppose $i$ and $j$ are relatives. It is of some interest to compute the covariance $\text{Cov}(X_i, X_j)$ between the trait values $X_i$ and $X_j$ of $i$ and $j$. Let us do this calculation under the simplifying assumption that neither $i$ nor $j$ is inbred. Conditioning on the various identity states and using the facts $\sum_k p_k = 1$, $\sum_k \alpha_k p_k = 0$, $\sum_l \delta_{kl} p_l = 0$, and $\alpha_k = \sum_l \mu_{kl} p_l$, we deduce

$$
\begin{aligned}
&\mathrm{E}(X_i X_j) \\
=\ & \Delta_{7ij} \sum_k \sum_l (\alpha_k + \alpha_l + \delta_{kl})^2 p_k p_l \\
& + \Delta_{8ij} \sum_k \sum_l \sum_m (\alpha_k + \alpha_l + \delta_{kl})(\alpha_k + \alpha_m + \delta_{km}) p_k p_l p_m \\
& + \Delta_{9ij} \sum_k \sum_l \sum_m \sum_n (\alpha_k + \alpha_l + \delta_{kl})(\alpha_m + \alpha_n + \delta_{mn}) p_k p_l p_m p_n \\
=\ & \Delta_{7ij} \left[ 2 \sum_k \alpha_k^2 p_k + \sum_k \sum_l \delta_{kl}^2 p_k p_l \right] + \Delta_{8ij} \sum_k \alpha_k^2 p_k \\
=\ & 2 \left[ \frac{1}{2} \Delta_{7ij} + \frac{1}{4} \Delta_{8ij} \right] 2 \sum_k \alpha_k^2 p_k + \Delta_{7ij} \sum_k \sum_l \delta_{kl}^2 p_k p_l \\
=\ & 2 \Phi_{ij} \sigma_a^2 + \Delta_{7ij} \sigma_d^2,
\end{aligned}
$$

where $\sigma_a^2 = 2 \sum_k \alpha_k^2 p_k$ and $\sigma_d^2 = \sum_k \sum_l \delta_{kl}^2 p_k p_l$ are explicit expressions for the additive and dominance genetic variances. Since $\mathrm{E}(X_i) = \mathrm{E}(X_j) = 0$, the desired covariance $\mathrm{Cov}(X_i, X_j) = \mathrm{E}(X_i X_j)$. When $i$ and $j$ represent the same person, $\mathrm{Var}(X_i) = \sigma_a^2 + \sigma_d^2$ is the total genetic variance. If $i$ is a parent of $j$, then $\mathrm{Cov}(X_i, X_j) = \frac{1}{2} \sigma_a^2$. If $i$ and $j$ are siblings, then $\mathrm{Cov}(X_i, X_j) = \frac{1}{2} \sigma_a^2 + \frac{1}{4} \sigma_d^2$.

The above arguments generalize to allow some environmental determination of the trait. Suppose that $W_i$ and $W_j$ are the random genotypes of two non-inbred relatives $i$ and $j$. If $X_i$ and $X_j$ are independent given $W_i$ and $W_j$, then the expression for $\mathrm{Cov}(X_i, X_j)$ continues to hold provided we define $\mu_{kl} = \mathrm{E}(X \mid W = a_k / a_l)$ for the trait value $X$ and genotype $W$ of a random person. Indeed, in view of our convention that $\mathrm{E}(X) = 0$, we find that

$$
\begin{aligned}
\mathrm{Cov}(X_i, X_j) &= \mathrm{E}(X_i X_j) \\
&= \mathrm{E}[\mathrm{E}(X_i X_j \mid W_i, W_j)] \\
&= \mathrm{E}[\mathrm{E}(X_i \mid W_i) \mathrm{E}(X_j \mid W_j)].
\end{aligned}
$$

However, the total trait variance of any person is inflated because

$$
\begin{aligned}
\mathrm{Var}(X) &= \mathrm{Var}[\mathrm{E}(X \mid W)] + \mathrm{E}[\mathrm{Var}(X \mid W)] \\
&= \sigma_a^2 + \sigma_d^2 + \mathrm{E}[\mathrm{Var}(X \mid W)].
\end{aligned}
$$

These simple variance and covariance expressions extend straightforwardly to polygenic traits, where many genes of small effect act additively to determine a quantitative trait. Many interesting statistical problems arise in this classical biometrical genetics setting.

## 6.4     Risk Ratios and Genetic Model Discrimination

The correlation patterns among relatives provide a simple yet powerful means of discriminating between genetic models for a trait. Following Risch [10], let us explore these patterns for a genetic disease characterized by two states, normal and affected. To any person in a population there corresponds an indicator random variable $X$ such that $X = 0$ if the person is normal and $X = 1$ if the person is affected. In this notation the prevalence of the disease is $K = \Pr(X = 1) = E(X)$.

The disease may have both genetic and environmental determinants. For the sake of simplicity, we assume that the disease indicators $X_i$ and $X_j$ of two relatives $i$ and $j$ are independent given their genotypes. We further suppose that the prevalence of the disease does not vary with age and that genetic equilibrium holds at the disease locus. These strong assumptions are apt to be violated in practice, but they may hold approximately. For instance, if selection is weak and mating is nearly random, then the assumption of genetic equilibrium may not be too damaging. Furthermore, if by a certain age every person definitely does or does not contract the disease, then we can restrict our attention to people beyond this cutoff age.

Now consider two non-inbred relatives $i$ and $j$ of type $R$. Given that person $i$ is affected, it is often possible to estimate empirically the conditional probability $K_R = \Pr(X_j = 1 \mid X_i = 1)$ that $j$ is affected also. The joint probability of both $i$ and $j$ being affected is

$$
\begin{aligned}
K K_R &= \Pr(X_i = 1, X_j = 1) \\
&= E(X_i X_j).
\end{aligned}
\tag{6.2}
$$

For a single-locus model, the covariance decomposition for two relatives gives

$$
\begin{aligned}
E(X_i X_j) &= \mathrm{Cov}(X_i, X_j) + K^2 \\
&= 2\Phi_{ij}\sigma_a^2 + \Delta_{7ij}\sigma_d^2 + K^2.
\end{aligned}
\tag{6.3}
$$

An important index for discriminating between genetic models is the risk ratio $\lambda_R = K_R/K$ for a relative of type $R$. $\lambda_R$ measures the increased risk of disease for the relative of an affected person compared to the population prevalence. It follows from equations (6.2) and (6.3) that

$$
\lambda_R - 1 = \Phi_R \frac{2\sigma_a^2}{K^2} + \Delta_{7R} \frac{\sigma_d^2}{K^2}.
\tag{6.4}
$$

In equation (6.4), $\Phi$ and $\Delta_7$ are subscripted by the relative type $R$. Table 6.1 lists some relative types and their corresponding values of $\lambda_R - 1$. Note that parent–offspring pairs are first-degree relatives; half-siblings, grandparent–grandchild, and uncle–niece pairs are typical second-degree relatives; and first cousins are typical third-degree relatives.

Evidently from the entries of the table for first, second, and third-degree relatives,

$$
\begin{aligned}
\lambda_1 - 1 &= 2(\lambda_2 - 1) \\
&= 4(\lambda_3 - 1),
\end{aligned}
\tag{6.5}
$$

TABLE 6.1. $\lambda_R$ for Different Relative Types $R$

| R | Relative Type | Adjusted Risk Ratios $\lambda_R - 1$ |
|---|---|---|
| M | Identical twin | $\dfrac{\sigma_a^2}{K^2} + \dfrac{\sigma_d^2}{K^2}$ |
| S | Sibling | $\dfrac{\sigma_a^2}{2K^2} + \dfrac{\sigma_d^2}{4K^2}$ |
| 1 | First-degree | $\dfrac{\sigma_a^2}{2K^2}$ |
| 2 | Second-degree | $\dfrac{\sigma_a^2}{4K^2}$ |
| 3 | Third-degree | $\dfrac{\sigma_a^2}{8K^2}$ |

and if $\sigma_d^2 = 0$, then for identical twins and siblings

$$\begin{aligned}
\lambda_M - 1 &= 2(\lambda_S - 1) \\
&= 2(\lambda_1 - 1).
\end{aligned}$$

More complicated multilocus models yield different patterns for the decline in $\lambda_R - 1$. For example, consider a two-locus multiplicative model. The disease indicator $X$ now satisfies $X = YZ$, where $Y$ and $Z$ are indicators for two independent loci. This model is appropriate for a double-dominant disease. In this case, if the alleles at the first locus are $A$ and $a$ and at the second locus $B$ and $b$, then people of unordered genotypes $\{A/A, B/B\}, \{A/a, B/B\}, \{A/A, B/b\},$ and $\{A/a, B/b\}$ are affected, and people of all other genotypes are normal.

As noted above, the population prevalence is

$$\begin{aligned}
K &= \mathrm{E}(X) \\
&= \mathrm{E}(Y)\mathrm{E}(Z) \\
&= K_1 K_2,
\end{aligned}$$

with $K_1 = \mathrm{E}(Y)$ and $K_2 = \mathrm{E}(Z)$. For two relatives of type $R$, the joint probability of both being affected is in obvious notation

$$\begin{aligned}
K K_R &= \mathrm{E}(X_i X_j) \\
&= \mathrm{E}(Y_i Z_i Y_j Z_j) \\
&= \mathrm{E}(Y_i Y_j)\mathrm{E}(Z_i Z_j) \\
&= K_1 K_{1R} K_2 K_{2R}.
\end{aligned}$$

This computation relies on the $Y$ random variables being independent of the $Z$ random variables. The risk ratio

$$
\begin{aligned}
\lambda_R &= \frac{K_R}{K} \\
&= \frac{K_{1R}}{K_1} \frac{K_{2R}}{K_2} \\
&= \lambda_{1R} \lambda_{2R}.
\end{aligned}
$$

Using the equations (6.5) for each locus separately, it follows that for second-degree relatives

$$
\begin{aligned}
\lambda_2 &= \lambda_{12} \lambda_{22} \\
&= \left( \frac{1}{2} \lambda_{11} + \frac{1}{2} \right) \left( \frac{1}{2} \lambda_{21} + \frac{1}{2} \right),
\end{aligned}
$$

and for third-degree relatives

$$
\begin{aligned}
\lambda_3 &= \lambda_{13} \lambda_{23} \\
&= \left( \frac{1}{4} \lambda_{11} + \frac{3}{4} \right) \left( \frac{1}{4} \lambda_{21} + \frac{3}{4} \right),
\end{aligned}
$$

again in more or less obvious notation. The simple formulas

$$
\begin{aligned}
\frac{\lambda_2 - 1}{\lambda_1 - 1} &= \frac{\lambda_3 - 1}{\lambda_2 - 1} \\
&= \frac{1}{2}
\end{aligned}
$$

no longer apply. For instance, when $\lambda_{11} = \lambda_{21} = 4$, we have

$$
\begin{aligned}
\lambda_1 &= 16, \\
\frac{\lambda_2 - 1}{\lambda_1 - 1} &= .35, \\
\frac{\lambda_3 - 1}{\lambda_2 - 1} &= .39.
\end{aligned}
$$

Thus, the ratio $(\lambda_n - 1)/(\lambda_{n-1} - 1)$ declines faster than for a single-locus model.

A possibly more realistic variant of the single-locus model is a two-locus genetic heterogeneity model. In this model either of two independent loci can cause the disease. Let $Y$ be the disease indicator random variable for the first locus, and let $Z$ be the disease indicator random variable for the second locus. Since the two forms of the disease are indistinguishable, $X = Y + Z - YZ$ is the indicator for the disease caused by either or both loci. For a moderately rare disease, the term $YZ$ will be 0 with probability nearly 1. Neglecting the term $YZ$, the approximate population prevalence of the disease under the heterogeneity model is

$$
\begin{aligned}
K &= E(Y) + E(Z) \\
&= K_1 + K_2.
\end{aligned}
$$

Again in obvious notation, the joint probability of $i$ and $j$ both being affected is approximately

$$
\begin{aligned}
KK_R &= \mathrm{E}[(Y_i + Z_i)(Y_j + Z_j)] \\
&= \mathrm{E}(Y_i Y_j) + \mathrm{E}(Y_i)\,\mathrm{E}(Z_j) + \mathrm{E}(Y_j)\,\mathrm{E}(Z_i) + \mathrm{E}(Z_i Z_j) \\
&= K_1 K_{1R} + 2K_1 K_2 + K_2 K_{2R}.
\end{aligned}
$$

The equations for $K$ and $KK_R$ can be combined to yield

$$
\begin{aligned}
KK_R - K^2 &= K_1 K_{1R} + 2K_1 K_2 + K_2 K_{2R} - (K_1 + K_2)^2 \\
&= K_1^2(\lambda_{1R} - 1) + K_2^2(\lambda_{2R} - 1), \tag{6.6}
\end{aligned}
$$

where $\lambda_{1R} = K_{1R}/K_1$ and $\lambda_{2R} = K_{2R}/K_2$. Dividing (6.6) by $K^2$ now gives

$$
\lambda_R - 1 = \left(\frac{K_1}{K}\right)^2 (\lambda_{1R} - 1) + \left(\frac{K_2}{K}\right)^2 (\lambda_{2R} - 1),
$$

with $\lambda_R = K_R/K$.

We conclude from this analysis that the pattern of decline of $\lambda_R - 1$ for the two-locus heterogeneity model is indistinguishable from that for the single-locus model. Risch [10] argues that the index $\lambda_R - 1$ declines too rapidly in schizophrenia to fit the pattern dictated by these two models. He reports a prevalence of $K = .0085$ and the risk ratios displayed in Table 6.2.

TABLE 6.2. Risk Ratios for Schizophrenia

| Relative Type $R$ | Risk Ratio $\lambda_R$ |
|---|---|
| Identical twin | 52.1 |
| Fraternal twin | 14.2 |
| Sibling | 8.6 |
| Offspring | 10.0 |
| Half-sibling | 3.5 |
| Niece or nephew | 3.1 |
| Grandchild | 3.3 |
| First cousin | 1.8 |

## 6.5    An Affecteds-Only Method of Linkage Analysis

Genetic epidemiologists are now actively attempting to map some of the genes contributing to common diseases. This task is complicated by the poorly understood inheritance patterns for many of these diseases. While major genes certainly contribute to some common diseases such as breast cancer and Alzheimer disease, the classical monogenic patterns of inheritance typically do not fit pedigree and population data. One strategy for identifying disease predisposing genes is to restrict mapping studies to pedigrees showing multiple affecteds with early age of onset. Such pedigrees are more apt to segregate major genes than pedigrees with isolated affecteds showing late onset. Even this enrichment strategy does not guarantee a single Mendelian pattern of inheritance in the ascertained pedigrees.

In the absence of a well-defined disease inheritance model, it is still profitable to pursue linkage analysis by robust methods. Robust linkage methods are predicated on the observation that a marker allele will track a closely linked disease allele as both descend from a founder through a pedigree. Only recombination can separate a pair of such alleles present in a pedigree founder. Thus, marker genes can be used as surrogates for disease genes. Robust linkage tests seek to assess the amount of marker allele sharing among affecteds. Excess sharing is taken as evidence that the marker locus is closely linked to a disease predisposing locus. The marker locus may be a candidate locus for the disease. In this case it is perhaps better to speak of association between the marker and the disease.

Our immediate goal is to examine one robust linkage statistic and to compute the mean and variance of this statistic using kinship coefficients [12]. These computations are valid under the null hypothesis of independent segregation of the marker locus and the disease. Beyond this independence assumption, nothing specific is assumed about disease causation.

Consider a pedigree and two affected individuals $i$ and $j$ in that pedigree who are typed at a given marker locus. We assume that the marker locus is in Hardy-Weinberg equilibrium and that its alleles are codominant with the $k$th allele having population frequency $p_k$. At the heart of our robust statistic is the pairwise statistic $Z_{ij}$ assessing the marker sharing between $i$ and $j$. It is desirable for $Z_{ij}$ to give greater weight to shared rare alleles than to shared common alleles. This weighting is accomplished via a weighting function $f(p)$ of the population frequency $p$ of the shared allele. Typical choices for $f(p)$ are $f(p) = 1$, $f(p) = 1/\sqrt{p}$, and $f(p) = 1/p$. Now let $M_i$ and $M_j$ be the observed marker genotypes of $i$ and $j$. Imagine drawing one marker gene $G_i$ at random from $i$ and one marker gene $G_j$ at random from $j$.

The statistic $Z_{ij}$ is defined as the conditional expectation

$$Z_{ij} = \mathrm{E}(1_{\{G_i = G_j\}} f(p_{G_i}) \mid M_i, M_j), \qquad (6.7)$$

where the indicator function $1_{\{G_i = G_j\}}$ is 1 when the sampled genes $G_i$ and $G_j$ match in state. Although substituting i.b.d. for identity by state might be attractive in this definition, the alternative statistic with i.b.d. matches counted would be considerably more difficult to evaluate. In any event, if person $i$ has observed genotype $M_i = a_k/a_l$ and person $j$ has genotype $M_j = a_m/a_n$, then definition (6.7) reduces to

$$Z_{ij} = \frac{1}{4} 1_{\{a_k = a_m\}} f(p_k) + \frac{1}{4} 1_{\{a_k = a_n\}} f(p_k)$$

$$+ \frac{1}{4} 1_{\{a_l = a_m\}} f(p_l) + \frac{1}{4} 1_{\{a_l = a_n\}} f(p_l).$$

From the pairwise statistics $Z_{ij}$, we form an overall statistic $Z = \sum_{\{i,j\}} Z_{ij}$ by summing over all affected pairs $\{i, j\}$ typed in the pedigree. In most applications we take $i \neq j$, but the contrary procedure of comparing an affected person to himself can be useful for inbred affecteds if the disease is thought to be caused by recessively acting genes.

Since the mean and variance of $Z$ obviously are

$$\mathrm{E}(Z) = \sum_{\{i,j\}} \mathrm{E}(Z_{ij})$$

$$\mathrm{Var}(Z) = \sum_{\{i,j\}} \sum_{\{k,l\}} \mathrm{Cov}(Z_{ij}, Z_{kl}),$$

it suffices to calculate $\mathrm{E}(Z_{ij})$ and $\mathrm{Cov}(Z_{ij}, Z_{kl})$. If we condition on whether the two sampled genes $G_i$ and $G_j$ are i.b.d., then it follows that

$$\mathrm{E}(Z_{ij}) = \mathrm{E}[1_{\{G_i = G_j\}} f(p_{G_i})]$$

$$= \Phi_{ij} \sum_k f(p_k) p_k + (1 - \Phi_{ij}) \sum_k f(p_k) p_k^2.$$

The covariance $\mathrm{Cov}(Z_{ij}, Z_{kl}) = \mathrm{E}(Z_{ij} Z_{kl}) - \mathrm{E}(Z_{ij}) \mathrm{E}(Z_{kl})$ can be computed by first noting that $1_{\{G_i = G_j\}} f(p_{G_i})$ depends only on the observed marker genotypes $M_i$ and $M_j$ and that $1_{\{G_k = G_l\}} f(p_{G_k})$ depends only on the observed marker genotypes $M_k$ and $M_l$. These two facts imply that

$$\mathrm{E}(Z_{ij} Z_{kl})$$
$$= \mathrm{E}[\mathrm{E}(1_{\{G_i = G_j\}} f(p_{G_i}) \mid M_i, M_j) \mathrm{E}(1_{\{G_k = G_l\}} f(p_{G_k}) \mid M_k, M_l)]$$
$$= \mathrm{E}[\mathrm{E}(1_{\{G_i = G_j\}} f(p_{G_i}) 1_{\{G_k = G_l\}} f(p_{G_k}) \mid M_i, M_j, M_k, M_l)]$$
$$= \mathrm{E}[1_{\{G_i = G_j\}} f(p_{G_i}) 1_{\{G_k = G_l\}} f(p_{G_k})].$$

To evaluate the last expectation, we condition on how the four sampled genes $G_i$, $G_j$, $G_k$, and $G_l$ are partitioned under identity by descent. Consider again the condensed identity states of Figure 5.3 of Chapter 5. In each state imagine genes $G_i$ and $G_j$ appearing on the top row in no particular order and genes $G_k$ and $G_l$ appearing on the bottom row in no particular order. Let $\Upsilon_r$ denote the probability of the condensed identity state $S_r$ under these conventions. Then

$$\mathrm{E}(Z_{ij} Z_{kl}) = \sum_r \mathrm{E}(1_{\{G_i = G_j\}} f(p_{G_i}) 1_{\{G_k = G_l\}} f(p_{G_k}) \mid S_r) \Upsilon_r.$$

Table 6.3 lists the necessary conditional expectations. The entries of the table are straightforward to verify. For instance, consider the entry for state $S_8$. In this condensed identity state, one of the two genes on the top row is i.b.d. with one of the two genes on the bottom row. Thus, $1_{\{G_i = G_j\}} 1_{\{G_k = G_l\}} = 1$ only when $G_i$, $G_j$,

$G_k$, and $G_l$ all agree in state. By independence, all four genes coincide with the $m$th allele with probability $p_m^3$.

To compute the probabilities $\Upsilon_r$, we reason as we did in Chapter 5 in passing between generalized kinship coefficients and condensed identity coefficients. Consider the 15 detailed identity states possible for 4 genes as depicted in Figure 5.2 of Chapter 5. Now imagine in all states that the sampled genes $G_i$ and $G_j$ occupy the top row in some particular order and that $G_k$ and $G_l$ occupy the bottom row in some particular order. The probability of any detailed identity state is just a generalized kinship coefficient involving the four sampled genes $G_i$, $G_j$, $G_k$, and $G_l$. Under the usual correspondence between detailed and condensed states, adding the appropriate generalized kinship coefficients yields each $\Upsilon_r$.

TABLE 6.3. Conditional Expectations for Marker Sharing

| State $r$ | $E(1_{\{G_i=G_j\}} f(p_{G_i}) 1_{\{G_k=G_l\}} f(p_{G_k}) \mid S_r)$ |
|:---:|:---:|
| 1 | $\sum_m p_m f(p_m)^2$ |
| 2 | $\{\sum_m p_m f(p_m)\}^2$ |
| 3, 5, 7 | $\sum_m p_m^2 f(p_m)^2$ |
| 4, 6 | $\{\sum_m p_m^2 f(p_m)\}\{\sum_m p_m f(p_m)\}$ |
| 8 | $\sum_m p_m^3 f(p_m)^2$ |
| 9 | $\{\sum_m p_m^2 f(p_m)\}^2$ |

Given a collection of pedigrees, it is helpful to combine the marker-sharing statistics from the individual pedigrees into one grand statistic. The grand statistic should reflect the information content available in the individual pedigrees and should lead to easily approximated $p$-values. If $Z_m$ is the statistic corresponding to the $m$th pedigree, then these goals can be achieved by defining

$$T = \frac{\sum_m w_m [Z_m - E(Z_m)]}{\sqrt{\sum_m w_m^2 \operatorname{Var}(Z_m)}},$$

where $w_m$ is a positive weight assigned to pedigree $m$. Under the null hypothesis of independent segregation of the disease phenotype and the marker alleles, the grand statistic $T$ has mean 0 and variance 1. For a moderately large number of pedigrees, $T$ should be approximately normally distributed as well. In practice, $p$-values can be computed by simulation, and normality need not be taken for granted. A one-sided test is appropriate because excess marker sharing increases the observed value of $T$.

Choice of the weights is bound to be somewhat arbitrary. With $r_m$ typed affecteds in a pedigree, results of Hodge [6] suggest

$$w_m = \sqrt{\frac{r_m - 1}{\text{Var}(Z_m)}}.$$

This weighting scheme represents a compromise between giving all pedigrees equal weight ($w_m = 1/\sqrt{\text{Var}(Z_m)}$) and overweighting large pedigrees with many affecteds ($w_m = 1$). Overweighting is a potential problem because the number of affected pairs $r_m(r_m - 1)/2$ is a quadratic rather than a linear function of $r_m$.

Applications of the statistic $T$ to pedigree data on Huntington disease, rheumatoid arthritis, breast cancer, and Alzheimer disease are discussed in the references [5, 9, 12]. Extension of the statistic to multiple linked markers is undertaken in [13].

## 6.6 Problems

1. Let the disease allele at a recessive disease locus have population frequency $q$. If a child has inbreeding coefficient $f$, argue that his or her disease risk is $fq+(1-f)q^2$. What assumptions does this formula entail? Now suppose that a fraction $\alpha$ of all marriages in the surrounding population are between first cousins [1]. Show that the fraction of affecteds due to first-cousin marriages is

$$\frac{\alpha(\frac{1}{16}q + \frac{15}{16}q^2)}{\bar{f}q + (1 - \bar{f})q^2} = \frac{\alpha(1 + 15q)}{16[\bar{f} + (1 - \bar{f})q]},$$

where $\bar{f}$ is the average inbreeding coefficient of the population. Compute this fraction for $\alpha = .02, \bar{f} = .002$, and for $q = .01$ and $q = .001$. What conclusions do you draw from your results?

2. Consider a disease trait partially determined by an autosomal locus with two alleles 1 and 2 having frequencies $p_1$ and $p_2$. Let $\phi_{k/l}$ be the probability that a person with genotype $k/l$ manifests the disease. For the sake of simplicity, assume that people mate at random and that the disease states of two relatives $i$ and $j$ are independent given their genotypes at the disease locus. Now let $X_i$ and $X_j$ be indicator random variables that assume the value 1 when $i$ or $j$ is affected, respectively. Show that

$$\Pr(X_j = 1 \mid X_i = 1) = \sum_{g_i} \sum_{g_j} \sum_{S_r} \Pr(X_j = 1 \mid g_j) \Pr(g_j \mid S_r, g_i)$$
$$\times \Pr(S_r \mid g_i) \Pr(g_i \mid X_i = 1), \qquad (6.8)$$

where $g_i$ and $g_j$ are the possible genotypes of $i$ and $j$ and $S_r$ is a condensed identity state. This gives an alternative to computing risks by multiplying the relative risk ratio $\lambda_R$ by the prevalence $K$. Explicitly evaluate the risk (6.8) for identical twins and parent–offspring pairs.

3. Suppose that marker loci on different chromosomes are typed on two putative relatives. At locus $i$, let $p_{ij}$ be the likelihood of the observed pair of phenotypes conditional on the relatives being in condensed identity state $S_j$. In the absence of inbreeding, only the states $S_7$, $S_8$, and $S_9$ are possible. If we want to estimate the true relationship between the pair, then we can write the likelihood of the observations as

$$L(\Delta) \;=\; \prod_i (\Delta_7 p_{i7} + \Delta_8 p_{i8} + \Delta_9 p_{i9})$$

and attempt to estimate the $\Delta$'s [11]. Describe an EM algorithm to find the maximum likelihood estimates. The value of $L(\Delta)$ can be compared under the maximum likelihood estimates and under choices for the $\Delta$'s characterizing typical relative pairs such as parent–offspring, siblings, first cousins, and so forth. Discuss the merits and demerits of this strategy. For one objection, see Problem 2 of Chapter 5.

4. Suppose that the two relatives $i$ and $j$ are inbred. Show that the covariance between their trait values $X_i$ and $X_j$ is

$$\begin{aligned}
\mathrm{Cov}(X_i, X_j) \;=\;& (4\Delta_1 + 2\Delta_3 + 2\Delta_5 + 2\Delta_7 + \Delta_8) \sum_k \alpha_k^2 p_k \\
&+ (4\Delta_1 + \Delta_3 + \Delta_5) \sum_k \alpha_k \delta_{kk} p_k \\
&+ \Delta_1 \sum_k \delta_{kk}^2 p_k + \Delta_7 \sum_k \sum_l \delta_{kl}^2 p_k p_l \\
&+ (\Delta_2 - f_i f_j) \left( \sum_k \delta_{kk} p_k \right)^2 .
\end{aligned}$$

What is $\mathrm{Cov}(X_i, X_j)$ when $\sigma_d^2 = 0$?

5. For a locus with two alleles, show that the additive genetic variance satisfies

$$\begin{aligned}
\sigma_a^2 \;=\;& 2p_1 p_2 (\alpha_1 - \alpha_2)^2 \\
=\;& 2p_1 p_2 [p_1(\mu_{11} - \mu_{12}) + p_2(\mu_{12} - \mu_{22})]^2 . \tag{6.9}
\end{aligned}$$

As a consequence of formula (6.9), $\sigma_a^2$ can be 0 only in the unlikely circumstance that $\mu_{12}$ lies outside the interval with endpoints $\mu_{11}$ and $\mu_{22}$. (Hint: Expand $0 = 2(\alpha_1 p_1 + \alpha_2 p_2)^2$ and subtract from the expression defining $\sigma_a^2$.)

Show that the dominance genetic variance satisfies

$$\sigma_d^2 \;=\; p_1^2 p_2^2 (\mu_{11} - 2\mu_{12} + \mu_{22})^2 .$$

It follows that if either $p_1$ or $p_2$ is small, then $\sigma_d^2$ will tend to be small compared to $\sigma_a^2$. Hint: Let $\overline{\mu} = p_1^2 \mu_{11} + 2p_1 p_2 \mu_{12} + p_2^2 \mu_{22}$. Since $\overline{\mu} = 0$, it follows that

$$
\begin{aligned}
\delta_{11} &= \mu_{11} - 2\alpha_1 + \overline{\mu} \\
&= p_2^2(\mu_{11} - 2\mu_{12} + \mu_{22}) \\
\delta_{12} &= -p_1 p_2(\mu_{11} - 2\mu_{12} + \mu_{22}) \\
\delta_{22} &= p_1^2(\mu_{11} - 2\mu_{12} + \mu_{22}).
\end{aligned}
$$

6. Prove that any pair of nonnegative numbers $(\sigma_a^2, \sigma_d^2)$ can be realized as additive and dominance genetic variances. The special pairs $(\frac{1}{2}, 0)$ and $(0, 1)$ show that the two matrices $\Phi = (\Phi_{ij})$ and $\Delta_7 = (\Delta_{7ij})$ defined for an arbitrary non-inbred pedigree are legitimate covariance matrices. (Hint: Based on the previous problem,

$$
\begin{aligned}
\sigma_a^2 &= 2p_1 p_2(p_1 u + p_2 v)^2 \\
\sigma_d^2 &= p_1^2 p_2^2(u - v)^2
\end{aligned}
$$

for $u = \mu_{11} - \mu_{12}$ and $v = \mu_{12} - \mu_{22}$. Solve for $u$ and $v$.)

7. Show that the matrices $\Phi$ and $\Delta_7$ of coefficients assigned to a pedigree do not necessarily commute. It is therefore pointless to attempt a simultaneous diagonalization of these two matrices. (Hint: Consider a nuclear family consisting of a mother, father, and two siblings.)

8. Let $(X_1, \ldots, X_n)$ and $(Y_1, \ldots, Y_n)$ be measured values for two different traits on a pedigree of $n$ people. Suppose that both traits are determined by the same locus. Show that there exist constants $\sigma_{axy}$ and $\sigma_{dxy}$ such that

$$
\text{Cov}(X_i, Y_j) = 2\Phi_{ij}\sigma_{axy} + \Delta_{7ij}\sigma_{dxy}
$$

for any two non-inbred relatives $i$ and $j$ [8]. Prove that the two matrices

$$
\begin{pmatrix} \sigma_{axx}^2 & \sigma_{axy} \\ \sigma_{axy} & \sigma_{ayy}^2 \end{pmatrix}
\qquad
\begin{pmatrix} \sigma_{dxx}^2 & \sigma_{dxy} \\ \sigma_{dxy} & \sigma_{dyy}^2 \end{pmatrix}
$$

are covariance matrices, where $\sigma_{axx}^2$, $\sigma_{dxx}^2$, $\sigma_{ayy}^2$, and $\sigma_{dyy}^2$ are the additive and dominance genetic variances of the $X$ and $Y$ traits, respectively. (Hints: For the first part, consider the artificial trait $W = X + Y$ for a typical person. For the second part, prove that

$$
\begin{aligned}
\sigma_{axy} &= 2\,\text{Cov}(A_1, B_1) \\
\sigma_{dxy} &= \text{Cov}(X - A_1 - A_2, Y - B_1 - B_2),
\end{aligned}
$$

where $A_k = E(X \mid Z_k)$ and $B_k = E(Y \mid Z_k)$, $Z_1$ and $Z_2$ being the maternal and paternal alleles at the common locus.)

9. In the two-locus heterogeneity model with $X = Y + Z - YZ$, carry through the computations retaining the product term $YZ$. In particular, let $K_m$ be the prevalence of the $m$th form of the disease, and let $K_{mR}$ be the recurrence risk for a relative of type $R$ under the $m$th form. If $K$ is the prevalence and $K_R$ is the recurrence risk to a relative of type $R$ under either form of the disease, then show that

$$K = K_1 + K_2 - K_1 K_2$$
$$K K_R = K_1 K_{1R} + K_1 K_2 - K_1 K_{1R} K_2 + K_1 K_2 + K_2 K_{2R}$$
$$- K_1 K_2 K_{2R} - K_1 K_{1R} K_2 - K_1 K_2 K_{2R} + K_1 K_{1R} K_2 K_{2R}.$$

Assuming that $K_1$, $K_2$, $K_{1R}$, and $K_{2R}$ are relatively small, verify the approximation

$$
\begin{aligned}
\lambda_R - 1 \\
= \left(\frac{K_1}{K}\right)^2 (\lambda_{1R} - 1) + \left(\frac{K_2}{K}\right)^2 (\lambda_{2R} - 1) \\
+ \frac{K_1 K_2}{K^2}[2K_1 + 2K_2 - K_1 K_2 - 2K_{1R} - 2K_{2R} + K_{1R} K_{2R}] \\
\approx \left(\frac{K_1}{K}\right)^2 (\lambda_{1R} - 1) + \left(\frac{K_2}{K}\right)^2 (\lambda_{2R} - 1),
\end{aligned}
$$

where $\lambda_{mR} = K_{mR}/K_m$ and $\lambda_R = K_R/K$.

10. In the pedigree depicted in Figure 6.1, compute the marker-sharing statistic $Z$ and its expectation $E(Z)$ for the three phenotyped affecteds 3, 4, and 6. Assume $f(p) = 1/p$, $p_a = 1/2$, $p_b = 1/4$, and for a third unobserved marker allele $p_c = 1/4$.

# References

[1] Crow JF, Kimura M (1970) *An Introduction to Population Genetics Theory*. Harper and Row, New York

[2] Elston RC, Lange K (1976) The genotypic distribution of relatives of homozygotes when consanguinity is present. *Ann Hum Genet* 39:493–496

[3] Fisher RA (1918) The correlation between relatives on the supposition of Mendelian inheritance. *Trans Roy Soc Edinb* 52:399–433

[4] Grimmett GR, Stirzaker DR (1992) *Probability and Stochastic Processes*, 2nd ed. Oxford University Press, Oxford

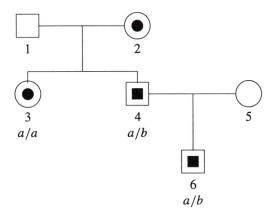

FIGURE 6.1. A Pedigree Illustrating Marker Sharing Among Affecteds

[5] Hall JM, Lee MK, Newman B, Morrow JE, Anderson LA, Huey B, King M-C (1990) Linkage of early-onset familial breast cancer to chromosome 17q21. *Science* 250:1684–1689

[6] Hodge SE (1984) The information contained in multiple sibling pairs. *Genet Epidemiology* 1:109–122

[7] Jacquard A (1974) *The Genetic Structure of Populations.* Springer-Verlag, New York

[8] Lange K, Boehnke M (1983) Extensions to pedigree analysis. IV. Covariance components models for multivariate traits. *Amer J Med Genet* 14:513–524

[9] Pericak-Vance MA, Bebout JL, Gaskell PC Jr, Yamaoka LH, Hung W-Y, Alberts MJ, Walker AP, Barlett RJ, Haynes CA, Welsh KA, Earl NL, Heyman A, Clark CM, Roses AD (1991) Linkage studies in familial Alzheimer disease: evidence for chromosome 19 linkage. *Amer J Hum Genet* 48:1034–1050

[10] Risch N (1990) Linkage strategies for genetically complex traits. I. Multilocus models. *Amer J Hum Genet* 46:22–228

[11] Thompson EA (1986) *Pedigree Analysis in Human Genetics.* Johns Hopkins University Press, Baltimore

[12] Weeks DE, Lange K (1988) The affected-pedigree-member method of linkage analysis. *Amer J Hum Genet* 42:315–326

[13] Weeks DE, Lange K (1992) A multilocus extension of the affected-pedigree-member method of linkage analysis. *Amer J Hum Genet* 50:859–868

# 7
# Computation of Mendelian Likelihoods

## 7.1 Introduction

Rigorous analysis of human pedigree data is a vital concern in genetic epidemiology, human gene mapping, and genetic counseling. In this chapter we investigate efficient algorithms for likelihood computation on pedigree data, placing particular stress on the pioneering algorithm of Elston and Stewart [5]. It is no accident that their research coincided with the introduction of modern computing. To analyze human pedigree data is tedious, if not impossible, without computers. Pedigrees lack symmetry, and all simple closed-form solutions in mathematics depend on symmetry. The achievement of Elston and Stewart [5] was to recognize that closed-form solutions are less relevant than good algorithms. However, the Elston-Stewart algorithm is not the end of the story. Evaluation of pedigree likelihoods remains a subject sorely in need of further theoretical improvement. Linkage calculations alone are among the most demanding computational tasks in modern biology.

## 7.2 Mendelian Models

Besides the raw materials of pedigree structure and observed phenotypes, a genetic model is a prerequisite for likelihood calculation. At its most elementary level, a model postulates the number of loci necessary to explain the phenotypes. Mendelian models, as opposed to polygenic models, involve only a finite number of loci. For purposes of discussion, it is convenient to use the term "genotype" when discussing the multilocus, ordered genotypes of an underlying model. Because ordered genotypes preserve phase, they are preferable to unordered genotypes for theoretical and computational purposes. Of course, observed genotypes are always unordered.

Any Mendelian model revolves around the three crucial notions of **priors, penetrances**, and **transmission probabilities** [5]. Prior probabilities pertain only to founders. If $G$ is a possible genotype for a founder, then in the absence of other knowledge, Prior($G$) is the probability that the founder carries genotype $G$. Almost all models postulate that prior probabilities conform to Hardy-Weinberg and linkage equilibrium.

Penetrance functions specify the likelihood of an observed phenotype $X$ given an unobserved genotype $G$. We denote a penetrance by Pen($X \mid G$). Penetrances apply to all people in a pedigree, founders and nonfounders alike. Implicit in the notion of penetrance is the assumption that the phenotypes of two or more people are independent given their genotypes. This restriction rules out complex models in which common environment influences phenotypes. It is easier to incorporate environmental effects in polygenic models. In Mendelian models, likelihood evaluation involves combinatorics; in polygenic models, it involves linear algebra.

In general, Pen($X \mid G$) can represent a conditional likelihood as well as a conditional probability. This would be the case, for instance, with a quantitative trait $X$ following a different Gaussian density for each genotype. For many genetic traits, Pen($X \mid G$) is either 0 or 1; in other words, each genotype leads to one and only one phenotype. When a phenotype is unobserved, it is natural to assume that the penetrance function is identically 1.

The third and last component probability of a likelihood summarizes the genetic transmission of the trait or traits observed. Let Tran($G_k \mid G_i, G_j$) denote the probability that a mother $i$ with genotype $G_i$ and a father $j$ with genotype $G_j$ produce a child $k$ with genotype $G_k$. For ordered genotypes, the child's genotype $G_k$ can be visualized as an ordered pair of gametes $(U_k, V_k)$, $U_k$ being maternal in origin and $V_k$ being paternal in origin. If all participating loci reside on the same chromosome, then $U_k$ and $V_k$ are haplotypes. Because any two parents create gametes independently, the transmission probability

$$\text{Tran}(G_k \mid G_i, G_j) = \text{Tran}(U_k \mid G_i)\text{Tran}(V_k \mid G_j)$$

factors into two **gamete transmission probabilities**. Unordered genotypes do not obey this gamete factorization rule.

Specification of gamete transmission probabilities is straightforward for single-locus models. For a single autosomal locus, Tran($H \mid G$) is either 1, $\frac{1}{2}$, or 0, depending on whether the single allele $H$ is identical in state to both, one, or neither of the two alleles of the parental genotype $G$, respectively. For multiple linked loci, Haldane's model [7] permits easy computation of gamete transmission probabilities, provided one is willing to neglect the phenomenon of chiasma interference. For the sake of computational simplicity, we now adopt Haldane's model, which postulates that recombination occurs independently on disjoint intervals.

To apply Haldane's model, one begins by discarding all homozygous loci in the parent. This entails no loss of information because recombination events can never be inferred between such loci. Between each remaining adjacent pair of heterozygous loci, gametes can be scored as recombinant or nonrecombinant. Once adjacent intervals have been consolidated to the point where all interval endpoints are marked by heterozygous loci, calculation of gamete transmission probabilities is straightforward. Invoking independence, the probability of a gamete is now $\frac{1}{2}$ times the product over all consolidated intervals of the corresponding recombination fractions $\theta$ or of their complements $1 - \theta$, depending on whether the gamete shows recombination on a given interval or not. The factor of $\frac{1}{2}$ accounts for the parental chromosome chosen for the first locus. In the exceptional case where there are no heterozygous loci, the gamete transmission probability is 1. If there is only one heterozygous locus, the gamete transmission probability is $\frac{1}{2}$. Recombination fractions for consolidated intervals can be computed via Trow's formula as described in Problem 1.

The likelihood $L$ of a pedigree with $n$ people can now be assembled from these component parts. Let the $i$th person have phenotype $X_i$ and possible genotype $G_i$. Conditioning on the genotypes of each of the $n$ people yields Ott's [21] representation of the likelihood

$$
\begin{aligned}
L &= \sum_{G_1} \cdots \sum_{G_n} \Pr(X_1, \ldots, X_n \mid G_1, \ldots, G_n) \Pr(G_1, \ldots, G_n) \\
&= \sum_{G_1} \cdots \sum_{G_n} \prod_i \mathrm{Pen}(X_i \mid G_i) \Pr(G_1, \ldots, G_n) \qquad (7.1) \\
&= \sum_{G_1} \cdots \sum_{G_n} \prod_i \mathrm{Pen}(X_i \mid G_i) \prod_j \mathrm{Prior}(G_j) \prod_{\{k,l,m\}} \mathrm{Tran}(G_m \mid G_k, G_l),
\end{aligned}
$$

where the product on $j$ is taken over all founders and the product on $\{k, l, m\}$ is taken over all parent–offspring triples.

Several comments are appropriate at this point concerning the explicit likelihood representation (7.1). First, ranges of summation for the genotypes are not specified. At the very least it is profitable to eliminate any genotype $G_i$ with $\mathrm{Pen}(X_i \mid G_i) = 0$. We will discuss later an algorithm for genotype elimination that performs much better than this naive tactic in most circumstances. Second, the notation in (7.1) does not make it clear whether the likelihood $L$ should be computed as a joint sum or as an iterated sum. One can argue rigorously that an iterated sum is always preferable to a joint sum if minimizing counts of additions and multiplications is taken as a criterion [12]. Viewing (7.1) as an iterated sum opens up the possibility of rearranging the order of summation so as to achieve the most efficient computation. Third, calculation of $L$ is numerically stable since only additions and multiplications of nonnegative numbers are involved. There will be no disastrous roundoff errors due to subtraction of quantities of similar magnitude. However, serious underflows can be encountered because all terms are usually probabilities and hence lie in the interval [0, 1]. Underflows can be successfully defused by repeated rescaling and reporting the final answer as a loglikelihood. Last of all, the various terms in (7.1) can be viewed as values taken on by arrays. For instance,

Pen$(X_i \mid G_i)$ is an array of rank 1 that depends on the possible genotypes $G_i$ for $i$. Similarly, Tran$(G_k \mid G_i, G_j)$ is an array of rank 3 depending on $G_i$, $G_j$, and $G_k$ jointly. Thus, computation of $L$ is inherently array-oriented.

## 7.3    Genotype Elimination and Allele Consolidation

As hinted above, systematic genotype elimination is a powerful technique for accelerating likelihood evaluation. This preprocessing step involves more than just using those genotypes compatible with a person's observed phenotype. The phenotypes of the person's relatives also impose rigid compatibility constraints on his or her possible genotypes. Although this fact has long been known informally, it is helpful to state a formal algorithm that mimics how geneticists reason. In doing so, we will focus on ordered genotypes at a single autosomal locus. Because ordered genotypes carry phase information, applying the algorithm separately to several linked loci automatically eliminates superfluous phases among the loci as well as superfluous genotypes within each locus.

Here then is the algorithm [12, 13]:

**(A)** For each pedigree member, list only those ordered genotypes compatible with his or her phenotype.

**(B)** For each nuclear family:

    **(1)** Consider each mother–father genotype pair.

        **(a)** Determine which zygotes can arise from the genotype pair.

        **(b)** If each child in the nuclear family has one or more of these zygote genotypes among his or her current list of genotypes, then save the parental genotypes and any child genotype matching one of the created zygote genotypes.

        **(c)** If any child has none of these zygote genotypes among his or her current list of genotypes—in other words, is incompatible with the current parental pair of genotypes—then take no action to save any genotypes.

    **(2)** For each person in the nuclear family, exclude any genotypes not saved during step (1) above.

**(C)** Repeat part (B) until no more genotypes can be excluded.

As an illustration of the algorithm, consider the pedigree of Figure 7.1 at the ABO locus, and suppose individuals 2, 3, and 5 alone are typed at this locus. Then applying part (A) of the algorithm leads to the genotype sets displayed in column 2 of Table 7.1. Applying (B) to the nuclear family {3, 4, 5} gives column 3, and finally, applying (B) to the nuclear family {1, 2, 3} gives column 4. Recall that the maternal allele is listed to the left of the paternal allele in an ordered genotype.

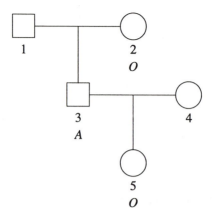

FIGURE 7.1. A Pedigree Partially Typed at the ABO Locus

This convention shows up in the genotype set for person 3. No further genotypes can be eliminated by repeated use of (B), and column 4 provides the minimal genotype sets. In more extensive pedigrees, genotype eliminations can ripple up and down through the pedigree, requiring multiple visits to each nuclear family. For pedigrees that are graphically trees, that is, have no loops or cycles, the algorithm is guaranteed to eliminate all superfluous genotypes for each person [13].

TABLE 7.1. Genotype Sets for a Genotype Elimination Example

| Person | After Applying (A) | After Applying (B) to {3, 4, 5} | After Applying (B) to {1, 2, 3} |
|--------|--------------------|----------------------------------|----------------------------------|
| 1 | All 9 genotypes | All 9 genotypes | $\{A/A, A/O, O/A, A/B, B/A\}$ |
| 2 | $\{O/O\}$ | $\{O/O\}$ | $\{O/O\}$ |
| 3 | $\{A/A, A/O, O/A\}$ | $\{A/O, O/A\}$ | $\{O/A\}$ |
| 4 | All 9 genotypes | $\{A/O, O/A, B/O, O/B, O/O\}$ | $\{A/O, O/A, B/O, O/B, O/O\}$ |
| 5 | $\{O/O\}$ | $\{O/O\}$ | $\{O/O\}$ |

Allele consolidation is another tactic that reduces ranges of summation. At a highly polymorphic, codominant marker locus, most pedigrees will segregate only a subset of the possible alleles. If a person $i$ is untyped, then the range of summation for his or her genotypes may involve genotypes composed of alleles not actually seen within the pedigree. These unseen genotypes can be consolidated by consolidating unseen alleles. In most applications, this action will not change the likelihood of the pedigree, provided the lumped allele is assigned the appropriate summed population frequency. For instance, suppose only the first three alleles from the set $\{A_i : 1 \leq i \leq 6\}$ of six possible alleles are seen among the typed people of a pedigree. Then one can consolidate alleles $A_4$, $A_5$, and $A_6$ into a single artificial allele $A_7$ with frequency $p_7 = p_4 + p_5 + p_6$ in obvious notation.

If allele consolidation is carried out on a pedigree-by-pedigree basis, then substantial computational savings can be realized. Even more dramatic savings can occur when allele consolidation is carried out locally within a pedigree. O'Connell and Weeks' [20] explanation of local consolidation is well worth reading but a little too lengthy to recite here. Finally, note that there are some problems such as allele frequency estimation from pedigree data [2] where allele consolidation is disastrous. A little common sense should be an adequate safeguard against these abuses.

## 7.4    Array Transformations and Iterated Sums

To elaborate on some of the comments made earlier about iterated sums and arrays, we now strip away the genetics overlay and concentrate on issues of numerical analysis. As an example [12], consider the problem of computing the sum of products

$$\sum_{G_1 \in S_1} \sum_{G_2 \in S_2} \sum_{G_3 \in S_3} A(G_1, G_2)B(G_2)C(G_2, G_3), \tag{7.2}$$

where $S_i$ is the finite range of summation for the index $G_i$, and where $A$, $B$, and $C$ are arrays of real numbers. Let $S_i$ have $m_i$ elements. Computing (7.2) as a joint sum requires $2m_1m_2m_3$ multiplications and $m_1m_2m_3 - 1$ additions. If we compute (7.2) as an iterated sum in the sequence $(3, 2, 1)$ specified, we first compute an array

$$D(G_2) \;=\; \sum_{G_3 \in S_3} C(G_2, G_3)$$

in $m_2(m_3 - 1)$ additions. Note that the arrays $A$ and $B$ do not depend on the index $G_3$ so it is uneconomical to involve them in the sum on $G_3$. Next we compute an array

$$E(G_1) \;=\; \sum_{G_2 \in S_2} A(G_1, G_2)B(G_2)D(G_2) \tag{7.3}$$

in $2m_1m_2$ multiplications and $m_1(m_2 - 1)$ additions. Last of all, we compute the sum $\sum_{G_1 \in S_1} E(G_1)$ in $m_1 - 1$ additions. The total arithmetic operations needed for the joint sum is $3m_1m_2m_3 - 1$; for the iterated sum the total is $m_2(3m_1 + m_3 - 1) - 1$. It is clear that the iterated sum requires the same number of operations when $m_3 = 1$

and strictly fewer operations when $m_3 > 1$. If we take the alternative order $(3, 1, 2)$, the total operations are $m_2(m_1 + m_3 + 1) - 1$. Thus, the order $(3, 1, 2)$ is even better than the original order $(3, 2, 1)$.

Note that when (7.2) is computed as an iterated sum, arrays are constantly being created and discarded. At each summation, those arrays depending on the current summation index are multiplied together, and the resulting product array is summed on this index. This process leads to a new array no longer depending on the eliminated index, and those arrays participating in the formation of the new array can now be discarded. Each summation therefore transforms the original computational problem into a problem of the same sort, except that the number of indices is reduced by one. Eventually, all indices are eliminated, and the original problem is solved.

Finding an optimal or nearly optimal summation sequence is highly nontrivial. In genetics problems, one can attempt to generate such sequences by working from the periphery of the pedigree inward. Such pruning of the pedigree succeeds for graphically simple pedigrees. However, in the presence of inbreeding, cycles or loops in the graphical structure of the pedigree impede this approach. Furthermore, a purely graphical treatment ignores the important differences in the number of genotypes per person. A detailed analysis of this problem is carried out in [6].

Greedy algorithms provide useful heuristics for choosing a nearly optimal summation sequence. For instance, we can always sum on that index requiring the fewest current arithmetic operations to eliminate. Thus, in our toy example, we would start with index 1 or 3 depending on whether $m_1 < m_3$ or $m_1 > m_3$. A tie $m_1 = m_3$ is broken arbitrarily. This greedy heuristic is not always optimal, as Problem 3 indicates.

Another context where greedy algorithms arise naturally is in the formation of array products. Consider, for instance, equation (7.3). If we first multiply array $B$ times array $D$ to get

$$F(G_2) = B(G_2)D(G_2),$$

and then multiply and sum to form

$$E(G_1) = \sum_{G_2 \in S_2} A(G_1, G_2)F(G_2),$$

we save $m_1 m_2 - m_2$ multiplications. This example illustrates that arrays should always be multiplied pairwise until only two arrays involving the current summation index survive. These last two arrays can then be multiplied and afterwards summed, or they can be simultaneously multiplied and summed. The latter method generalizes matrix multiplication; it entails the same amount of arithmetic but requires less storage than the former method. In forming pairwise products of intermediate arrays, we face the question of what two arrays to multiply at any given step. In equation (7.3) the answer is obvious. In more complex examples, we can resort to a greedy approach; namely, at each stage we always pick the two arrays that cost the least to multiply. Ties at any stage are broken by arbitrarily choosing one of the best pairs of arrays.

## 7.5   Array Factoring

The calculation of pedigree likelihoods involving many linked markers has raised interesting challenges. Even with complete phenotyping of all pedigree members, phase ambiguities pose a problem. Lathrop et al. [16] show that for many fully typed nuclear families (with or without grandparents appended), the likelihood factors into a product of likelihoods involving subsets of the loci. These multiplicand likelihoods can be quickly evaluated. Lander and Green [11] take a different approach. They redefine the likelihood expression (7.1) so that the sums extend over loci rather than people. In other words, their algorithm steps through the likelihood calculation locus by locus while considering all people simultaneously at each locus. This tactic has the consequence of radically displacing the source of computational complexity. Instead of scaling exponentially in the number of loci, their algorithm scales linearly. However, since all pedigree members are taken simultaneously, it scales exponentially in the number of pedigree members. Although the clever speedups proposed by Kruglyak et al. [9, 10] help, very large pedigrees are simply beyond the reach of the Lander and Green algorithm.

A synthesis of these two methods is possible [6]. On one hand, the factorization method of Lathrop et al. [16] ultimately depends on being able to factor the prior, penetrance, and transmission arrays. On the other hand, the method of Lander and Green [11] shifts summations from people to loci. It is possible to decompose on both people and loci in such a manner that the prior, penetrance, and transmission arrays factor. This suggestion entails viewing the multilocus ordered genotypes of a given person as originating from a Cartesian product of his or her single-locus ordered genotypes. A negative consequence of this synthesis is the substitution of a swarm of small arrays where a few large ones formerly sufficed. In compensation for this complication is the potential benefit of encountering much smaller initial and intermediate arrays in the likelihood calculation.

To elaborate on this synthesis, consider again a typical person $i$ in a pedigree with $n$ members. Suppose that $i$'s phenotype $X_i$ is determined by $m$ loci $1, \ldots, m$ taken in their natural order along a chromosome. A multilocus ordered genotype $G_i$ of $i$ decomposes into an ordered sequence $G_i = (G_{i1}, \ldots, G_{im})$ of single-locus ordered genotypes $G_{ij}$. Under Hardy-Weinberg and linkage equilibrium, the prior $\text{Prior}(G_i)$ factors as

$$\text{Prior}(G_i) \quad = \quad \prod_{j=1}^{m} \text{Prior}(G_{ij}). \tag{7.4}$$

If $i$'s phenotype $X_i$ also decomposes into separate observations $X_{ij}$ at each locus, then most penetrance functions exhibit the factorization

$$\text{Pen}(X_i \mid G_i) \quad = \quad \prod_{j=1}^{m} \text{Pen}(X_{ij} \mid G_{ij}). \tag{7.5}$$

Failures of assumption (7.5) are rare in linkage studies and represent **epistasis** among loci. Both equations (7.4) and (7.5) are forms of probabilistic independence.

Factorization of transmission arrays is more subtle. According to Haldane's model, a gamete transmission probability $\mathrm{Tran}(H_k \mid G_i)$ factors into terms encompassing blocks of loci, with each block delimited by two heterozygous loci in the parent $i$. For example, suppose $r$ and $s$, $1 < r < s < m$, are the only heterozygous loci in the parental genotype $G_i$. Then the transmission probability for the haplotype $H_k$ factors as

$$
\begin{aligned}
\mathrm{Tran}(H_k \mid G_i) \;=\; & \mathrm{Tran}[(H_{k1}, \ldots, H_{kr}) \mid (G_{i1}, \ldots, G_{ir})] \\
& \times \mathrm{Tran}[(H_{k,r+1}, \ldots, H_{ks}) \mid (G_{ir}, \ldots, G_{is}), H_{kr}] \\
& \times \mathrm{Tran}[(H_{k,s+1}, \ldots, H_{km}) \mid (G_{is}, \ldots, G_{im}), H_{ks}],
\end{aligned}
$$

where the block $(r, \ldots, s)$ spans the only interval on which recombination can be counted. Traversing the haplotype from locus 1 to locus $m$, a factor of $\frac{1}{2}$ accounts for which parental allele is encountered at the first heterozygous locus $r$. Thus,

$$
\mathrm{Tran}[(H_{k1}, \ldots, H_{kr}) \mid (G_{i1}, \ldots, G_{ir})] \;=\; \frac{1}{2}.
$$

Recombination or nonrecombination between loci $r$ and $s$ is summarized by

$$
\begin{aligned}
& \mathrm{Tran}[(H_{k,r+1}, \ldots, H_{ks}) \mid (G_{ir}, \ldots, G_{is}), H_{kr}] \\
& = \begin{cases} \theta_{rs} & \text{for recombination on interval } [r, s] \\ 1 - \theta_{rs} & \text{for nonrecombination on interval } [r, s], \end{cases}
\end{aligned}
$$

where $\theta_{rs}$ is the recombination fraction between loci $r$ and $s$. Finally, because recombination cannot be scored between loci $s$ and $m$,

$$
\mathrm{Tran}[(H_{k,s+1}, \ldots, H_{km}) \mid (G_{is}, \ldots, G_{im}), H_{ks}] \;=\; 1.
$$

For transmission array factorization to be useful, it must take the same form for all possible multilocus genotypes $G_i$. Clearly, a transmission array can be uniformly factored into two terms involving a given locus and loci to the right and left of it, respectively, only if the contributing parent is an obligate heterozygote at the locus. A combination of inspection and genotype elimination quickly identifies all obligate heterozygous loci in parents.

In this reformulated model, the summations in the likelihood representation (7.1) are replaced by analogous summations over person–locus combinations $G_{ij}$. Although this substitution increases the complexity of finding a good summation sequence, most other features of likelihood evaluation remain unchanged. For instance, genotype elimination is already carried out one locus at a time. Array creation and annihilation are handled similarly in both likelihood formulations, except that more numerous but smaller arrays are encountered in the person–locus mode of calculation.

## 7.6    Examples of Pedigree Analysis

**Example 7.1** *Paternity Testing*

Paternity testing confirms or eliminates a putative father as the actual father of a child. Phenotyping of the mother, child, and putative father is done at a number of different marker loci. If a genetic inconsistency is found, then the putative father is absolved. On the other hand, if the trio is consistent at all loci typed, then either a rare event has occurred or the putative father is the actual father. There are two ways of quantifying the rarity of this event. The Bayesian approach is to compute a likelihood ratio of the trio with the putative father as real father versus the trio with the real father as a random male. This likelihood ratio or **paternity index** can be transformed into a posterior probability if a prior probability of paternity is supplied.

A strictly frequentist approach to the problem is to compute the probability that a random male would be excluded by at least one of the tests based on the phenotypes of the mother and child. This **exclusion probability** relieves a judge or jury from the necessity of quantifying their prior probabilities of paternity. Both posterior and exclusion probabilities can be computed for each locus separately and then cumulated over all loci jointly. The locus-by-locus statistics are useful in determining which loci are critically important in confirming paternity. The cumulative statistics are the ones quoted in court.

To compute the paternity index, imagine two pedigrees. The first pedigree, $\text{Ped}_1$, contains the mother and child and the putative father as actual father. The second pedigree, $\text{Ped}_2$, substitutes a random male with all phenotypes unknown for the actual father. The putative father is present as an isolated individual unrelated to the child in $\text{Ped}_2$. Suppose the vector $X^j$ denotes the observed phenotypes for the trio of mother, child, and putative father at the $j$th locus of a set of marker loci in Hardy-Weinberg and linkage equilibrium. The paternity index for the $j$th locus is $\Pr(X^j \mid \text{Ped}_1)/\Pr(X^j \mid \text{Ped}_2)$. Over all loci it is

$$\frac{\prod_j \Pr(X^j \mid \text{Ped}_1)}{\prod_j \Pr(X^j \mid \text{Ped}_2)}. \tag{7.6}$$

Let $\alpha$ be the prior probability that the putative father is the actual father based on the nongenetic evidence; let $\beta$ be the posterior probability that the putative father is the actual father based on both the nongenetic and the genetic evidence. Then a convenient form of Bayes' theorem is

$$\frac{\beta}{1-\beta} = \frac{\alpha \prod_j \Pr(X^j \mid \text{Ped}_1)}{(1-\alpha) \prod_j \Pr(X^j \mid \text{Ped}_2)}.$$

The exclusion probability for the $j$th locus can be found by carrying out the genotype elimination algorithm on $\text{Ped}_2$ for this locus. Let $S_j$ be the set of non-excluded genotypes for the random male. Then the exclusion probability for locus

TABLE 7.2. Phenotypes for a Paternity-Testing Problem

| Person | ABO Phenotype | ADA Phenotype |
|--------|---------------|---------------|
| Mother | $AB$ | 1/1 |
| Child | $B$ | 1/2 |
| Putative Father | $B$ | 1/2 |

$j$ is $1 - \sum_{G_j \in S_j} \Pr(G_j)$. The exclusion probability over all loci typed is clearly $1 - \prod_j [\sum_{G_j \in S_j} \Pr(G_j)]$.

As a simple numerical example, consider the phenotype data in Table 7.2. At the ABO locus, suppose the three alleles $A$, $B$, and $O$ have population frequencies of .28, .06, and .66, respectively. At the ADA locus, suppose the two codominant alleles 1 and 2 have population frequencies of .934 and .066, respectively. It is evident in this case that the only excluded genotype for the father at the ABO locus is $A/A$; at the ADA locus, the only excluded genotype is 1/1. Table 7.3 lists the computed paternity indices and exclusion probabilities. Although the ADA locus is less polymorphic than the ABO locus, in this situation it yields the more decisive statistics. In practice, a larger number of individually more polymorphic loci would be used.  ■

TABLE 7.3. Statistics for the Paternity-Testing Example

| Locus | Paternity Index | Exclusion Probability |
|-------|-----------------|------------------------|
| ABO | 1.39 | .078 |
| ADA | 7.58 | .872 |
| Both Loci | 10.5 | .882 |

**Example 7.2** *Multiple Allele Segregation Analysis*

When a new locus is investigated, one of the first statistical tasks is to check whether its proposed alleles and genotypes conform to Mendelian segregation ratios. The typical way of doing this is to subdivide all available nuclear families into different mating types. For any given pair of parental genotypes, there are from one to four possible offspring genotypes. The numbers of offspring observed in the various genotypic categories are used to compute an approximate $\chi^2$ statistic. These approximate $\chi^2$ statistics are then added over the various mating types to give a grand $\chi^2$. This classical procedure suffers from the fact that the component $\chi^2$ statistics often lack adequate numbers for large sample theory to apply. If the alleles at the proposed locus involve dominance, then defining mating types is also problematic.

An alternative procedure is to assign each allele $A_i$ a segregation parameter $\tau_i$. These parameters can be estimated by maximum likelihood from the available pedigree data. The parameters enter the likelihood calculations at the level of gamete transmission probabilities. For two different alleles $A_i$ and $A_j$, take $\Pr(A_i \mid A_i/A_j) = \tau_i/(\tau_i + \tau_j)$. For a homozygous parental genotype $A_i/A_i$, take $\Pr(A_i \mid A_i/A_i) = 1$. Under the hypothesis of Mendelian segregation, all the $\tau$'s are equal. Because multiplying the $\tau$'s by the same constant preserves segregation ratios, one should arbitrarily constrain one $\tau_i = 1$ and test by a likelihood ratio statistic whether all other $\tau_j = 1$.

As a simple numerical example, consider the four alleles $1+$, $1-$, $2+$, and $2-$ of the PGM1 marker locus on chromosome 1. The PGM1 data of Lewis et al. [17] lists 93 people in 5 pedigrees. For these data, the maximum likelihood estimates are $\hat{\tau}_{1+} = 1$, $\hat{\tau}_{1-} = .79$, $\hat{\tau}_{2+} = .84$, and $\hat{\tau}_{2-} = 1.26$. The likelihood ratio statistic is approximately distributed as a $\chi^2$ with three degrees of freedom. The observed value of this statistic,

$$2 \times [(-114.70) - (-115.64)] \; = \; 1.88,$$

suggests that the alleles of the PGM1 locus do conform to Mendelian inheritance. Note that this analysis safely ignores the issue of ascertainment.    ∎

**Example 7.3** *Risk Prediction*

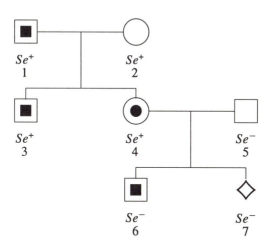

FIGURE 7.2. Risk Prediction for a Pedigree Segregating Myotonic Dystrophy

Risk prediction in genetic counseling reduces to an exercise in computing conditional probabilities. Figure 7.2 depicts a typical risk prediction problem. The fetus 7 in Figure 7.2 has been tested for the marker gene secretor linked to the autosomal dominant disease myotonic dystrophy. Assuming that myotonic dystrophy cannot be clinically diagnosed at the fetal stage, what is the risk that the fetus will eventually develop the disease? At the myotonic dystrophy locus, affected individuals in the pedigree are denoted by partially darkened circles or squares. The disease allele $Dm^+$ is so rare that the $Dm^+/Dm^+$ disease genotype is virtually nonexistent. The secretor locus also exhibits dominance, with the $Se^+$ allele being dominant to the $Se^-$ allele. Thus, phenotypes at either locus convey whether the dominant allele is present.

To compute the risk to the fetus, we must form the ratio of two probabilities. The denominator probability is just the probability of the observed phenotypes within the pedigree. These phenotypes include the fetus's secretor phenotype, but not its unknown disease phenotype. The numerator probability is the probability of the observed phenotypes within the pedigree plus an assigned phenotype of affected for the fetus at the myotonic dystrophy locus. Given Hardy-Weinberg and linkage equilibrium, allele frequencies of $p_{Se^+} = .52$ and $p_{Dm^+} = .0001$, and a recombination fraction of $\theta = .08$ between the two loci, the risk to the fetus can be computed as .84. It takes great patience to carry out these calculations by hand, but a computer does them in less than a twinkling of an eye [14]. This example is mainly of historical interest since the gene for myotonic dystrophy has been cloned [1]. ∎

**Example 7.4** *Lod Scores and Location Scores*

Geneticists are keenly interested in mapping genes to particular regions of particular chromosomes. Classically they have defined linkage groups in plants and non-human species by testing for reduced recombination between two loci in a breeding experiment. In humans, planned matings are ethically objectionable, and geneticists must rely on the random recombination data provided by human pedigrees. During the past decade a vigorous effort has been made to map large numbers of marker loci using a common group of specially chosen pedigrees [3]. These CEPH (Centre d'Etude du Polymorphisme Humain) pedigrees are large nuclear families. Most of them include all four associated grandparents; this helps determine phase relations in the parents. With the advent of physical mapping techniques such as somatic cell hybrids, in situ hybridization, and radiation hybrids, pedigree analysis has diminished in importance, but it still is the only method for mapping clinically important diseases of unknown etiology.

In mapping a disease locus, the CEPH pedigrees are useless. Only pedigrees segregating the disease trait of interest contain linkage information on that trait. In a typical clinical genetics study, the likelihood of the trait and a single marker is computed over one or more relevant pedigrees. This likelihood $L(\theta)$ is a function of the recombination fraction $\theta$ between the trait locus and the marker locus. The standardized loglikelihood $Z(\theta) = \log_{10}[L(\theta)/L(\frac{1}{2})]$ is referred to as a **lod score**. Here "lod" is an abbreviation for "logarithm of the odds." A lod score permits easy visualization of linkage evidence. As a rule of thumb, most geneticists provisionally accept linkage if $Z(\hat{\theta}) \geq 3$ at its maximum $\hat{\theta}$ on the interval $[0, \frac{1}{2}]$; they provisionally reject linkage at a particular $\theta$ if $Z(\theta) \leq -2$ . Acceptance and rejection are treated asymmetrically because with 22 pairs of human autosomes it is unlikely that a random marker even falls on the same chromosome as a trait locus.

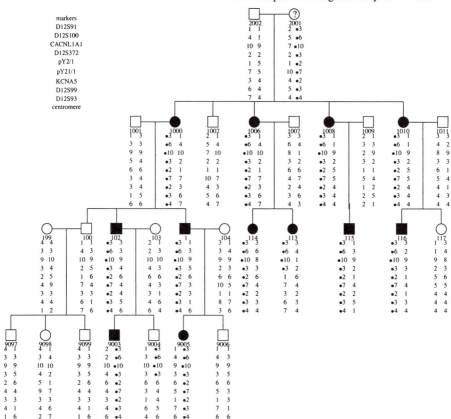

FIGURE 7.3. An Episodic Ataxia Pedigree with Reconstructed Haplotypes

Figure 7.3 depicts an updated version of pedigree 4 from an article by Litt et al. mapping the autosomal dominant disease episodic ataxia to chromosome 12p [18]. Presumably the great-grandmother 2001 is the source of the disease gene in this pedigree. Her affected descendants are indicated by black circles and squares. Except for 2001 and the spouse 1011 of 1010, all of the remaining 29 members of the pedigree were available for typing with the nine 12p markers shown in the figure. Figure 7.4 plots the lod score between episodic ataxia and the marker D12S372. This pedigree strongly suggests but does not prove linkage. Note that individual 9004 is a definite recombinant between the disease locus and D12S372; this fact explains the limiting behavior $\lim_{\theta \to 0} Z(\theta) = -\infty$ of the lod score curve. In this example as in all examples, $Z(\frac{1}{2}) = 0$ by definition.

Once a disease gene is mapped to a particular chromosome region, geneticists saturate the region by typing many nearby markers in the disease pedigrees. Because the order and separation of the markers are usually known from the CEPH families, the goal now becomes one of positioning the disease locus on the known marker map. In the method of **location scores**, we accomplish this task by evaluating and plotting the joint likelihood of the disease and marker phenotypes as

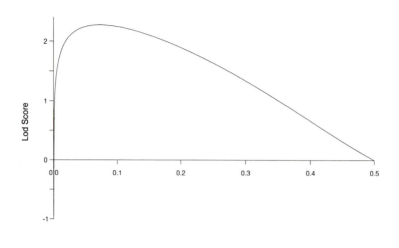

FIGURE 7.4. Lod Score Curve for Episodic Ataxia Versus Marker D12S372

a function of the position of the disease locus [15, 22, 23]. This necessitates converting recombination fractions into map distances, a process we will consider in
Chapter 12. For the moment, let us only mention Haldane's map function

$$d = -\frac{1}{2}\ln(1 - 2\theta) \tag{7.7}$$

and its inverse

$$\theta = \frac{1}{2}(1 - e^{-2d}), \tag{7.8}$$

which is certainly reminiscent of a Poisson process with intensity 2. The map distance $d$ featured in these formulas represents the expected number of crossovers
between the two loci per gamete. The unit of distance is the Morgan (or the cen-
tiMorgan, which equals $10^{-2}$ Morgans), in honor of one of the pioneers of gene
mapping. Map distances have the advantages of being additive over large distances
and approximately equaling recombination fractions for small distances.

A location score is analogous to a lod score. An origin is arbitrarily fixed
and map distances $d$ are now measured relative to it. If $L(d)$ denotes the likelihood of the trait and marker data when the trait locus is at position $d$, then
$Z(d) = \log_{10}[L(d)/L(\infty)]$ defines the location score. In effect, one standardizes
the loglikelihood by moving the trait off the marker chromosome. This extreme
position entails independent segregation of the trait gene relative to the marker
genes. One defect of location scores is that recombination fractions do not depend
on sex. If we postulate a common ratio of female to male map distances, then even
this defect can be remedied.

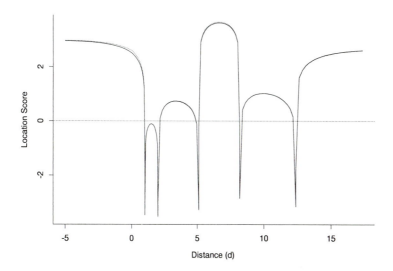

FIGURE 7.5. Location Score Curve for Episodic Ataxia Versus 12p Markers

Figure 7.5 plots a location score curve for the episodic ataxia pedigree drawn in Figure 7.3. Owing to the computational difficulty of this problem, only the representative marker pY2/1 from the tight cluster pY2/1, pY21/1, KCNA5, and D12S99 was used in the calculations. The map for the six participating chromosome 12p markers can be summarized as follows:

**Recombination Fractions Between Adjacent 12p Markers**

$$S91 \xrightarrow{1cM} S100 \xrightarrow{1cM} CACNL1A1 \xrightarrow{3cM} S372 \xrightarrow{3cM} pY2/1 \xrightarrow{4cM} S93$$

In Figure 7.5, the origin occurs at locus D12S91, and distances are given in units of centiMorgans (cM).

Inspection of the location score curve shows that it rises above the magical level of 3 and that the episodic ataxia gene probably resides on the interval from D12S372 to pY2/1. Where the marker S372 is uninformative for linkage, other markers fill the information gap. Thus, location scores make better use of scarce disease pedigrees than lod scores do. In Chapter 9 we will revisit this problem and demonstrate how the haplotypes displayed in Figure 7.3 are reconstructed and how one can compute location scores using an almost arbitrary number of markers. ∎

## 7.7 Problems

1. Under Haldane's model of independent recombination on disjoint intervals, it is possible to compute the recombination fraction $\theta_{ij}$ between two loci

$i < j$ by Trow's formula

$$1 - 2\theta_{ij} = \prod_{k=i}^{j-1}(1 - 2\theta_{k,k+1}), \qquad (7.9)$$

where the loci occur in numerical order along the chromosome, and where $\theta_{k,k+1}$ is the recombination fraction between the adjacent loci $k$ and $k + 1$. Verify Trow's formula first for three loci ($i = 1$ and $j = 3$) and then by induction for an arbitrary number of loci. (Hint: For three loci, recombination occurs between loci 1 and 3 if and only if it occurs between loci 1 and 2 and not between loci 2 and 3, or vice versa.)

2. Consider the partially typed, inbred pedigree depicted in Figure 7.6. The phenotypes displayed in the figure are unordered genotypes at a single codominant locus with three alleles. Show that the genotype elimination algorithm fails to eliminate some superfluous genotypes in this pedigree.

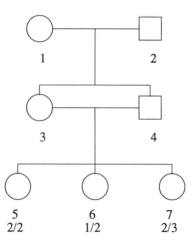

FIGURE 7.6. A Genotype Elimination Counterexample

3. The sum of array products

$$\sum_{G_1 \in S_1} \cdots \sum_{G_9 \in S_9} A(G_1, G_2, G_3, G_4)B(G_4, G_5)$$

$$\times C(G_5, G_6)D(G_6, G_7, G_8, G_9)$$

can be evaluated as an iterated sum by the greedy algorithm. If all range sets $S_i$ have the same number of elements $m > 2$, then show that one greedy summation sequence is $(5, 1, 2, 3, 4, 7, 8, 9, 6)$. Prove that the alternative nongreedy sequence $(1, 2, 3, 4, 5, 7, 8, 9, 6)$ requires fewer arithmetic operations (additions plus multiplications) [12].

4. Consider the array product

$$E(G_1, G_2, G_3, G_4, G_5)$$
$$= A(G_1, G_2, G_3)B(G_1)C(G_2, G_3, G_4)D(G_5),$$

where the range set $S_i$ for the index $G_i$ has 4, 2, or 3 elements according as $i = 1$, $i \in \{2, 3, 4\}$, or $i = 5$. Show that the greedy tactic of assembling the product array from the pairwise products of the multiplicand arrays first multiplies $B$ times $D$, then $A$ times $C$, and finally the product $BD$ times the product $AC$. Demonstrate that the alternative of multiplying $A$ times $B$, then $C$ times $D$, and finally the product $AB$ times the product $CD$ requires fewer total multiplications [12].

5. Verify the numerical entries in Table 7.3.

6. Do by hand the risk prediction calculation for myotonic dystrophy, showing the various steps of the computations in detail. Neglect the extremely rare $Dm + /Dm+$ genotype at the myotonic dystrophy locus. Using two-locus genotypes, evaluate the two required likelihoods as seven-fold iterated sums.

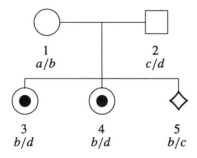

FIGURE 7.7. Risk Prediction for a Recessive Disease via a Linked Marker

7. Figure 7.7 gives a pedigree for an autosomal recessive disease and a linked marker. The four marker genes $a$, $b$, $c$, and $d$ are assumed distinct. If the recombination fraction between the two loci is $\theta$, then show that the risk of the fetus 5 being affected is

$$\frac{(1 - \theta)^5\theta + (1 - \theta)^4\theta^2 + (1 - \theta)^2\theta^4 + (1 - \theta)\theta^5}{(1 - \theta)^4 + 2(1 - \theta)^2\theta^2 + \theta^4}.$$

8. A healthy male had a sister with cystic fibrosis (CF), but she and his parents are dead. What is his risk of being a carrier for this recessive disease? About 75 percent of all disease alleles at the CF locus are accounted for by the $\Delta F_{508}$ mutation. If he tests negative for the $\Delta F_{508}$ mutation, what is his risk of being a carrier [22]?

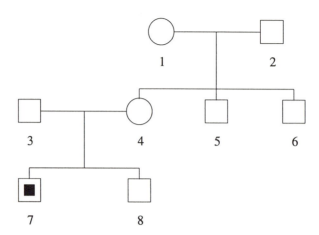

FIGURE 7.8. Risk Prediction for an X-linked Recessive Disease

9. The grandson 7 depicted in the pedigree of Figure 7.8 is afflicted by a lethal, X-linked recessive disease [19]. Problem 11 of Chapter 1 notes that if the carrier females for such a disease are fully fit, then they have a population frequency $4\mu$, where $\mu$ is the mutation rate to the disease allele. In view of this fact, demonstrate that the mother 4 has a chance of approximately $5/13$ of carrying the disease allele. Consequently, her next son has a chance of $5/26$ of being affected. (Hints: Either the grandmother 1 is a carrier, or the mother 4 is a new mutation and passes the disease allele to the grandson 7, or the grandson 7 is a new mutation. The presence of unaffected uncles 5 and 6 and an unaffected brother 8 modifies the probabilities of these contingencies. Because $\mu$ is very small, you may approximate the probability of a carrier female passing either the normal or disease allele as $1/2$. You may also approximate the prior probability of a normal female or normal male as 1.)

10. Consider a nuclear family in which one parent is affected by an autosomal dominant disease [8]. If the affected parent is heterozygous at a codominant marker locus, the normal parent is homozygous at the marker locus, and the number of children $n \geq 2$, then the family is informative for linkage. Because of the phase ambiguity in the affected parent, we can split the children of the family into two disjoint sets of size $k$ and $n - k$, the first set consisting of recombinant children and the second set consisting of nonrecombinant children, or vice versa. Show that the likelihood of the family is

$$L(\theta) = \frac{1}{2}\theta^k(1 - \theta)^{n-k} + \frac{1}{2}\theta^{n-k}(1 - \theta)^k,$$

where $\theta$ is the recombination fraction between the disease and marker loci. A harder problem is to characterize the maximum of $L(\theta)$ on the interval

$[0, \frac{1}{2}]$. Without loss of generality, take $k \leq \frac{n}{2}$. Then demonstrate that the likelihood curve is unimodal with maximum at $\theta = 0$ when $k = 0$, at $\theta = \frac{1}{2}$ when $(n - 2k)^2 \leq n$, and at $\theta \in (0, \frac{1}{2})$ otherwise. (Hints: The case $k = 0$ can be resolved straightforwardly by inspecting the derivative $L'(\theta)$. For the remaining two cases, write $L'(\theta) = \theta^{k-1}(1 - \theta)^{n-k}g(\tau)$, where $g(\tau)$ is a polynomial in $\tau = \frac{\theta}{1-\theta}$. From this representation check that $\theta = 0$ is a local minimum of $L(\theta)$ and that $\theta = \frac{1}{2}$ is a stationary point of $L(\theta)$. The maximum of $L(\theta)$ must therefore occur at $\theta = \frac{1}{2}$ or some other positive root of $g(\tau)$. Use Descartes' rule of signs [4] and symmetry to limit the number of positive roots of $g(\tau)$ on $\tau \in (0, 1]$, that is, $\theta \in (0, \frac{1}{2}]$. Compute $L''(\frac{1}{2})$ to determine the nature of the stationary point $\theta = \frac{1}{2}$.)

# References

[1] Aslanidis C, Jansen G, Amemiya C, Shutler G, Mahadevan M, Tsilfidis C, Chen C, Alleman J, Wormskamp NGM, Vooijs M, Buxton J, Johnson K, Smeets HJM, Lennon GG, Carrano AV, Korneluk RG, Wieringa B, deJong PJ (1992) Cloning of the essential myotonic dystrophy region and mapping of the putative defect. *Nature* 355:548–551

[2] Boehnke M (1991) Allele frequency estimation from data on relatives. *Amer J Hum Genet* 48:22–25

[3] Dausset J, Cann H, Cohen D, Lathrop M, Lalouel J-M, White R (1990) Centre d'Etude du Polymorphisme Humain (CEPH): Collaborative genetic mapping of the human genome. *Genomics* 6:575–577

[4] Dickson LE (1939) *New First Course in the Theory of Equations*. Wiley, New York

[5] Elston RC, Stewart J (1971) A general model for the genetic analysis of pedigree data. *Hum Hered* 21:523–542

[6] Goradia TM, Lange K, Miller PL, Nadkarni PM (1992) Fast computation of genetic likelihoods on human pedigree data. *Hum Hered* 42:42–62

[7] Haldane JBS (1919) The combination of linkage values, and the calculation of distance between the loci of linked factors. *J Genet* 8:299–309

[8] Hulbert-Shearon T, Boehnke M, Lange K (1995) Lod score curves for phase-unknown matings. Hum Hered 46:55–57

[9] Kruglyak L, Daly MJ, Lander ES (1995) Rapid multipoint linkage analysis of recessive traits in nuclear families, including homozygosity mapping. *Amer J Hum Genet* 56:519–527

[10] Kruglyak L, Daly MJ, Reeve-Daly MP, Lander ES (1996) Parametric and nonparametric linkage analysis: a unified multipoint approach. *Amer J Hum Genet* 58:1347–1363

[11] Lander ES, Green P (1987) Construction of multilocus genetic linkage maps in humans. *Proc Natl Acad Sci USA* 84:2363–2367

[12] Lange K, Boehnke M (1983) Extensions to pedigree analysis. V. Optimal calculation of Mendelian likelihoods. *Hum Hered* 33:291–301

[13] Lange K, Goradia TM (1987) An algorithm for automatic genotype elimination. *Amer J Hum Genet* 40:250–256

[14] Lange K, Weeks D, Boehnke M (1988) Programs for pedigree analysis: MENDEL, FISHER, and dGENE. *Genet Epidemiology* 5:471–472

[15] Lathrop GM, Lalouel JM, Julier C, Ott J (1984) Strategies for multilocus linkage analysis in humans. *Proc Natl Acad Sci* 81:3443–3446

[16] Lathrop GM, Lalouel J-M, White RL (1986) Construction of human linkage maps: Likelihood calculations for multilocus linkage analysis. *Genet Epidemiology* 3:39–52

[17] Lewis M, Kaita H, Philipps S, Giblet E, Anderson JE, McAlpine PJ, Nickel B (1980) The position of the Radin blood group locus in relation to other chromosome 1 loci. *Ann Hum Gent* 44:179–184

[18] Litt M, Kramer P, Browne D, Gancher S, Brunt ERP, Root D, Phromchotikul T, Dubay CJ, Nutt J (1994) A gene for Episodic Ataxia/Myokymia maps to chromosome 12p13. *Amer J Hum Genet* 55:702–709

[19] Murphy EA, Chase GA (1975) *Principles of Genetic Counseling*. Year Book Medical Publishers, Chicago

[20] O'Connell JR, Weeks DE (1995) The VITESSE algorithm for rapid exact multilocus linkage analysis via genotype set-recoding and fuzzy inheritance. *Nature Genet* 11:402–408

[21] Ott J (1974) Estimation of the recombination fraction in human pedigrees: efficient computation of the likelihood for human linkage studies. *Amer J Hum Genet* 26:588–597

[22] Ott J (1991) *Analysis of Human Genetic Linkage*, revised ed. Johns Hopkins University Press, Baltimore

[23] Terwilliger JD, Ott J (1994) *Handbook of Human Genetic Linkage*. Johns Hopkins University Press, Baltimore

# 8

# The Polygenic Model

## 8.1 Introduction

The standard polygenic model of biometrical genetics can be motivated by considering a quantitative trait determined by a large number of loci acting independently and additively [9]. In a pedigree of $m$ people, let $X_i^k$ be the contribution of locus $k$ to person $i$. The trait value $X_i = \sum_k X_i^k$ for person $i$ forms part of a vector $X = (X_1, \ldots, X_m)^t$ of trait values for the pedigree. If the effects of the various loci are comparable, then the central limit theorem implies that $X$ follows an approximate multivariate normal distribution [13, 15]. Furthermore, independence of the various loci implies $\text{Cov}(X_i, X_j) = \sum_k \text{Cov}(X_i^k, X_j^k)$. From our covariance decomposition for two non-inbred relatives at a single locus, it follows that

$$\text{Cov}(X_i, X_j) \quad = \quad 2\Phi_{ij}\sigma_a^2 + \Delta_{7ij}\sigma_d^2,$$

where $\sigma_a^2$ and $\sigma_d^2$ are the additive and dominance genetic variances summed over all participating loci. These covariances can be expressed collectively in matrix notation as $\text{Var}(X) = 2\sigma_a^2\Phi + \sigma_d^2\Delta_7$. Again it is convenient to assume that $X$ has mean $E(X) = \mathbf{0}$. Although it is an article of faith that the assumptions necessary for the central limit theorem actually hold for any given trait, one can check multivariate normality empirically.

Environmental effects can be incorporated in this simple model by supposing that the observed trait value for person $i$ is the sum $Y_i = X_i + Z_i$ of a genetic contribution $X_i$ and an environmental contribution $Z_i$. If we assume that the random vector $Z = (Z_1, \ldots, Z_m)^t$ is uncorrelated with $X$ and follows a multivariate normal distribution with mean vector $\nu$ and covariance matrix $\Upsilon$, then the trait vector

$Y = (Y_1, \ldots, Y_m)'$ is multivariate normal with mean $E(Y) = \nu$ and covariance $\text{Var}(Y) = 2\sigma_a^2 \Phi + \sigma_d^2 \Delta_7 + \Upsilon$.

Different levels of environmental sophistication can be incorporated by appropriately choosing $\nu$ and $\Upsilon$. Typically $\nu$ is defined to be a linear function $\nu = A\mu$ of a parameter vector $\mu$ of $p$ **mean components**. The $m \times p$ **design matrix** $A$ specifies the measured covariates determining this linear function. For instance, $\mu$ might be $(\mu_f, \mu_m, \alpha)'$, where $\mu_f$ is the female population mean at birth, $\mu_m$ is the male population mean at birth, and $\alpha$ is a regression coefficient on age. A row of $A$ is then either $(1, 0, \text{age})$ or $(0, 1, \text{age})$, depending on whether the corresponding person is female or male.

Among the simplest possibilities for $\Upsilon$ is $\Upsilon = \sigma_e^2 I$, where $I$ is the $m \times m$ identity matrix. The parameter $\sigma_e^2$ is referred to as a **variance component**. For this choice of $\Upsilon$, environmental contributions are uncorrelated among pedigree members. Note that environmental contributions include trait measurement errors. To represent shared environments within a pedigree, it is useful to define a household indicator matrix $H = (h_{ij})$ with entries

$$h_{ij} = \begin{cases} 1 & i \text{ and } j \text{ are in the same household} \\ 0 & \text{otherwise.} \end{cases}$$

A reasonable covariance model incorporating both household and random effects is $\Upsilon = \sigma_h^2 H + \sigma_e^2 I$, giving an overall covariance matrix $\Omega$ for $Y$ of

$$\Omega = 2\sigma_a^2 \Phi + \sigma_d^2 \Delta_7 + \sigma_h^2 H + \sigma_e^2 I.$$

This last representation suggests studying the general model

$$\Omega = \sum_{k=1}^{r} \sigma_k^2 \Gamma_k, \tag{8.1}$$

where the variance components $\sigma_k^2$ are nonnegative and the matrices $\Gamma_k$ are known covariance matrices. Since measurement error will enter almost all models, it is convenient to assume $\Gamma_r = I$.

## 8.2   Maximum Likelihood Estimation by Scoring

The mean components $\mu_1, \ldots, \mu_p$ and the variance components $\sigma_1^2, \ldots, \sigma_r^2$ appear as parameters in the multivariate normal loglikelihood

$$L(\gamma) = -\frac{m}{2} \ln 2\pi - \frac{1}{2} \ln \det \Omega - \frac{1}{2}(y - A\mu)' \Omega^{-1} (y - A\mu) \tag{8.2}$$

for the observed data $Y = y$ [11, 12, 15, 17]. In equation (8.2), $\det \Omega$ denotes the determinant of $\Omega$, and $\gamma = (\mu_1, \ldots, \mu_p, \sigma_1^2, \ldots, \sigma_r^2)'$ denotes the parameters collected into a column vector. Because $\Gamma_r = I$, $\Omega$ is nonsingular whenever $\sigma_r^2 > 0$.

To implement the scoring algorithm for maximum likelihood estimation of $\gamma$, we need the loglikelihood $L(\gamma)$, score $dL(\gamma)$, and expected information $J(\gamma)$ over all the pedigrees in a sample. Because these quantities add for independent pedigrees, it suffices to consider a single pedigree. In deriving the score and expected information for a single pedigree, we could use the general results presented in Chapter 3 for exponential families. It is more illuminating to proceed directly after reviewing the following facts from linear algebra and calculus:

(a) If $B = (b_{ij})$ is a symmetric matrix with cofactor $B_{ij}$ corresponding to entry $b_{ij}$, then the determinant $\det B = \sum_j b_{ij} B_{ij}$ is expandable on any row $i$. If $B$ is invertible as well, then its inverse $B^{-1} = \frac{1}{\det B}(B_{ij})$.

(b) If $B = (b_{ij})$ is a square matrix, then the trace $\mathrm{tr}(B)$ of $B$ is defined by the formula $\mathrm{tr}(B) = \sum_i b_{ii}$. The trace function satisfies $\mathrm{tr}(BC) = \mathrm{tr}(CB)$ for any two conforming matrices $B$ and $C$.

(c) The matrix transpose operation satisfies $(BC)^t = C^t B^t$.

(d) If $B$ is a matrix and $W$ is a random vector, then the vector $BW$ has expectation $\mathrm{E}(BW) = B\,\mathrm{E}(W)$. The quadratic form $W^t BW$ has expectation $\mathrm{E}(W^t BW) = \mathrm{tr}[B\,\mathrm{Var}(W)] + \mathrm{E}(W)^t B\,\mathrm{E}(W)$. To verify this last assertion, observe that

$$
\begin{aligned}
\mathrm{E}(W^t BW) &= \mathrm{E}\left(\sum_{ij} W_i b_{ij} W_j\right) \\
&= \sum_{ij} b_{ij}\, \mathrm{E}(W_i W_j) \\
&= \sum_{ij} b_{ij}[\mathrm{Cov}(W_i, W_j) + \mathrm{E}(W_i)\,\mathrm{E}(W_j)] \\
&= \mathrm{tr}[B\,\mathrm{Var}(W)] + \mathrm{E}(W)^t B\,\mathrm{E}(W).
\end{aligned}
$$

(e) The partial derivative of a matrix $B = (b_{ij})$ with respect to a scalar parameter $\theta$ is the matrix with entries $(\frac{\partial}{\partial \theta} b_{ij})$. Because the trace function is linear, $\frac{\partial}{\partial \theta}\mathrm{tr}(B) = \mathrm{tr}(\frac{\partial}{\partial \theta}B)$. The product rule of differentiation yields the formula $\frac{\partial}{\partial \theta}(BC) = (\frac{\partial}{\partial \theta}B)C + B\frac{\partial}{\partial \theta}C$.

(f) The derivative of a matrix inverse is $\frac{\partial}{\partial \theta}B^{-1} = -B^{-1}(\frac{\partial}{\partial \theta}B)B^{-1}$. To derive this formula, solve for $\frac{\partial}{\partial \theta}B^{-1}$ in

$$
\begin{aligned}
0 &= \frac{\partial}{\partial \theta}I \\
&= \frac{\partial}{\partial \theta}(B^{-1}B) \\
&= \left(\frac{\partial}{\partial \theta}B^{-1}\right)B + B^{-1}\frac{\partial}{\partial \theta}B.
\end{aligned}
$$

**(g)** If $B$ is a symmetric matrix, then $\frac{\partial}{\partial\theta}\ln\det B = \operatorname{tr}(B^{-1}\frac{\partial}{\partial\theta}B)$. This formula is validated by

$$
\begin{aligned}
\frac{\partial}{\partial\theta}\ln\det B &= \sum_{ij}\Big(\frac{\partial}{\partial b_{ij}}\ln\det B\Big)\frac{\partial}{\partial\theta}b_{ij} \\
&= \sum_{ij}\frac{B_{ij}}{\det B}\frac{\partial}{\partial\theta}b_{ij} \\
&= \operatorname{tr}\Big(B^{-1}\frac{\partial}{\partial\theta}B\Big)
\end{aligned}
$$

using property (a).

Applying the above facts to the loglikelihood (8.2) leads to the score. With respect to the mean component $\mu_i$, we have

$$
\begin{aligned}
\frac{\partial}{\partial\mu_i}L &= \frac{1}{2}\Big(A\frac{\partial}{\partial\mu_i}\mu\Big)^t\Omega^{-1}(y-A\mu)+\frac{1}{2}(y-A\mu)^t\Omega^{-1}A\frac{\partial}{\partial\mu_i}\mu \\
&= \Big(\frac{\partial}{\partial\mu_i}\mu\Big)^t A^t\Omega^{-1}(y-A\mu).
\end{aligned}
$$

With respect to the variance component $\sigma_i^2$, we have

$$
\begin{aligned}
\frac{\partial}{\partial\sigma_i^2}L &= -\frac{1}{2}\frac{\partial}{\partial\sigma_i^2}\ln\det\Omega-\frac{1}{2}(y-A\mu)^t\frac{\partial}{\partial\sigma_i^2}\Omega^{-1}(y-A\mu) \\
&= -\frac{1}{2}\operatorname{tr}(\Omega^{-1}\Gamma_i)+\frac{1}{2}(y-A\mu)^t\Omega^{-1}\Gamma_i\Omega^{-1}(y-A\mu).
\end{aligned}
$$

In similar fashion, the elements of the observed information matrix are

$$
\begin{aligned}
-\frac{\partial^2}{\partial\mu_i\partial\mu_j}L &= \Big(\frac{\partial}{\partial\mu_j}\mu\Big)^t A^t\Omega^{-1}A\frac{\partial}{\partial\mu_i}\mu \\
-\frac{\partial^2}{\partial\sigma_i^2\partial\mu_j}L &= \Big(\frac{\partial}{\partial\mu_j}\mu\Big)^t A^t\Omega^{-1}\Gamma_i\Omega^{-1}(y-A\mu) \\
&= -\frac{\partial^2}{\partial\mu_j\partial\sigma_i^2}L \\
-\frac{\partial^2}{\partial\sigma_i^2\partial\sigma_j^2}L &= -\frac{1}{2}\operatorname{tr}(\Omega^{-1}\Gamma_i\Omega^{-1}\Gamma_j) \\
&\quad +\frac{1}{2}(y-A\mu)^t\Omega^{-1}\Gamma_i\Omega^{-1}\Gamma_j\Omega^{-1}(y-A\mu) \\
&\quad +\frac{1}{2}(y-A\mu)^t\Omega^{-1}\Gamma_j\Omega^{-1}\Gamma_i\Omega^{-1}(y-A\mu) \\
&= -\frac{1}{2}\operatorname{tr}(\Omega^{-1}\Gamma_i\Omega^{-1}\Gamma_j) \\
&\quad +(y-A\mu)^t\Omega^{-1}\Gamma_i\Omega^{-1}\Gamma_j\Omega^{-1}(y-A\mu).
\end{aligned}
$$

Note that $\frac{\partial}{\partial \mu_i}\mu$ and $\frac{\partial}{\partial \sigma_i^2}\Omega = \Gamma_i$ are treated as constants in these derivations.

Since $E(Y) = A\mu$, the expected information matrix has entries

$$E\left(-\frac{\partial^2}{\partial \mu_i \partial \mu_j}L\right) = \left(\frac{\partial}{\partial \mu_j}\mu\right)' A'\Omega^{-1}A\frac{\partial}{\partial \mu_i}\mu$$

$$E\left(-\frac{\partial^2}{\partial \sigma_i^2 \partial \mu_j}L\right) = \left(\frac{\partial}{\partial \mu_j}\mu\right)' A'\Omega^{-1}\Gamma_i\Omega^{-1}[E(Y) - A\mu]$$

$$= 0$$

$$E\left(-\frac{\partial^2}{\partial \sigma_i^2 \partial \sigma_j^2}L\right) = -\frac{1}{2}\text{tr}(\Omega^{-1}\Gamma_i\Omega^{-1}\Gamma_j) + \text{tr}(\Omega^{-1}\Gamma_i\Omega^{-1}\Gamma_j\Omega^{-1}\Omega)$$

$$= \frac{1}{2}\text{tr}(\Omega^{-1}\Gamma_i\Omega^{-1}\Gamma_j).$$

Some simplification in the above formulas can be achieved by defining the partial score $d_\mu L = (\frac{\partial}{\partial \mu_1}L, \ldots, \frac{\partial}{\partial \mu_p}L)$ and the corresponding partial observed information matrix $-d_\mu^2 L$. Since $(\frac{\partial}{\partial \mu_1}\mu, \ldots, \frac{\partial}{\partial \mu_p}\mu)$ is the identity matrix $I$,

$$d_\mu L' = A'\Omega^{-1}(y - A\mu)$$
$$-d_\mu^2 L = A'\Omega^{-1}A.$$

The expected information matrix $J$ evidently has the block diagonal form

$$J = \begin{pmatrix} E[-d_\mu^2 L] & 0 \\ 0 & E[-d_{\sigma^2}^2 L] \end{pmatrix},$$

where $-d_{\sigma^2}^2 L$ is the observed information matrix on $\sigma^2 = (\sigma_1^2, \ldots, \sigma_r^2)'$.

Since the block form of $J$ is retained under matrix inversion, the current $\mu$ is updated in the scoring algorithm by

$$\mu + (A'\Omega^{-1}A)^{-1}A'\Omega^{-1}(y - A\mu) = (A'\Omega^{-1}A)^{-1}A'\Omega^{-1}y. \qquad (8.3)$$

If there is more than one pedigree, the quantities $A'\Omega^{-1}A$ and $A'\Omega^{-1}y$ must be summed over all pedigrees before matrix inversion and multiplication. The scoring increment $\Delta\sigma^2$ to $\sigma^2$ is similarly expressed as

$$\Delta\sigma^2 = E(-d_{\sigma^2}^2 L)^{-1}d_{\sigma^2}L'.$$

It is convenient to initialize $\sigma^2$ at $(0, \ldots, 0, 1)'$. Since $\Omega = I$ in this case, the first iteration of the scoring algorithm (8.3) produces the standard linear regression estimate of $\mu$. Of course, this estimate does not take into account the correlational architecture of the pedigree. In those unlikely circumstances when $\mu$ is known exactly, the initial value $\sigma^2 = (0, \ldots, 0, 1)'$ leads to a least-squares estimate of $\sigma^2$ after a single iteration of scoring. (See Problem 3.) Computation of the score $dL$ and expected information matrix $E(-d^2 L)$ also simplifies drastically when $\Omega = I$. This fact can be used to good advantage in a quasi-Newton search of the loglikelihood [20]. Although quasi-Newton methods explicitly require neither the observed nor the expected information matrix, a good initial approximation to the observed information matrix is crucial. Starting a search at $\sigma^2 = (0, \ldots, 0, 1)^t$ and approximating the observed information matrix by the expected information matrix fits in well with the quasi-Newton strategy.

The basic model just presented can be generalized in many useful ways [15]. First, missing observations can be handled by deleting the appropriate rows (and columns) of the observation vector $Y$, the design matrix $A$, and the covariance matrix $\Omega$ of a pedigree. Second, theoretical means and covariances that depend nonlinearly on the parameters can be accommodated. To be precise, suppose $A(\mu)$ is the mean vector and $\Omega(\sigma^2)$ is the covariance matrix of a given pedigree. Then in the scoring algorithm any appearance of $A\frac{\partial}{\partial\mu_i}\mu$ is replaced by $\frac{\partial}{\partial\mu_i}A(\mu)$, and any appearance of $\Gamma_i$ is replaced by $\frac{\partial}{\partial\sigma_i^2}\Omega(\sigma^2)$. Third, as we shall see later in this chapter, covariance models for multivariate traits can be devised.

## 8.3    Application to Gc Measured Genotype Data

Human group specific component (Gc) is a transport protein for vitamin D. The Gc locus determines qualitative variation in the Gc protein. An interesting question is whether the genotypes at the Gc locus also influence quantitative differences in plasma concentrations of the Gc protein. Daiger et al. [3] collected relevant data on 31 identical twin pairs, 13 fraternal twin pairs, and 45 unrelated controls. Gc concentrations and Gc genotypes are available on all individuals. The three genotypes 1/1, 1/2, and 2/2 at the Gc locus are distinguishable.

A reasonable model for these data involves $p = 5$ mean components and $r = 3$ variance components [1]. The covariates are Gc genotype, sex, and age. To accommodate these covariates requires mean components $\mu_{1/1}$, $\mu_{1/2}$, and $\mu_{2/2}$ for the three genotypes, a male offset $\mu_{male}$ to distinguish males from females, and a regression coefficient $\mu_{age}$ on age. With these components, the expected trait value for a 35-year-old female with Gc genotype 1/2 is $\mu_{1/2} + 35\mu_{age}$, for instance. For a 15-year-old male with genotype 2/2, the expected trait value is $\mu_{2/2} + \mu_{male} + 15\mu_{age}$.

Instead of choosing the variance components $\sigma_a^2$, $\sigma_d^2$, and $\sigma_e^2$ parameterizing the additive genetic variance, the dominance genetic variance, and the random environmental variance, we can proceed somewhat differently for these data. Let $\sigma_{tot}^2$ be the total trait variance, $\rho_{ident}$ the correlation between identical twins, and $\rho_{frat}$ the correlation between fraternal twins. Mathematically the two sets of parameters give the same model. Biologically, the interpretation of the second set of parameters is less rigid. Note that the hypothesis $\sigma_d^2 = 0$ is equivalent to the hypothesis $\rho_{frat} = \frac{1}{2}\rho_{ident}$.

Table 8.1 summarizes maximum likelihood output from the computer program FISHER [16] for these data. In the first analysis conducted, all eight parameters were estimated under the model just described. The second analysis was performed under the constraint $\rho_{frat} = \frac{1}{2}\rho_{ident}$, and the third analysis was performed under

TABLE 8.1. Maximum Loglikelihoods for the Gc Data

| Model | Loglikelihood | Parameters |
|---|---|---|
| Full Model | -217.610 | 8 |
| $\rho_{\text{frat}} = \frac{1}{2}\rho_{\text{ident}}$ | -217.695 | 7 |
| $\mu_{1/1} = \mu_{1/2} = \mu_{2/2}$ | -230.252 | 6 |

the constraints $\mu_{1/1} = \mu_{1/2} = \mu_{2/2}$. A likelihood ratio test shows that there is virtually no evidence against the assumption $\rho_{\text{frat}} = \frac{1}{2}\rho_{\text{ident}}$. Furthermore, under the model $\rho_{\text{frat}} = \frac{1}{2}\rho_{\text{ident}}$, the estimated correlation between identical twins is .80, indicating a highly heritable trait. High heritability is also suggested by the extremely significant likelihood ratio test for the equality of the three Gc genotype means. Although further test statistics do detect modest departures from normality in these data, it is safe to say that Gc genotypes have a major impact on plasma concentrations of the Gc protein.

## 8.4   Multivariate Traits

Often geneticists collect pedigree data on more than one quantitative trait. To understand the common genetic and environmental determinants of two traits, let $X = (X_1, \ldots, X_n)^t$ and $Y = (Y_1, \ldots, Y_n)^t$ be their measured values on the $n$ members of a non-inbred pedigree [15]. If both traits are determined by the same locus, then in the absence of environmental effects, we know that

$$\text{Cov}(X_i, X_j) = 2\Phi_{ij}\sigma_{ax}^2 + \Delta_{7ij}\sigma_{dx}^2 \tag{8.4}$$
$$\text{Cov}(Y_i, Y_j) = 2\Phi_{ij}\sigma_{ay}^2 + \Delta_{7ij}\sigma_{dy}^2, \tag{8.5}$$

where $\sigma_{ax}^2$ and $\sigma_{dx}^2$ are the additive and dominance genetic variances of the $X$ trait, and $\sigma_{ay}^2$ and $\sigma_{dy}^2$ are the additive and dominance genetic variances of the $Y$ trait. If we consider the sum $Z_i = X_i + Y_i$, then we can likewise write the decomposition

$$\text{Cov}(Z_i, Z_j) = 2\Phi_{ij}\sigma_{az}^2 + \Delta_{7ij}\sigma_{dz}^2 \tag{8.6}$$

in obvious notation. Subtracting equations (8.4) and (8.5) from equation (8.6), dividing by 2, and invoking symmetry and the bilinearity of the covariance operator, we deduce that

$$\text{Cov}(X_i, Y_j) = \text{Cov}(Y_i, X_j)$$
$$= 2\Phi_{ij}\sigma_{axy} + \Delta_{7ij}\sigma_{dxy},$$

where

$$\sigma_{axy} = \frac{1}{2}(\sigma_{az}^2 - \sigma_{ax}^2 - \sigma_{ay}^2)$$

$$\sigma_{dxy} = \frac{1}{2}(\sigma_{dz}^2 - \sigma_{dx}^2 - \sigma_{dy}^2)$$

are additive and dominance **cross covariances**, respectively. In partitioned matrix and Kronecker product notation [21], these covariances can be collectively expressed as

$$
\mathrm{Var}\left[\begin{pmatrix} X \\ Y \end{pmatrix}\right]
$$

$$
= \sigma_{ax}^2 \begin{pmatrix} 2\Phi & 0 \\ 0 & 0 \end{pmatrix} + \sigma_{axy} \begin{pmatrix} 0 & 2\Phi \\ 2\Phi & 0 \end{pmatrix} + \sigma_{ay}^2 \begin{pmatrix} 0 & 0 \\ 0 & 2\Phi \end{pmatrix}
$$

$$
+ \sigma_{dx}^2 \begin{pmatrix} \Delta_7 & 0 \\ 0 & 0 \end{pmatrix} + \sigma_{dxy} \begin{pmatrix} 0 & \Delta_7 \\ \Delta_7 & 0 \end{pmatrix} + \sigma_{dy}^2 \begin{pmatrix} 0 & 0 \\ 0 & \Delta_7 \end{pmatrix} \qquad (8.7)
$$

$$
= 2 \begin{pmatrix} \sigma_{ax}^2 & \sigma_{axy} \\ \sigma_{axy} & \sigma_{ay}^2 \end{pmatrix} \otimes \Phi + \begin{pmatrix} \sigma_{dx}^2 & \sigma_{dxy} \\ \sigma_{dxy} & \sigma_{dy}^2 \end{pmatrix} \otimes \Delta_7.
$$

The covariance representation (8.7) carries over to two traits determined by multiple loci if each locus contributes additively to each trait. Random measurement error can also be incorporated in this scheme if the covariance matrix (8.7) is amended to include the further terms

$$
\sigma_{ex}^2 \begin{pmatrix} I & 0 \\ 0 & 0 \end{pmatrix} + \sigma_{exy} \begin{pmatrix} 0 & I \\ I & 0 \end{pmatrix} + \sigma_{ey}^2 \begin{pmatrix} 0 & 0 \\ 0 & I \end{pmatrix} = \begin{pmatrix} \sigma_{ex}^2 & \sigma_{exy} \\ \sigma_{exy} & \sigma_{ey}^2 \end{pmatrix} \otimes I.
$$

Finally, if we desire to model common household effects, then we tack on the additional terms

$$
\sigma_{hx}^2 \begin{pmatrix} H & 0 \\ 0 & 0 \end{pmatrix} + \sigma_{hxy} \begin{pmatrix} 0 & H \\ H & 0 \end{pmatrix} + \sigma_{hy}^2 \begin{pmatrix} 0 & 0 \\ 0 & H \end{pmatrix} = \begin{pmatrix} \sigma_{hx}^2 & \sigma_{hxy} \\ \sigma_{hxy} & \sigma_{hy}^2 \end{pmatrix} \otimes H,
$$

where $H$ is the household indicator matrix.

We are now in a position to perform scoring for a bivariate trait. The analog of the covariance decomposition (8.1) for a univariate trait holds provided that we do not insist on covariance parameters such as $\sigma_{axy}$ being positive and symmetric matrices such as $\begin{pmatrix} 0 & 2\Phi \\ 2\Phi & 0 \end{pmatrix}$ being positive definite. Of course, the separate Kronecker products appearing in (8.7) and the following expressions are legitimate covariance matrices. Means can be linearly parameterized as

$$
\mathrm{E}\left[\begin{pmatrix} X \\ Y \end{pmatrix}\right] = \begin{pmatrix} A \\ B \end{pmatrix} \mu
$$

for given design matrices $A$ and $B$.

## 8.5  Left and Right-Hand Finger Ridge Counts

Total finger ridge count is a highly heritable trait for which an abundance of pedigree data exists. In her Tables 1 and 3, Holt [10] records left and right-hand ridge counts on 48 nuclear families and 18 pairs of identical twins. To assess the degree

TABLE 8.2. Maximum Likelihood Estimates for the Ridge Count Data

| Parameter | Estimate | Parameter | Estimate |
|-----------|----------|-----------|----------|
| $\mu_{ml}$ | $66.6 \pm 2.8$ | $\rho_{alr}$ | $.992 \pm .008$ |
| $\mu_{fl}$ | $59.0 \pm 2.8$ | $\sigma_{ar}^2$ | $657.5 \pm 69.2$ |
| $\mu_{mr}$ | $68.9 \pm 2.9$ | $\sigma_{el}^2$ | $30.3 \pm 7.5$ |
| $\mu_{fr}$ | $62.7 \pm 2.8$ | $\rho_{elr}$ | $-.146 \pm .178$ |
| $\sigma_{al}^2$ | $638.8 \pm 65.7$ | $\sigma_{er}^2$ | $35.6 \pm 9.9$ |

to which the left and right-hand counts are under common genetic and environmental control, we can treat these counts as bivariate traits and estimate mean and covariance components by maximum likelihood. Because it is well known that dominance effects are small for ridge counts, we postulate an additive genetic variance for each hand ($\sigma_{al}^2$ and $\sigma_{ar}^2$), a random environmental variance for each hand ($\sigma_{el}^2$ and $\sigma_{er}^2$), an additive genetic correlation between hands ($\rho_{alr}$), and a random environmental correlation between hands ($\rho_{elr}$). We also postulate for each hand a separate mean for males ($m$) and females ($f$). This gives the four mean parameters $\mu_{ml}$, $\mu_{fl}$, $\mu_{mr}$, and $\mu_{fr}$ in addition to the six covariance parameters.

The maximum likelihood estimates and corresponding asymptotic standard errors for Holt's data appear in Table 8.2. From this table we can draw several tentative conclusions. First, ridge counts tend to be higher for males than females and for right hands than left hands. Second, left and right-hand ridge counts are highly heritable traits, as reflected in the ratio of the additive genetic variances to the corresponding random environmental variances. Third, the additive genetic correlation is surprisingly strong and the environmental correlation is surprisingly weak. If these data are credible, then ridge counts on the left and right hands are basically determined by the same set of genes. Furthermore, the environmental determinants for the two hands may act independently; indeed, the estimate of environmental correlation is less than one standard deviation from 0. Although we omit them here, overall goodness of fit statistics suggest that the model is reasonably accurate [15].

## 8.6    The Hypergeometric Polygenic Model

Two elaborations of the polygenic model present substantial computational difficulties. In the **polygenic threshold model**, a qualitative trait such as the presence or absence of a birth defect is determined by an underlying polygenically determined **liability**. If a person's liability falls above a fixed threshold, then the person possesses the trait [6, 7]; otherwise, he or she does not. Likelihood evaluation under the polygenic threshold model involves multivariate normal distribution functions and nasty numerical integrations [17]. In the **mixed model**, a quantitative trait is determined as the sum of a polygenic contribution plus a major gene contribution [5, 18]. In this case, computational problems arise because the likelihood is a mixture of numerous multivariate normal densities [19].

One strategy to overcome these computational barriers is to approximate polygenic inheritance by segregation at a large, but finite, number of additive loci. In the **finite polygenic model**, the alleles at $n$ symmetric loci are termed **polygenes** and are categorized as positive or negative [8, 22]. Positive polygenes contribute $+1$ and negative polygenes $-1$ to a trait. If positive and negative polygenes are equally frequent at each locus, then the trait mean and variance for a random noninbred person are 0 and $2n$, respectively. An arbitrary mean $\mu$ and variance $\sigma^2$ for the trait $X$ can be achieved by transforming $X$ to $\frac{\sigma}{\sqrt{2n}} X + \mu$. When the number of loci $n$ is moderately large, $X$ appropriately standardized is approximately normal. Although the finite polygenic model is superficially attractive, it is defeated by the $3^n$ multilocus genotypes per person necessary to implement it.

If one is willing to allow nongenetic transmission, then the situation can be salvaged by employing the **hypergeometric polygenic model** of Cannings et al. [2]. In this model the $2n$ polygenes of a person exist in a common pool that ignores separate loci. If we equate a person's genotype to the number of positive polygenes within it, then there are only $2n + 1$ possible genotypes. This is a major reduction from $3^n$. A gamete is generated in this model by randomly sampling without replacement $n$ polygenes from a parental pool of $2n$ polygenes. Thus, a person having $i$ positive polygenes transmits a gamete having $j$ positive polygenes with hypergeometric probability

$$\tau_{i \to j} \quad = \quad \frac{\binom{i}{j}\binom{2n-i}{n-j}}{\binom{2n}{n}}.$$

Two independently generated gametes unite to form a child. To make this hypergeometric polygenic model as similar as possible to the finite polygenic model, we finally postulate that all pedigree founders independently share the binomial distribution

$$\binom{2n}{i}\left(\frac{1}{2}\right)^{2n} \tag{8.8}$$

for their number of positive polygenes $i$.

The hypergeometric polygenic model mimics the polygenic model well in two regards. First, both models entail the same pattern of variances and covariances among the relatives of a non-inbred pedigree. Second, as $n \to \infty$ in the hypergeometric polygenic model, appropriately standardized trait values within a pedigree tend to multivariate normality. We will verify the first of these assertions, leaving the second for interested readers to glean from the reference [14].

To compute the means, variances, and covariances of the trait values within a pedigree, let $X_i$ denote the trait value of pedigree member $i$. When $i$ is a pedigree founder, $E(X_i) = 0$ by virtue of the binomial distribution (8.8). If $i$ has parents $k$ and $l$ in the pedigree, then we can decompose

$$X_i \quad = \quad Y_{k \to i} + Y_{l \to i}$$

into a gamete contribution from $k$ plus a gamete contribution from $l$. Assuming that the parental trait means vanish, we infer that

$$
\begin{aligned}
\mathrm{E}(X_i) &= \mathrm{E}[\mathrm{E}(X_i \mid X_k, X_l)] \\
&= \mathrm{E}[\mathrm{E}(Y_{k\to i} \mid X_k)] + \mathrm{E}[\mathrm{E}(Y_{l\to i} \mid X_l)] \\
&= \mathrm{E}\left(\frac{1}{2}X_k\right) + \mathrm{E}\left(\frac{1}{2}X_l\right) \\
&= 0
\end{aligned}
$$

and inductively conclude that all trait means in the pedigree vanish.

To compute trait covariances, let $j$ be another member of the pedigree who is not a descendant of $i$. If $i$ is a founder, then $X_i$ and $X_j$ are independent and consequently uncorrelated. If $i$ has parents $k$ and $l$, then

$$
\begin{aligned}
\mathrm{Cov}(X_i, X_j) &= \mathrm{E}[\mathrm{Cov}(X_i, X_j \mid X_k, X_l)] \\
&\quad + \mathrm{Cov}[\mathrm{E}(X_i \mid X_k, X_l), \mathrm{E}(X_j \mid X_k, X_l)] \\
&= 0 + \mathrm{Cov}[\frac{1}{2}(X_k + X_l), \mathrm{E}(X_j \mid X_k, X_l)] \\
&= \frac{1}{2}\,\mathrm{Cov}[X_k, \mathrm{E}(X_j \mid X_k, X_l)] \\
&\quad + \frac{1}{2}\,\mathrm{Cov}[X_l, \mathrm{E}(X_j \mid X_k, X_l)] \\
&= \frac{1}{2}\,\mathrm{Cov}[\mathrm{E}(X_k \mid X_k, X_l), \mathrm{E}(X_j \mid X_k, X_l)] \\
&\quad + \frac{1}{2}\,\mathrm{Cov}[\mathrm{E}(X_l \mid X_k, X_l), \mathrm{E}(X_j \mid X_k, X_l)] \\
&= \frac{1}{2}\,\mathrm{Cov}(X_k, X_j) + \frac{1}{2}\,\mathrm{Cov}(X_l, X_j).
\end{aligned}
$$

Note that $\mathrm{E}[\mathrm{Cov}(X_i, X_j \mid X_k, X_l)] = 0$ in this calculation because $X_i$ and $X_j$ are independent conditional on the parental values $X_k$ and $X_l$. The recurrence

$$
\mathrm{Cov}(X_i, X_j) = \frac{1}{2}\,\mathrm{Cov}(X_k, X_j) + \frac{1}{2}\,\mathrm{Cov}(X_l, X_j) \tag{8.9}
$$

is precisely the recurrence obeyed by ordinary kinship coefficients.

Calculation of variances is a little more complicated. If $i$ is a founder, then the binomial distribution (8.8) implies $\mathrm{Var}(X_i) = 2n$. If $i$ has parents $k$ and $l$, then

$$
\begin{aligned}
\mathrm{Var}(X_i) &= \mathrm{E}[\mathrm{Var}(X_i \mid X_k, X_l)] + \mathrm{Var}[\mathrm{E}(X_i \mid X_k, X_l)] \\
&= \mathrm{E}[\mathrm{Var}(Y_{k\to i} \mid X_k, X_l)] + \mathrm{E}[\mathrm{Var}(Y_{l\to i} \mid X_k, X_l)] \\
&\quad + \mathrm{Var}[\frac{1}{2}(X_k + X_l)] \\
&= \mathrm{E}[\mathrm{Var}(Y_{k\to i} \mid X_k)] + \mathrm{E}[\mathrm{Var}(Y_{l\to i} \mid X_l)] \tag{8.10} \\
&\quad + \frac{1}{4}\,\mathrm{Var}(X_k) + \frac{1}{2}\,\mathrm{Cov}(X_k, X_l) + \frac{1}{4}\,\mathrm{Var}(X_l).
\end{aligned}
$$

To make further progress, we must compute $E[\text{Var}(Y_{k\to i} \mid X_k)]$. With this end in mind, suppose we label the $2n$ polygenes of $k$ by the numbers $1, \ldots, 2n$, and let $W_m$ be $+1$ or $-1$ according as the $m$th polygene of $k$ is positive or negative. If we also let $A_m$ be the event that the $m$th polygene of $k$ is sampled in forming the gamete contribution $Y_{k\to i}$, then $X_k = \sum_{m=1}^{2n} W_m$ and $Y_{k\to i} = \sum_{m=1}^{2n} 1_{A_m} W_m$. A moment's reflection shows that $\text{Var}(1_{A_r}) = \frac{1}{4}$ and that

$$
\text{Cov}(1_{A_r}, 1_{A_s}) = \frac{\binom{2n-2}{n-2}}{\binom{2n}{n}} - \frac{1}{4}
$$

$$
= -\frac{1}{4(2n-1)}.
$$

Conditional on $W = (W_1, \ldots, W_{2n})^t$, we therefore calculate that

$$
\begin{aligned}
\text{Var}(Y_{k\to i} \mid X_k) &= \text{Var}(Y_{k\to i} \mid W) \\
&= \left[\frac{1}{4} + \frac{1}{4(2n-1)}\right] \sum_{m=1}^{2n} W_m^2 - \frac{1}{4(2n-1)} \left(\sum_{m=1}^{2n} W_m\right)^2 \\
&= \left[\frac{1}{4} + \frac{1}{4(2n-1)}\right] 2n - \frac{1}{4(2n-1)} X_k^2
\end{aligned}
$$

and consequently that

$$
E[\text{Var}(Y_{k\to i} \mid X_k)] = \left[\frac{1}{4} + \frac{1}{4(2n-1)}\right] 2n - \frac{1}{4(2n-1)} \text{Var}(X_k).
$$

Substituting this and a similar expression for $E[\text{Var}(Y_{l\to i} \mid X_l)]$ in equation (8.10) produces the recurrence

$$
\begin{aligned}
\text{Var}(X_i) &= \left[\frac{1}{2} + \frac{1}{2(2n-1)}\right] 2n + \frac{1}{4}\left(1 - \frac{1}{2n-1}\right) \text{Var}(X_k) \\
&\quad + \frac{1}{4}\left(1 - \frac{1}{2n-1}\right) \text{Var}(X_l) + \frac{1}{2} \text{Cov}(X_k, X_l). \quad (8.11)
\end{aligned}
$$

In the absence of inbreeding, $\text{Cov}(X_k, X_l) = 0$, and one can argue inductively that $\text{Var}(X_i) = 2n$. Indeed, the induction hypothesis $\text{Var}(X_k) = 2n$ and $\text{Var}(X_l) = 2n$ and the recurrence (8.11) imply that $\text{Var}(X_i) = 2n$.

Summarizing the situation for a non-inbred pedigree, all expectations reduce to $E(X_i) = 0$ and all variances to $\text{Var}(X_i) = 2n$. All covariances either are 0 or are governed by the recurrence (8.9). Thus, insofar as first and second moments are concerned, the hypergeometric polygenic model exactly mimics the inheritance of a fully additive polygenic trait ($\sigma_d^2 = 0$) with mean 0 and variance $2n$. This resemblance and the empirical calculations carried out by Elston, Fernando, and Stricker [8, 22] for the mixed model suggest that the hypergeometric polygenic model is a computationally efficient substitute for the polygenic model. A pleasing aspect of this substitution is that all computations can be performed via a version of the Elston-Stewart algorithm featured in Chapter 7.

## 8.7  Application to Risk Prediction

For a simple numerical application to the polygenic threshold model, consider the pedigree of Figure 8.1. In this pedigree, darkened individuals are afflicted by a hypothetical disease with a prevalence of .01 and a **heritability** of .75. We approximate the polygenic liability to disease of person $i$ in the pedigree by the sum $Z_i = \sigma_a(\frac{1}{\sqrt{2n}}X_i) + \sigma_e Y_i$, where $X_i$ is determined by the hypergeometric polygenic model with $2n$ polygenes; the $Y_i$ are independent, standard normal deviates; $\sigma_a^2 = .75$; and $\sigma_e^2 = .25$. The ratio $\sigma_a^2/(\sigma_a^2 + \sigma_e^2)$ is by definition the heritability of each $Z_i$. Given that each $Z_i$ follows an approximate standard normal distribution, the liability threshold of 2.326 is determined by the prevalence condition $\frac{1}{\sqrt{2\pi}}\int_{2.326}^{\infty} e^{-z^2/2}dz = .01$.

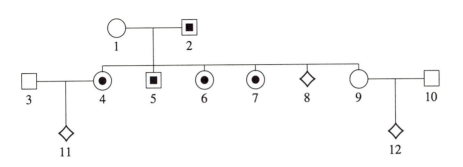

FIGURE 8.1. Risk Prediction Under the Polygenic Threshold Model

The individuals represented by ◇ marks in Figure 8.1 are unborn, potential children. Table 8.3 gives the conditional probabilities that these children will be afflicted with the disease. The recurrence risks recorded evidently stabilize at about 35 percent, 12 percent, and 6 percent as the number of polygenes $2n \rightarrow \infty$. Under the alternative hypothesis of an autosomal dominant mode of disease, these risks are 1/2, 1/2, and 0, respectively. In counseling families such as this one where risks are strongly model dependent, one should obviously exercise caution.

TABLE 8.3. Recurrence Risks for the Unborn Children in Figure 8.1

| Polygenes $2n$ | Child 8 | Child 11 | Child 12 |
|---|---|---|---|
| 10 | .326 | .081 | .054 |
| 20 | .349 | .104 | .057 |
| 30 | .354 | .111 | .058 |
| 40 | .357 | .115 | .058 |
| 50 | .358 | .117 | .058 |

## 8.8   Problems

1. Suppose that $A_i \hat{\mu}$ and $\hat{\Omega}_i$ are the mean vector and covariance matrix for the $i$th of $s$ pedigrees evaluated at the maximum likelihood estimates. Under the multivariate normal model (8.1), show that

$$\sum_{i=1}^{s} (Y^i - A_i \hat{\mu})' \hat{\Omega}_i^{-1} (Y^i - A_i \hat{\mu}) \; = \; \sum_{i=1}^{s} m_i,$$

where $m_i$ is the number of entries of the trait vector $Y^i$ [11]. Hint:

$$\sum_{k=1}^{r} \hat{\sigma}_k^2 \frac{\partial}{\partial \sigma_k^2} L(\hat{\gamma}) \; = \; 0.$$

2. In the notation of Problem 1, prove that the pedigree statistic

$$(Y^i - A_i \mu)' \Omega_i^{-1} (Y^i - A_i \mu)$$

has a $\chi_{m_i}^2$ distribution when evaluated at the true values of $\mu$ and $\sigma^2$ [21]. This $\chi_{m_i}^2$ distribution holds approximately when the maximum likelihood estimates $\hat{\mu}$ and $\hat{\sigma}^2$ are substituted for their true values. There is a slight dependence among the statistics

$$(Y^i - A_i \hat{\mu})' \hat{\Omega}_i^{-1} (Y^i - A_i \hat{\mu})$$

because of the functional relationship featured in Problem 1. (Hint: What is the distribution of $\Omega_i^{-\frac{1}{2}} (Y^i - A_i \mu)$? Recall that a linear transformation of a multivariate normal variate is multivariate normal.)

3. Suppose all pedigrees from a sample have been amalgamated into a single pedigree. For a trait vector $Y$ with $E(Y) = \mathbf{0}$, consider the covariance components model

$$Y_i Y_j \; = \; \sum_{k=1}^{r} \sigma_k^2 \Gamma_{kij} + e_{ij}, \tag{8.12}$$

where the $e_{ij}$ are independent, identically distributed random errors. Let $U$ be the matrix $YY'$, $W_k$ be the matrix $\Gamma_k$, and $e$ be the matrix $(e_{ij})$, all written in lexicographical order as column vectors. Then the model (8.12) can be written as

$$U \; = \; W\sigma^2 + e, \tag{8.13}$$

where $W = (W_1, \ldots, W_r)$. Show that the normal equations for estimating $\sigma^2$ reduce to one step of scoring starting from $(0, \ldots, 0, 1)'$. This result is due to Robert Jennrich.

4. As an alternative to scoring in the polygenic model, one can implement the EM algorithm [4]. In the notation of the text, consider a multivariate normal random vector $Y$ with mean $\nu = A\mu$ and covariance $\Omega = \sum_{k=1}^{r} \sigma_k^2 \Gamma_k$, where $A$ is a fixed design matrix, the $\sigma_k^2 > 0$, the $\Gamma_k$ are positive definite covariance matrices, and $\Gamma_r = I$. Let the complete data consist of independent, multivariate normal random vectors $X^1, \ldots, X^r$ designed so that $Y = \sum_{k=1}^{r} X^k$ and so that $X^k$ has mean $1_{\{k=r\}} A\mu$ and covariance $\sigma_k^2 \Gamma_k$. If $\gamma = (\mu_1, \ldots, \mu_p, \sigma_1^2, \ldots, \sigma_r^2)^t$, and the observed data are amalgamated into a single pedigree with $m$ people, then prove the following assertions:

(a) The complete data loglikelihood is

$$
\ln f(X \mid \gamma) = -\frac{1}{2} \sum_{k=1}^{r} \{ \ln \det \Gamma_k + m \ln \sigma_k^2
$$

$$
+ \frac{1}{\sigma_k^2} [X^k - E(X^k)]^t \Gamma_k^{-1} [X^k - E(X^k)] \}.
$$

(b) Omitting irrelevant constants, the $Q(\gamma \mid \gamma_n)$ function of the EM algorithm is

$$
Q(\gamma \mid \gamma_n)
$$
$$
= -\frac{1}{2} \sum_{k=1}^{r-1} \{ m \ln \sigma_k^2 + \frac{1}{\sigma_k^2} [\operatorname{tr}(\Gamma_k^{-1} \Upsilon_{nk}) + \nu_{nk}^t \Gamma_k^{-1} \nu_{nk}] \}
$$
$$
- \frac{m}{2} \ln \sigma_r^2 - \frac{1}{2\sigma_r^2} [\operatorname{tr}(\Upsilon_{nr}) + (\nu_{nr} - A\mu)^t (\nu_{nr} - A\mu)],
$$

where $\nu_{nk}$ is the conditional mean vector

$$
\nu_k = 1_{\{k=r\}} A\mu + \sigma_k^2 \Gamma_k \Omega^{-1} [y - A\mu]
$$

and $\Upsilon_{nk}$ is the conditional covariance matrix

$$
\Upsilon_k = \sigma_k^2 \Gamma_k - \sigma_k^2 \Gamma_k \Omega^{-1} \sigma_k^2 \Gamma_k
$$

of $X^k$ given $Y = y$ evaluated at the current iterate $\gamma_n$. (Hint: Consider the concatenated random normal vector $\begin{pmatrix} X^k \\ Y \end{pmatrix}$ and use fact (d) proved in the text.)

(c) The solution of the M step is

$$
\mu_{n+1} = (A^t A)^{-1} A^t \nu_{nr}
$$
$$
\sigma_{n+1,k}^2 = \frac{1}{m} [\operatorname{tr}(\Gamma_k^{-1} \Upsilon_{nk}) + \nu_{nk}^t \Gamma_k^{-1} \nu_{nk}], \quad 1 \le k \le r - 1
$$
$$
\sigma_{n+1,r}^2 = \frac{1}{m} [\operatorname{tr}(\Upsilon_{nr}) + (\nu_{nr} - A\mu_{n+1})^t (\nu_{nr} - A\mu_{n+1})].
$$

In the above update, $\mu_{n+1}$ is the next iterate of the mean vector $\mu$ and not a component of $\mu$.

5. Continuing Problem 4, show that $\sigma_{nk}^2 \geq 0$ holds for all $k$ and $n$ if $\sigma_{1k}^2 \geq 0$ holds initially for all $k$. If all $\sigma_{1k}^2 > 0$, show that all $\sigma_{nk}^2 > 0$.

6. Continuing Problem 4, suppose that one or more of the covariance matrices $\Gamma_k$ is singular. For instance, in modeling common household effects, the corresponding missing data $X^k$ can be represented as $X^k = \sigma_k M_k W^k$, where $M_k$ is a constant $m \times s$ matrix having exactly one entry 1 and the remaining entries 0 in each row, and where $W^k$ has $s$ independent, standard normal components. Each component of $W^k$ corresponds to a different household; each row of $M_k$ chooses the correct household for a given person. It follows from this description that

$$\begin{aligned} \sigma_k^2 \Gamma_k &= \mathrm{Var}(X^k) \\ &= \sigma_k^2 M_k M_k^t. \end{aligned}$$

The matrix $H_k = M_k M_k^t$ is the household indicator matrix described in the text. When $X^k$ has the representation $\sigma_k M_k W^k$, one should replace $X^k$ in the complete data by $\sigma_k W^k$. With this change, show that the EM update for $\sigma_k^2$ is

$$\sigma_{n+1,k}^2 = \frac{1}{s}[\mathrm{tr}(\Upsilon_{nk}) + v_{nk}^t v_{nk}],$$

where $v_{nk}$ and $\Upsilon_{nk}$ are

$$\begin{aligned} v_k &= \sigma_k^2 M_k^t \Omega^{-1}(y - A\mu) \\ \Upsilon_k &= \sigma_k^2 I - \sigma_k^4 M_k^t \Omega^{-1} M_k \end{aligned}$$

evaluated at the current parameter vector $\gamma_n$.

7. In the hypergeometric polygenic model, verify that the number of positive polygenes a non-inbred person possesses follows the binomial distribution (8.8). Do this by a qualitative argument and by checking analytically the reproductive property

$$\sum_{g_1} \sum_{g_2} \binom{2n}{g_1}\left(\frac{1}{2}\right)^{2n} \binom{2n}{g_2}\left(\frac{1}{2}\right)^{2n} \tau_{g_1 \times g_2 \to g_3} = \binom{2n}{g_3}\left(\frac{1}{2}\right)^{2n}$$

for polygene transmission under sampling without replacement.

8. In the hypergeometric polygenic model, $\mathrm{Var}(X_i) = 2n$ holds for each person $i$ in a non-inbred pedigree. In the presence of inbreeding, give a counterexample to this formula. However, prove that

$$0 \leq \mathrm{Cov}(X_i, X_j) \leq (2 + q)n$$

for all pairs $i$ and $j$ from a pedigree with $q$ people. Note that the special case $i = j$ gives an upper bound on trait variances. (Hint: Argue by induction using the recurrence formulas for variances and covariances.)

9. In the hypergeometric polygenic model, suppose that one randomly samples each of the $n$ polygenes transmitted to a gamete with replacement rather than without replacement. If $j \neq i$ is not a descendant of $i$, and $i$ has parents $k$ and $l$, then show that this altered model entails

$$E(X_i) = 0$$

$$\mathrm{Cov}(X_i, X_j) = \frac{1}{2}\mathrm{Cov}(X_k, X_j) + \frac{1}{2}\mathrm{Cov}(X_l, X_j)$$

$$\mathrm{Var}(X_i) = 2n + \frac{1}{4}\left(1 - \frac{1}{n}\right)\mathrm{Var}(X_k) + \frac{1}{4}\left(1 - \frac{1}{n}\right)\mathrm{Var}(X_l)$$

$$+ \frac{1}{2}\mathrm{Cov}(X_k, X_l).$$

10. Continuing Problem 9, let $v_m$ be the trait variance of a person $m$ generations removed from his or her relevant pedigree founders in a non-inbred pedigree. Verify that $v_m$ satisfies the difference equation

$$v_m = 2n + \frac{1}{2}\left(1 - \frac{1}{n}\right)v_{m-1}$$

with solution

$$v_m = \frac{4n}{1 + \frac{1}{n}} + \left[\frac{1}{2}\left(1 - \frac{1}{n}\right)\right]^m \left(v_0 - \frac{4n}{1 + \frac{1}{n}}\right).$$

Check that $v_m$ steadily increases from $v_0 = 2n$ to the limit $v_\infty = \frac{4n}{1 + \frac{1}{n}}$.

# References

[1] Boerwinkle E, Chakraborty R, Sing CF (1986) The use of measured genotype information in the analysis of quantitative phenotypes in man. I. Models and analytical methods. *Ann Hum Genet* 50:181–194

[2] Cannings C, Thompson EA, Skolnick MH (1978) Probability functions on complex pedigrees. *Adv Appl Prob* 10:26–61

[3] Daiger SP, Miller M, Chakraborty R (1984) Heritability of quantitative variation at the group-specific component (Gc) locus. *Amer J Hum Genet* 36:663–676

[4] Dempster AP, Laird NM, Rubin DB (1977) Maximum likelihood estimation with incomplete data via the EM algorithm (with discussion). *J Roy Stat Soc B* 39:1–38

[5] Elston RC, Stewart J (1971) A general model for the genetic analysis of pedigree data. *Hum Hered* 21:523–542

[6] Falconer DS (1965) The inheritance of liability to certain diseases, estimated from the incidences among relatives. *Ann Hum Genet* 29:51–79

[7] Falconer DS (1967) The inheritance of liability to diseases with variable age of onset, with particular reference to diabetes mellitus. *Ann Hum Genet* 31:1–20

[8] Fernando RL, Stricker C, Elston RC (1994) The finite polygenic mixed model: An alternative formulation for the mixed model of inheritance. *Theor Appl Genet* 88:573–580

[9] Fisher RA (1918) The correlation between relatives on the supposition of Mendelian inheritance. *Trans Roy Soc Edinburgh* 52:399–433

[10] Holt SB (1954) Genetics of dermal ridges: Bilateral asymmetry in finger ridge-counts. *Ann Eugenics* 18:211–231

[11] Hopper JL, Mathews JD (1982) Extensions to multivariate normal models for pedigree analysis. *Ann Hum Genet* 46:373–383.

[12] Jennrich RI, Sampson PF (1976) Newton-Raphson and related algorithms for maximum likelihood variance component estimation. *Technometrics* 18:11–17

[13] Lange K (1978) Central limit theorems for pedigrees. *J Math Biol* 6:59–66

[14] Lange K (1997) An approximate model of polygenic inheritance.

[15] Lange K, Boehnke M (1983) Extensions to pedigree analysis. IV. Covariance component models for multivariate traits. *Amer J Med Genet* 14:513–524

[16] Lange K, Boehnke M, Weeks D (1987) Programs for pedigree analysis: MENDEL, FISHER, and dGENE. *Genet Epidemiology* 5:473–476

[17] Lange K, Westlake J, Spence MA (1976) Extensions to pedigree analysis. III. Variance components by the scoring method. *Ann Hum Genet* 39:484–491

[18] Morton NE, MacLean CJ (1974) Analysis of family resemblance. III. Complex segregation analysis of quantitative traits. *Amer J Hum Genet* 26:489–503

[19] Ott J (1979) Maximum likelihood estimation by counting methods under polygenic and mixed models in human pedigrees. *Amer J Hum Genet* 31:161–175

[20] Peressini AL, Sullivan FE, Uhl JJ Jr (1988) *The Mathematics of Nonlinear Programming*. Springer-Verlag, New York

[21] Rao CR (1973) *Linear Statistical Inference and its Applications*, 2nd ed. Wiley, New York

[22] Stricker C, Fernando RL, Elston RC (1995) Linkage analysis with an alternative formulation for the mixed model of inheritance: The finite polygenic mixed model. *Genetics* 141:1651–1656

# 9
# Markov Chain Monte Carlo Methods

## 9.1  Introduction

Mapping disease and marker loci from pedigree phenotypes is one of the most computationally onerous tasks in modern biology. Even tightly optimized software can be quickly overwhelmed by the synergistic obstructions of missing data, multiple marker loci, multiple alleles per marker locus, and inbreeding. This unhappy situation has prompted mathematical and statistical geneticists to adapt recent stochastic methods for numerical integration to the demands of pedigree analysis [9, 15, 16, 21, 25, 26, 28]. The current chapter explains in a concrete genetic setting how these powerful stochastic methods operate.

The Markov chain Monte Carlo (MCMC) revolution sweeping statistics is drastically changing how statisticians perform integration and summation. In particular, the Metropolis algorithm and Gibbs sampling make it straightforward to construct a Markov chain sampling from a complicated conditional distribution [4, 5, 7, 10, 19, 24]. Once a sample is available, then any conditional expectation can be approximated by forming its corresponding sample average. The implications of this insight are profound for both classical and Bayesian statistics. As a bonus, trivial changes to the Metropolis algorithm yield simulated annealing, a general-purpose algorithm for solving difficult combinatorial optimization problems [13, 20].

The agenda for the chapter is to (a) review the existing theory of finite-state Markov chains, (b) briefly explain the Metropolis algorithm and simulated annealing, and (c) apply these ideas to the computation of location scores and the reconstruction of haplotypes. These last two applications hardly exhaust the possibilities for stochastic sampling in genetics. Readers are urged to consult the references for further applications.

## 9.2    Review of Discrete-Time Markov Chains

For the sake of simplicity, we will only consider chains with a finite state space [3, 8]. The movement of such a chain from **epoch** (or generation) to epoch is governed by its **transition probability matrix** $P = (p_{ij})$. If $Z_n$ denotes the state of the chain at epoch $n$, then $p_{ij} = \Pr(Z_n = j \mid Z_{n-1} = i)$. As a consequence, every entry of $P$ satisfies $p_{ij} \geq 0$, and every row of $P$ satisfies $\sum_j p_{ij} = 1$. Implicit in the definition of $p_{ij}$ is the fact that the future of the chain is determined by its present without regard to its past. This Markovian property is expressed formally by the equation

$$\Pr(Z_n = i_n \mid Z_{n-1} = i_{n-1}, \ldots, Z_0 = i_0) \quad = \quad \Pr(Z_n = i_n \mid Z_{n-1} = i_{n-1}).$$

The $n$-step transition probability $p_{ij}^{(n)} = \Pr(Z_n = j \mid Z_0 = i)$ is given by the entry in row $i$ and column $j$ of the matrix power $P^n$. This follows because the decomposition

$$p_{ij}^{(n)} \quad = \quad \sum_{i_1} \cdots \sum_{i_{n-1}} p_{ii_1} \cdots p_{i_{n-1}j}$$

over all paths $i \to i_1 \to \cdots \to i_{n-1} \to j$ corresponds to matrix multiplication. A question of fundamental theoretical importance is whether the matrix powers $P^n$ converge. If the chain eventually forgets its starting state, then the limit should have identical rows. Denoting the common limiting row by $\pi$, we deduce that $\pi = \pi P$ from the calculation

$$\begin{pmatrix} \pi \\ \vdots \\ \pi \end{pmatrix} \quad = \quad \lim_{n \to \infty} P^{n+1}$$

$$= \quad \left( \lim_{n \to \infty} P^n \right) P$$

$$= \quad \begin{pmatrix} \pi \\ \vdots \\ \pi \end{pmatrix} P.$$

Any probability distribution $\pi$ on the states of the chain satisfying the condition $\pi = \pi P$ is termed an **equilibrium** or **stationary** distribution of the chain. For finite-state chains, equilibrium distributions always exist [3, 8]. The real issue is uniqueness.

Mathematicians have attacked the uniqueness problem by defining appropriate **ergodic** conditions. For a finite-state chain, two ergodic assumptions are invoked.

The first is **aperiodicity**; this means that the greatest common divisor of the set $\{n \geq 1 : p_{ii}^{(n)} > 0\}$ is 1 for every state $i$. Aperiodicity trivially holds when $p_{ii} > 0$ for all $i$. The second ergodic assumption is **irreducibility**; this means that for every pair of states $(i, j)$ there exists a positive integer $n_{ij}$ such that $p_{ij}^{(n_{ij})} > 0$. In other words, every state is reachable from every other state. Said yet another way, all states **communicate**. For an irreducible chain, Problem 5 states that the integer $n_{ij}$ can be chosen independently of the particular pair $(i, j)$ if and only if the chain is also aperiodic. Thus, we can merge the two ergodic assumptions into the single assumption that some power $P^n$ has all entries positive. Under this single ergodic condition, a unique equilibrium distribution $\pi$ exists with all entries positive.

Equally important is the ergodic theorem [3, 8]. This theorem permits one to run a chain and approximate theoretical means by sample means. More precisely, let $f(z)$ be some function defined on the states of an ergodic chain. Then $\lim_{n \to \infty} \frac{1}{n} \sum_{i=0}^{n-1} f(Z_i)$ exists and equals the theoretical mean

$$\mathrm{E}_\pi[f(Z)] \;=\; \sum_z \pi_z f(z)$$

of $f(Z)$ under the equilibrium distribution $\pi$. This result generalizes the law of large numbers for independent sampling.

The equilibrium condition $\pi = \pi P$ can be restated as the system of equations

$$\pi_j \;=\; \sum_i \pi_i p_{ij} \tag{9.1}$$

for all $j$. In many Markov chain models, the stronger condition

$$\pi_j p_{ji} \;=\; \pi_i p_{ij} \tag{9.2}$$

holds for all pairs $(i, j)$. If this is the case, then the probability distribution $\pi$ is said to satisfy **detailed balance**. Summing equation (9.2) over $i$ yields the equilibrium condition (9.1). An irreducible Markov chain with equilibrium distribution $\pi$ satisfying detailed balance is said to be **reversible**. Irreducibility is imposed to guarantee that all entries of $\pi$ are positive.

If $i_1, \ldots, i_m$ is any sequence of states in a reversible chain, then detailed balance implies

$$\pi_{i_1} p_{i_1 i_2} \;=\; \pi_{i_2} p_{i_2 i_1}$$
$$\pi_{i_2} p_{i_2 i_3} \;=\; \pi_{i_3} p_{i_3 i_2}$$
$$\vdots$$
$$\pi_{i_{m-1}} p_{i_{m-1} i_m} \;=\; \pi_{i_m} p_{i_m i_{m-1}}$$
$$\pi_{i_m} p_{i_m i_1} \;=\; \pi_{i_1} p_{i_1 i_m}.$$

Multiplying these equations together and cancelling the common positive factor $\pi_{i_1} \cdots \pi_{i_m}$ from both sides of the resulting equality gives **Kolmogorov's circulation criterion** [12]

$$p_{i_1 i_2} p_{i_2 i_3} \cdots p_{i_{m-1} i_m} p_{i_m i_1} \;=\; p_{i_1 i_m} p_{i_m i_{m-1}} \cdots p_{i_3 i_2} p_{i_2 i_1}. \tag{9.3}$$

Conversely, if an irreducible Markov chain satisfies Kolmogorov's criterion, then the chain is reversible. This fact can be demonstrated by explicitly constructing the equilibrium distribution and showing that it satisfies detailed balance. The idea behind the construction is to choose some arbitrary reference state $i$ and to pretend that $\pi_i$ is given. If $j$ is another state, let $i \to i_1 \to \cdots \to i_m \to j$ be any path leading from $i$ to $j$. Then the formula

$$\pi_j = \pi_i \frac{p_{ii_1} p_{i_1 i_2} \cdots p_{i_m j}}{p_{j i_m} p_{i_m i_{m-1}} \cdots p_{i_1 i}} \tag{9.4}$$

defines $\pi_j$. A straightforward application of Kolmogorov's criterion (9.3) shows that the definition (9.4) does not depend on the particular path chosen from $i$ to $j$. To validate detailed balance, suppose that $k$ is adjacent to $j$. Then the sequence $i \to i_1 \to \cdots \to i_m \to j \to k$ furnishes a path from $i$ to $k$ through $j$. It follows from (9.4) that $\pi_k = \pi_j p_{jk}/p_{kj}$, which is obviously equivalent to detailed balance. In general, the value of $\pi_i$ is not known beforehand. Setting $\pi_i = 1$ produces the equilibrium distribution up to a normalizing constant.

**Example 9.1** *Two Different Markov Chains on a DNA Strand*

As explained in the Appendix, a DNA strand is constructed from the four bases A (adenine), G (guanine), C (cytosine), and T (thymine). The strand has a directionality so that we can imagine starting at its $5'$ end and walking toward its $3'$ end. As one proceeds along the strand, the bases encountered are not independent. To a first approximation [23], the successive bases conform to a Markov chain with transition matrix

$$P = \begin{array}{c} \\ A \\ C \\ G \\ T \end{array} \begin{pmatrix} \overset{\text{A}}{.32} & \overset{\text{C}}{.18} & \overset{\text{G}}{.23} & \overset{\text{T}}{.27} \\ .37 & .23 & .05 & .35 \\ .30 & .21 & .25 & .24 \\ .23 & .19 & .25 & .33 \end{pmatrix}.$$

It is straightforward to check that the equilibrium distribution of this aperiodic chain is $\pi = (.30, .20, .20, .30)$.

Bishop et al. [1] use the above chain to construct a more complicated Markov chain capturing the random distances between restriction sites. **Restriction enzymes** recognize certain specific sequences of bases along a DNA strand and cut the DNA at these restriction sites. For instance, the enzyme AluI recognizes the sequence AGCT. To investigate the random distance between restriction sites for AluI, it is helpful to construct a chain with states A, C, G, T, AG, AGC, and AGCT. The first four states are interpreted as in the chain above. AG is the state where the current DNA base is G and the previous base is A. Here part of the desired restriction site pattern has been achieved. The state AGC is even further along on the way to the desired full restriction site AGCT. Once AGCT is attained, the next base encountered will completely disrupt the pattern, and we are back again at A,

C, G, or T. This second chain has transition matrix

$$P = \begin{array}{c} \\ A \\ C \\ G \\ T \\ AG \\ AGC \\ AGCT \end{array} \begin{array}{ccccccc} A & C & G & T & AG & AGC & AGCT \\ \begin{pmatrix} .32 & .18 & 0 & .27 & .23 & 0 & 0 \\ .37 & .23 & .05 & .35 & 0 & 0 & 0 \\ .30 & .21 & .25 & .24 & 0 & 0 & 0 \\ .23 & .19 & .25 & .33 & 0 & 0 & 0 \\ .30 & 0 & .25 & .24 & 0 & .21 & 0 \\ .37 & .23 & .05 & 0 & 0 & 0 & .35 \\ .23 & .19 & .25 & .33 & 0 & 0 & 0 \end{pmatrix} \end{array}.$$

## 9.3   The Hastings-Metropolis Algorithm and Simulated Annealing

The Hastings-Metropolis algorithm is a device for constructing a Markov chain with a prescribed equilibrium distribution $\pi$ on a given state space [10, 19]. Each step of the chain is broken into two stages, a proposal stage and an acceptance stage. If the chain is currently in state $i$, then in the proposal stage a new destination state $j$ is proposed according to a probability density $q_{ij} = q(j \mid i)$. In the subsequent acceptance stage, a random number is drawn uniformly from [0, 1] to determine whether the proposed step is actually taken. If this number is less than the Hastings-Metropolis acceptance probability

$$a_{ij} = \min\left\{1, \frac{\pi_j q_{ji}}{\pi_i q_{ij}}\right\}, \tag{9.5}$$

then the proposed step is taken. Otherwise, the proposed step is declined, and the chain remains in place.

Historically, Metropolis et al. [19] considered only symmetric proposal densities with $q_{ij} = q_{ji}$. In this case the acceptance probability reduces to

$$a_{ij} = \min\left\{1, \frac{\pi_j}{\pi_i}\right\}. \tag{9.6}$$

It is also noteworthy that in applying either formula (9.5) or formula (9.6), the $\pi_i$ need only be known up to a multiplicative constant.

To prove that $\pi$ is the equilibrium distribution of the chain constructed from the Hastings-Metropolis scheme (9.5), it suffices to check that detailed balance holds. If $\pi$ puts positive weight on all points of the state space, it is clear that we must impose the requirement that the inequalities $q_{ij} > 0$ and $q_{ji} > 0$ are simultaneously true or simultaneously false. This requirement is also implicit in definition (9.5). Now suppose without loss of generality that the fraction

$$\frac{\pi_j q_{ji}}{\pi_i q_{ij}} \leq 1$$

for some $j \neq i$. Then detailed balance follows immediately from

$$
\begin{aligned}
\pi_i q_{ij} a_{ij} &= \pi_i q_{ij} \frac{\pi_j q_{ji}}{\pi_i q_{ij}} \\
&= \pi_j q_{ji} \\
&= \pi_j q_{ji} a_{ji}.
\end{aligned}
$$

The Gibbs sampler is a special case of the Hastings-Metropolis algorithm for Cartesian product state spaces [4, 5, 7]. In the Gibbs sampler, we suppose that each sample point $i = (i_1, \ldots, i_m)$ has $m$ components. For instance, $i_c$ could represent the genotype of person $c$ in a pedigree of $m$ people. The Gibbs sampler updates one component of $i$ at a time. If the component is chosen randomly and resampled conditional on the remaining components, then the acceptance probability is 1. To prove this assertion, let $i_c$ be the uniformly chosen component, and denote the remaining components by $i_{-c} = (i_1, \ldots, i_{c-1}, i_{c+1}, \ldots, i_m)$. If $j$ is a neighbor of $i$ reachable by changing only component $i_c$, then $j_{-c} = i_{-c}$. Hence, the proposal probability

$$
q_{ij} = \frac{1}{m} \frac{\pi_j}{\sum_{\{k : k_{-c} = i_{-c}\}} \pi_k}
$$

satisfies $\pi_i q_{ij} = \pi_j q_{ji}$, and the ratio appearing in the acceptance probability (9.5) is 1. In the location score application discussed in this chapter, the Gibbs sampler leads to chains that either mix too slowly or are reducible. For this reason, the general Metropolis algorithm is preferable.

In simulated annealing we are interested in finding the most probable state of a Markov chain [13, 20]. If this state is $k$, then $\pi_k > \pi_i$ for all $i \neq k$. To accentuate the weight given to state $k$, we can replace the equilibrium distribution $\pi$ by a distribution putting probability

$$
\pi_i^{(\tau)} = \frac{\pi_i^{\frac{1}{\tau}}}{\sum_j \pi_j^{\frac{1}{\tau}}}
$$

on state $i$. Here $\tau$ is a small, positive parameter traditionally called temperature. For a chain with symmetric proposal density, the distribution $\pi_i^{(\tau)}$ can be attained by running the chain with acceptance probability

$$
a_{ij} = \min \left\{ 1, \left( \frac{\pi_j}{\pi_i} \right)^{\frac{1}{\tau}} \right\}. \tag{9.7}
$$

In fact, what is done in simulated annealing is that the chain is run with $\tau$ gradually decreasing to 0. If $\tau$ starts out large, then in the early steps of simulated annealing, almost all proposed steps are accepted, and the chain broadly samples the state space. As $\tau$ declines, fewer unfavorable steps are taken, and the chain eventually settles on some nearly optimal state. With luck this state is $k$ or a state equivalent to $k$ if several states are optimal. Simulated annealing is designed to mimic the gradual freezing of a substance into a crystalline state of perfect symmetry and hence minimum energy.

# 9.4    Descent States and Descent Graphs

To apply the MCMC method to the analysis of human pedigree data, we must choose an appropriate state space and a mechanism for moving between neighboring states of the space. Given our goal of computing location scores, the state space must capture gene flow at multiple marker loci. For the sake of simplicity and in keeping with most genetic practice, only codominant alleles will be allowed at the marker loci. The state space will also omit mention of the trait locus. This locus is handled somewhat differently and appears later in our discussion.

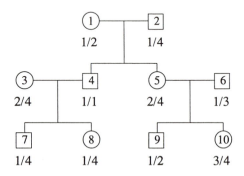

(a) Conventional Pedigree Representation

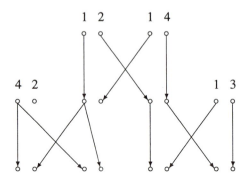

(b) Descent State Description of Gene Flow

FIGURE 9.1. Gene Flow in a Fully Typed Pedigree

The states of our state space are rather complicated graphs describing gene flow in a pedigree at the participating marker loci. It suffices to focus on a single pedigree because location scores are computed pedigree by pedigree. Figure 9.1 (a) depicts a typical pedigree with marker phenotypes noted at a codominant marker locus. Figure 9.1 (b) conveys more detailed, but consistent, information about the gene flow in the pedigree. Each person is replaced by two nodes; the left node is a place holder for his maternally inherited gene, and the right node is a place holder for his paternally inherited gene. Arcs connect parent nodes to child nodes and determine which grandparental genes children inherit. For example, the granddaughter 8 inherits from her father 4 the maternal gene of her grandmother 1. The maternal gene of the grandmother is labeled allele 1, which is consistent with the observed phenotype 1/4 of the granddaughter. The combination of the gene flow graph and the assigned founder alleles in Figure 9.1 (b) constitutes a **descent state** at the locus. The gene flow graph alone is called a **descent graph** at the locus. An assignment of one descent state (respectively, descent graph) to each participating locus constitutes a descent state (respectively, descent graph) of the pedigree.

Several comments are in order at this point. First, a descent state at a locus determines an ordered genotype for each and every person in the pedigree. Some descent states are consistent with the observed phenotypes of the pedigree, and some descent states are not. Those that are consistent are said to be **legal**; those that are not are illegal. Second, if a descent graph is consistent with at least one legal descent state, then the descent graph is legal; otherwise, it is illegal. Obviously, the collection of descent graphs is much smaller than the collection of descent states. This is the reason for preferring descent graphs to descent states as points of the state space [26, 27]. The size of the state space is further diminished by allowing only legal descent graphs.

The equilibrium distribution $\pi$ of our Markov chain should match the distribution of legal descent graphs $\widehat{G}$ conditioned on the observed marker phenotypes $M$ of the pedigree. Because the normalizing factor $\Pr(M)$ is irrelevant in applying the Metropolis acceptance formula (9.6), it suffices to calculate joint probabilities $\Pr(\widehat{G} \cap M)$ rather than the conditional probabilities $\pi_{\widehat{G}} = \Pr(\widehat{G} \mid M)$. If we let $G$ be an arbitrary descent state, then

$$\Pr(\widehat{G} \cap M) = \sum_{G \mapsto \widehat{G} \cap M} \Pr(G), \qquad (9.8)$$

where $G \mapsto \widehat{G} \cap M$ denotes consistency between $G$ and both $\widehat{G}$ and $M$. The descent state probability $\Pr(G)$ is the product

$$\Pr(G) = \mathrm{Prior}(G)\,\mathrm{Trans}(G)$$

of a prior probability and a transmission probability.

Under the usual assumptions of genetic equilibrium, $\mathrm{Prior}(G)$ is the product of the population frequencies of the founder alleles involved in $G$. Since a descent state entails no ambiguities about recombination, $\mathrm{Trans}(G)$ reduces under Haldane's model of recombination to a product of a power of $\frac{1}{2}$ and relevant powers of the recombination fractions and their complements for the adjacent intervals separating the markers. Finally, owing to the fact that all compatible descent states $G \mapsto \widehat{G}$ exhibit the same transmission pattern, we can reexpress the likelihood (9.8) as

$$\Pr(\widehat{G} \cap M) = \mathrm{Trans}(\widehat{G}) \sum_{G \mapsto \widehat{G} \cap M} \mathrm{Prior}(G). \qquad (9.9)$$

In the next section we tackle the subtle problem of quick computation of the sum of priors $\sum_{G \mapsto \widehat{G} \cap M} \text{Prior}(G)$ [14, 22].

## 9.5    Descent Trees and the Founder Tree Graph

Given $l$ loci in a pedigree with $p$ people and $f$ founders, there are $2lp$ nodes in a descent graph. These nodes are grouped in $2lf$ **descent trees**. The descent tree rooted at a particular founder node contains that founder node and those non-founder nodes inheriting the corresponding founder gene. All nodes of a descent tree involve the same locus. When a founder gene is not passed to any descendant of the founder, then the descent tree exists but is degenerate.

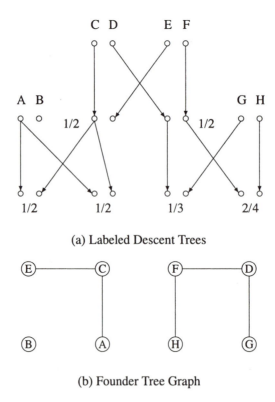

(a) Labeled Descent Trees

(b) Founder Tree Graph

FIGURE 9.2. Construction of a Founder Tree Graph

It is convenient to proceed to a higher level of abstraction and make the founder trees into an undirected graph. This abstraction serves to keep track of how founder alleles are constrained in a coupled manner by the observed marker phenotypes in the pedigree. The nodes of the **founder tree graph** are the descent trees of the descent graph. Two nodes of the founder tree graph are connected by an edge if and only if the two corresponding descent trees pass through the same typed locus of some person in the pedigree. This definition precludes connecting two descent trees associated with different loci. Part (a) of Figure 9.2 shows a descent graph for a single marker locus in which each descent tree is labeled above its rooting founder gene. Do not confuse these labels with the allele symbols used in descent states. Part (b) of Figure 9.2 shows the founder tree graph corresponding to this descent graph, assuming all nonfounders and no founders are typed at the locus.

It is possible for two descent trees at the same locus to mutually impinge on more than one person typed at the locus. Although this information is relevant to discerning whether the two trees are genetically compatible with the observed phenotypes in the pedigree, for the sake of simplicity, we will still view the descent trees as connected by just a single edge. When a descent tree intersects no one typed at its associated locus, the descent tree is isolated from all other descent trees in the founder tree graph.

As suggested above, if we assign an allele to each descent tree via the founder gene at its root, then the fates of two connected descent trees are coupled by the common, typed people through which they pass. For example, if both descent trees pass through an individual having heterozygous genotype $a_i/a_j$, then one of the descent trees must carry allele $a_i$ and the other allele $a_j$. They cannot produce a legal descent state if they both carry allele $a_i$ or allele $a_j$, or one descent tree carries a completely different allele. Refinement of these simple ideas involving the founder tree graph will permit us to compute the prior sum $\sum_{G \mapsto \hat{G} \cap M} \text{Prior}(G)$ associated with a descent graph.

One can subdivide the nodes of the founder tree graph into **connected components**. These components are sets of descent trees and should not be confused with the components of the descent graph, which are single descent trees. In the founder tree graph, two nodes belong to the same component if and only if one can travel from one node to the other by a finite sequence of edges. A component is said to be **singleton** if it consists of a single node. In a nonconsanguineous pedigree, a descent tree forms a singleton component if it passes through no one typed at its associated locus. In this case, all alleles can be legally assigned to the founder gene at the top of the descent tree. In a consanguineous pedigree, a descent tree can form a singleton component even if it descends through a typed child. However, the descent tree must descend to the child via both of its parents. If the typed child has homozygous genotype $a_i/a_i$, then $a_i$ is the only allele permitted for the founder gene. If the typed child is heterozygous, then no legal allele exists for the founder gene.

The situation for a multinode (or nonsingleton) component of the founder tree graph is equally simple. If we label the nodes of the component as $t_1, \ldots, t_k$, then the founder gene of node $t_1$ is transmitted to some typed person who is either homozygous or heterozygous. If the person is homozygous, then there is only one legal choice for the founder gene of $t_1$. Because this founder gene is connected to another founder gene through the current typed person or another typed person, the connected founder gene is also completely determined. This second founder gene is in turn connected to a third founder gene through some typed person.

TABLE 9.1. Allele Vectors for the Components of the Founder Tree Graph

| Descent Trees in Component | Legal Allele Vectors |
|---|---|
| $\{B\}$ | All singleton allele vectors |
| $\{A, C, E\}$ | (1, 2, 1) and (2, 1, 2) |
| $\{D, F, G, H\}$ | (1, 2, 3, 4) |

Hence, the third founder gene is also uniquely determined. In general, a cascade of connecting edges completely determines the permissible alleles for each of the founder genes of the component, unless, of course, an inconsistency is encountered at some step. If descent tree $t_1$ passes through a typed heterozygote, then the founder gene of $t_1$ may be either observed allele. Once one of these two alleles is chosen for $t_1$, then the alleles of all other founder genes in the component are determined by the argument just given. Thus, we can summarize the situation for a multinode component $t_1, \ldots, t_k$ by noting that either two, one, or no allele vectors $\mathbf{a} = (a_{t_1}, \ldots, a_{t_k})$ can be legally assigned to the founder genes of the component. Table 9.1 displays all legal allele vectors for each component of the founder tree graph shown in part (b) of Figure 9.2 based on the genotypes shown in part (a) of the same figure.

To simplify $\sum_{G \mapsto \widehat{G} \cap M} \text{Prior}(G)$, label the connected components of the founder tree graph $C_1, \ldots, C_m$, and let $G \mapsto \widehat{G} \cap M$ be a consistent descent state. As just noted, there is an allele vector $\mathbf{a}_i$ with constituent alleles $a_{ij}$ assigned to each component $C_i$ of the descent state $G$. Under conditions of genetic equilibrium, each founder gene is sampled independently; therefore,

$$\text{Prior}(G) = \prod_{i=1}^{m} \text{Pr}(\mathbf{a}_i)$$

$$= \prod_{i=1}^{m} \prod_{j} \text{Pr}(a_{ij}). \tag{9.10}$$

By construction, the founder genes assigned to different components do not impinge on one another. In other words, the set of founder genes consistent with $\widehat{G}$ and $M$ is drawn from the Cartesian product of the sets $S_1, \ldots, S_m$ of legal allele vectors for the components $C_1, \ldots, C_m$, respectively. Applying the distributive rule to equation (9.10) consequently yields

$$\sum_{G \mapsto \widehat{G} \cap M} \text{Prior}(G) = \prod_{i=1}^{m} \text{Pr}(C_i), \tag{9.11}$$

where

$$\text{Pr}(C_i) = \sum_{\mathbf{a}_i \in S_i} \prod_{j} \text{Pr}(a_{ij}).$$

As mentioned earlier, an allele vector set $S_i$ contains either all allele vectors or just two, one, or none. In the first case, $\Pr(C_i) = \sum_{\mathbf{a}_i \in S_i} \Pr(\mathbf{a}_i) = 1$, and in the remaining three cases, $\Pr(C_i) = \sum_{\mathbf{a}_i \in S_i} \Pr(\mathbf{a}_i)$ contains only two, one, or no terms. Hence, calculation of $\sum_{G \mapsto \widehat{G} \cap M} \Pr{ior}(G)$ reduces to easy component-by-component calculations.

## 9.6    The Descent Graph Markov Chain

The set of descent graphs over a pedigree becomes a Markov chain if we incorporate transition rules for moving between descent graphs. The most basic transition rule, which we call rule $T_0$, switches the origin of an arc descending from a parent to a child from the parental maternal node to the parental paternal node or vice versa [15, 16, 21, 26, 27]. The arbitrary arc chosen is determined by a combination of child, locus, and maternal or paternal source. Figure 9.3 illustrates rule $T_0$ at the black node.

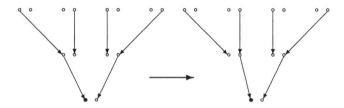

FIGURE 9.3. Example of Transition Rule $T_0$

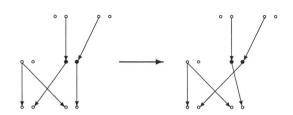

FIGURE 9.4. Example of Transition Rule $T_1$

From the basic rule $T_0$ we can design composite transition rules that make more radical changes in an existing descent graph and consequently speed up the circulation of the chain. For example, the composite transition rule $T_1$ illustrated in Figure 9.4 operates on the two subtrees descending from the person with black nodes at the given locus. One of these subtrees is rooted at the maternal node, and the other is rooted at the paternal node. The two subtrees are detached from their rooting nodes and rerooted at the opposite nodes. More formally, transition rule $T_1$ begins by choosing a person $i$ and a locus $l$. It then performs a $T_0$ transition at each node determined by a child of $i$, the given locus $l$, and the sex of $i$. Thus, every child of $i$ who previously inherited $i$'s maternal gene now inherits $i$'s paternal gene and vice versa.

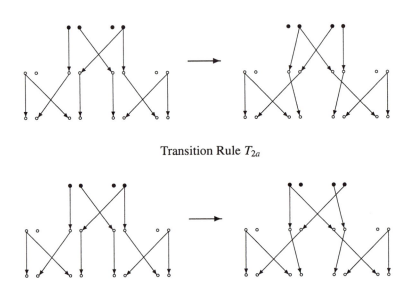

Transition Rule $T_{2a}$

Transition Rule $T_{2b}$

FIGURE 9.5. Examples of Transition Rules $T_{2a}$ and $T_{2b}$

Our second composite transition rule has the two variants $T_{2a}$ and $T_{2b}$ illustrated in Figure 9.5. Each variant begins by choosing a locus $l$ and a couple $i$ and $j$ with common children. Four different descent subtrees are rooted at the parents $i$ and $j$. In Figure 9.5 these start at the black nodes. Rule $T_{2a}$ exchanges the subtree rooted at the maternal node of $i$ with the subtree rooted at the maternal node of $j$; it likewise exchanges the paternally rooted subtrees of $i$ and $j$. In contrast, rule $T_{2b}$ exchanges the maternally rooted subtree of $i$ with the paternally rooted subtree of $j$ and vice versa. Two subtle points of this process are worth stressing. After swapping subtrees, we have paternally derived genes flowing to maternal nodes and vice versa. The obvious adjustments must be made in the children and grand-children to correct these forbidden patterns of gene flow. Also, if either parent has children with another spouse, then that parent's relevant subtrees are reduced. Only the paths descending through the children shared with the chosen spouse are pertinent. Problem 11 asks readers to provide a formal description of the sequence of $T_0$ and $T_1$ transitions invoked in executing a $T_{2a}$ or $T_{2b}$ transition.

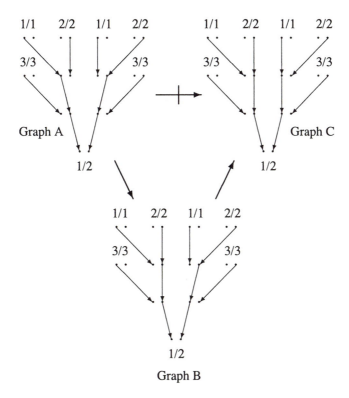

FIGURE 9.6. Failure of Descent Graphs A and C to Communicate

One of the complications in constructing a Markov chain on legal descent graphs is that two states may not communicate in the presence of three or more alleles per marker. Figure 9.6 gives a counterexample involving a single marker locus. In the pedigree depicted in Figure 9.6, all founders are typed and homozygous; the great-grandchild is typed and heterozygous. This great-grandchild must receive his allele 1 from one pair of great-grandparents and his allele 2 from the other pair. The two possibilities are labeled descent graph A and descent graph C. However, it is impossible to move in a finite number of transitions from descent graph A to descent graph C without passing through an illegal descent graph such as descent graph B, where the great-grandchild inherits a homozygous genotype.

The remedy to this dilemma is to "tunnel through" illegal descent graphs by taking multiple transitions per step of the Markov chain [21]. In practice, we employ a random number of transitions per step of the chain. This permits the chain to pass through illegal descent graphs on its way between legal descent graphs. Among the many devices for selecting the random number of transitions per step, one of the most natural is to sample from a geometric distribution with mean 2. This procedure entails taking a single transition with probability $\frac{1}{2}$, two transitions with probability $\frac{1}{4}$, three transitions with probability $\frac{1}{8}$, and so forth. For each transition one randomly selects a transition rule and a person and locus. If the transition rule selected is $T_0$, then one also randomly selects a maternal or paternal node to switch. If one of the $T_2$ transitions is selected, then one also randomly selects a spouse of the selected person.

Although selections are random, they need not be uniform. For example, it is probably wise to select transition $T_0$ more often than $T_1$, and $T_1$ more often than $T_2$, and to target untyped people more often than typed people. It makes sense to make other choices uniformly, such as the selection of a spouse for a $T_2$ transition or of a maternal or paternal node for a $T_0$ transition. In implementing the Metropolis algorithm, it simplifies matters to keep the proposal distribution symmetric and use equation (9.6). This is possible if independent choices are made at each transition. Indeed, because each transition is its own inverse, taking a given sequence of transitions in reverse order leads back from a proposed descent graph to the current descent graph.

The Metropolis algorithm always takes steps to more favorable descent graphs but never allows steps to illegal descent graphs. It is perfectly possible for the Markov chain to remain in place if a step is rejected or the step consists of a double application of the same transition. This feature forces the chain to be aperiodic. Finally, the chain is also irreducible since the tunneling mechanism permits the chain to move from any legal descent graph to any other legal descent graph in a single step.

## 9.7  Computing Location Scores

Location scores can be computed by a hybrid of stochastic sampling of marker descent graphs and deterministic likelihood evaluation. Denote the unknown trait position by $d$ and the observed trait phenotypes on a pedigree by $T$. Since it is trivial to compute the likelihood $\Pr(T)$ of the trait phenotypes in the absence of the marker phenotypes, the key ingredient in computing a location score $\log_{10}\left[\frac{\Pr_d(T \mid M)}{\Pr(T)}\right]$ is the conditional probability $\Pr_d(T \mid M)$.

If we can sample from marker descent graphs $\widehat{G}$ given the marker types $M$, then we can employ standard pedigree likelihood programs such as MENDEL [17] to estimate $\Pr_d(T \mid M)$. The basis for this computation is the obvious decomposition

$$\Pr_d(T \mid M) \;=\; \sum_{\widehat{G}} \Pr_d(T \mid \widehat{G}) \Pr(\widehat{G} \mid M), \tag{9.12}$$

which relies implicitly on the assumption of linkage equilibrium between the trait and marker loci. To evaluate (9.12), we run a Metropolis-coupled Markov chain on marker descent graphs $\widehat{G}$. This chain has equilibrium distribution matching the conditional distribution $\Pr(\widehat{G} \mid M)$. If a sequence of descent graphs $\widehat{G}_0, \ldots, \widehat{G}_{n-1}$ is generated by running the chain, then the sample average $\frac{1}{n}\sum_{i=0}^{n-1}\Pr_d(T \mid \widehat{G}_i)$ will approximate $\Pr_d(T \mid M)$ accurately for $n$ sufficiently large.

Deterministic computation of $\text{Pr}_d(T \cap \widehat{G}_i)$ can be done by MENDEL if it is alerted to recognize a mixture of the marker descent graph $\widehat{G}_i$ and the trait phenotypes $T$ as legitimate input [22]. Division of the joint likelihood $\text{Pr}_d(T \cap \widehat{G}_i)$ output by MENDEL by the marginal likelihood $\text{Pr}(\widehat{G}_i)$ then gives the requisite conditional likelihood $\text{Pr}_d(T \mid \widehat{G}_i)$ used in computing the sample average approximation. Since the trait locus is usually biallelic, and since sampling from the Markov chain fills in all of the missing information on marker gene flow, the deterministic part of a location score calculation is generally quick.

# 9.8   Finding a Legal Descent Graph

The MCMC method of location scores must start with a legal descent graph. Finding such a descent graph is harder than it first appears, but fortunately the problem yields to a randomized version of genotype elimination. The successful strategy proceeds locus by locus and constructs a legal vector of ordered genotypes for a pedigree. From this vector a descent state and corresponding descent graph are then assembled. Based on the genotype elimination method of Chapter 7, the following algorithm applies [22]:

1. Perform step (A) of genotype elimination on the pedigree.

2. Perform steps (B) and (C) of genotype elimination.

3. Consider each individual's genotype list:

   (a) If all people possess exactly one ordered genotype, then use these genotypes to construct a descent state, assigning sources in the process. If a parent is homozygous at a locus, then randomly assign sources to all genes contributed by the parent to his or her children. Exit the algorithm with success.

   (b) If any genotype list is empty, then either there is an inconsistency in the pedigree data, or one of the rare counterexamples to the optimality of genotype elimination has occurred. In either case, exit the algorithm with failure.

   (c) Otherwise, choose one of the people with multiple genotypes currently listed and randomly eliminate all but one of his or her ordered genotypes. Now return to step 2.

If the algorithm fails, then one should check the pedigree for phenotyping errors and nonpaternity. One of these two alternatives is certain for a graphically simple pedigree. If no errors are found, and the pedigree has cycles or loops—for instance, if it is inbred—then the algorithm should be retried with different random choices in step 3, part (c).

## 9.9    Haplotyping

In haplotyping one attempts to find the most likely descent state for a selected group of markers typed on a pedigree [14, 22]. Simulated annealing is designed to solve combinatorial optimization problems of just this sort. Because the space of descent states is very large, it is again advantageous to work on the much smaller state of descent graphs. This entails maximizing a different function than the conditional likelihood $\pi_{\widehat{G}}$ = Pr($\widehat{G}$ | $M$) of a descent graph $\widehat{G}$ given the marker phenotypes $M$. Here we assign to $\widehat{G}$ the joint likelihood Pr($G$) = Pr($G \cap M$) of the most likely descent state $G$ consistent with both $\widehat{G}$ and $M$. This modified objective function is substituted for $\pi_{\widehat{G}}$ in the simulated annealing acceptance probability (9.7).

Recall that the transmission probability Trans(G) in the joint likelihood Pr($G \cap M$) does not depend on $G$. A best descent state corresponding to the descent graph $\widehat{G}$ therefore maximizes the product

$$\text{Prior}(G) \quad = \quad \prod_{i=1}^{m} \text{Pr}(\mathbf{a}_i),$$

where $\mathbf{a}_i$ is any legal allele vector assigned to component $C_i$ of the founder tree graph associated with $\widehat{G}$. To maximize Prior($G$), one simply maximizes each factor Pr($\mathbf{a}_i$) over its set $S_i$ of legal allele vectors. When the set $S_i$ has one or two members, then it is trivial to choose the best member. If $S_i$ consists of more than two members, then $C_i$ must consist of a single descent tree, and $S_i$ contains all possible alleles for the corresponding founder gene. In this case, one chooses the allele with maximum population frequency. Except for the gradual lowering of temperature and the above indicated revision of the acceptance probability, the remaining details of simulated annealing exactly parallel the Markov chain simulations employed in calculating location scores.

## 9.10    Application to Episodic Ataxia

We now apply the preceding theory to the pedigree of episodic ataxia shown in Figure 7.3. After manually haplotyping the pedigree, Litt et al. [18] reject the standard CEPH marker map [2] because it "would result in an obligate triple crossover, within a 3-cM region, in individual 113." Accordingly, their Figure 2A presents a haplotype vector for the pedigree using the alternative order that shifts locus D12S99 two positions distal (toward the telomere) to its CEPH position. They claim that this alternative order reduces the apparent triple crossover to a single crossover.

The descent graph method improves on their manual haplotyping of the nine marker loci and produces the haplotypes shown in Figure 7.3. The original disease-bearing chromosome passed from affected to affected is flagged by • signs. This chromosome is disrupted twice by recombination events. Close inspection of our computer-generated reconstruction shows that it eliminates the triple crossover and a total of three superfluous recombination events postulated in the Litt et al. reconstruction [18]. Thus, there is no reason to question the CEPH map. Fortunately, these revisions do not affect the conclusion drawn by Litt et al. that the episodic ataxia locus lies between the marker D12S372 and the pY2/1–pY21/1–KCNA5–D12S99 marker cluster.

The episodic ataxia pedigree also illustrates MCMC calculation of location scores. As mentioned in Chapter 7, this pedigree is near the limit of what is computable by deterministic likelihood algorithms. Eliminating the three loci pY21/1, KCNA5, and D12S99, MENDEL calculates the exact location scores given by the continuous curve in Figure 7.5. The difference between the exact scores and the MCMC location scores (the dotted curve in Figure 7.5) is always less than 0.1 and usually less than 0.04. It is noteworthy that the deterministic calculations take 11 times longer than the MCMC calculations on one desktop computer —- 2 hours versus 22 hours. Even more impressive is that scaling up to larger pedigrees and a denser marker map is straightforward for the MCMC method but impractical for deterministic methods.

## 9.11    Problems

1. Numerically find the equilibrium distribution of the Markov chain corresponding to the AluI restriction site model. Is this chain reversible?

2. The restriction enzyme HhaI has the recognition site GCGC. Formulate a Markov chain for the attainment of this restriction site when moving along a DNA strand. What are the states and what are the transition probabilities?

3. "Selfing" is a plant breeding scheme that mates an organism with itself, selects one of the progeny randomly and mates it with itself, and so forth from generation to generation. Suppose at some genetic locus there are two alleles $A$ and $a$. A plant can have any of the three genotypes $A/A$, $A/a$, or $a/a$. Define a Markov chain with three states giving the genotype of the current plant in the selfing scheme. Show that the $n$th power of the transition matrix is

$$P^n = \begin{pmatrix} 1 & 0 & 0 \\ \frac{1}{2} - (\frac{1}{2})^{n+1} & (\frac{1}{2})^n & \frac{1}{2} - (\frac{1}{2})^{n+1} \\ 0 & 0 & 1 \end{pmatrix}.$$

What is $\lim_{n\to\infty} P^n$? Demonstrate that this Markov chain has multiple equilibrium distributions and characterize them.

4. Find a transition matrix $P$ such that $\lim_{n\to\infty} P^n$ does not exist.

5. For an irreducible chain, demonstrate that aperiodicity is a necessary and sufficient condition for some power $P^n$ of the transition matrix $P$ to have all entries positive.

6. Let $Z_0, Z_1, Z_2, \ldots$ be a realization of an ergodic chain. If we sample every $k$th epoch, then show (a) that the sampled chain $Z_0, Z_k, Z_{2k}, \ldots$ is ergodic, (b) that it possesses the same equilibrium distribution as the original chain, and (c) that it is reversible if the original chain is. Thus, we can estimate theoretical means by sample averages using only every $k$th epoch of the original chain.

7. The Metropolis acceptance mechanism (9.6) ordinarily implies aperiodicity of the underlying Markov chain. Show that if the proposal distribution is symmetric and if some state $i$ has a neighboring state $j$ such that $\pi_i > \pi_j$, then the period of state $i$ is 1, and the chain, if irreducible, is aperiodic. For a counterexample, assign probability $\pi_i = \frac{1}{4}$ to each vertex $i$ of a square. If the two vertices adjacent to a given vertex $i$ are each proposed with probability $\frac{1}{2}$, then show that all proposed steps are accepted by the Metropolis criterion and that the chain is periodic with period 2.

8. If the component updated in Gibbs sampling depends probabilistically on the current state of the chain, how must the Hastings-Metropolis acceptance probability be modified to preserve detailed balance? Under the appropriate modification, the acceptance probability is no longer always 1.

9. Importance sampling is one remedy when the states of a Markov chain communicate poorly [10]. Suppose that $\pi$ is the equilibrium distribution of the chain. If we sample from a chain whose distribution is $\nu$, then we can recover approximate expectations with respect to $\pi$ by taking weighted averages. In this scheme the state $z$ is given weight $w_z = \pi_z/\nu_z$. If $Z_0, Z_1, Z_2 \ldots$ is a run from the chain with equilibrium distribution $\nu$, then under the appropriate ergodic assumptions, prove that

$$\lim_{n \to \infty} \frac{\sum_{i=0}^{n-1} w_{Z_i} f(Z_i)}{\sum_{i=0}^{n-1} w_{Z_i}} = E_\pi[f(X)].$$

The choice $\nu_z \propto \pi_z^{1/\tau}$ for $\tau > 1$ lowers the peaks and raises the valleys of $\pi$ [11]. Unfortunately in practice, if $\nu$ differs too much from $\pi$, then the estimator

$$\frac{\sum_{i=0}^{n-1} w_{Z_i} f(Z_i)}{\sum_{i=0}^{n-1} w_{Z_i}}$$

of the expectation $E_\pi[f(X)]$ will have a large variance for $n$ of moderate size.

10. Another device to improve mixing of a Markov chain is to run several parallel chains on the same state space and occasionally swap their states [6]. If $\pi$ is the distribution of the chain we wish to sample from, then let $\pi^{(1)}$ equal

$\pi$, and define $m - 1$ additional distributions $\pi^{(2)}, \ldots, \pi^{(m)}$. For instance, incremental heating can be achieved by taking

$$\pi_z^{(k)} \propto \pi_z^{\frac{1}{1+(k-1)\tau}}$$

for $\tau > 0$. At epoch $n$ we sample for each chain $k$ a state $Z_{nk}$ given the chain's previous state $Z_{n-1,k}$. We then randomly select chain $i$ with probability $\frac{1}{m}$ and consider swapping states between it and chain $j = i + 1$. (When $i = m$ no swap is performed.) Under appropriate ergodic assumptions on the $m$ participating chains, show that if the acceptance probability for the proposed swap is

$$\min\left\{1, \frac{\pi_{Z_{nj}}^{(i)} \pi_{Z_{ni}}^{(j)}}{\pi_{Z_{ni}}^{(i)} \pi_{Z_{nj}}^{(j)}}\right\},$$

then the product chain is ergodic with equilibrium distribution given by the product distribution $\pi^{(1)} \otimes \pi^{(2)} \otimes \cdots \otimes \pi^{(m)}$. The marginal distribution of this distribution for chain 1 is just $\pi$. Therefore, we can throw away the outcomes of chains 2 through $m$, and estimate expectations with respect to $\pi$ by forming sample averages from the embedded run of chain 1. (Hint: The fact that no swap is possible at each step allows the chains to run independently for an arbitrary number of steps.)

11. Formally describe the transition rules $T_{2a}$ and $T_{2b}$ on descent graphs in terms of the transition rules $T_0$ and $T_1$.

# References

[1] Bishop DT, Williamson JA, Skolnick MH (1983) A model for restriction fragment length distributions. *Amer J Hum Genet* 35:795–815

[2] Dausset J, Cann H, Cohen D, Lathrop M, Lalouel J-M, White R (1990) Centre d'Etude du Polymorphisme Humain (CEPH): Collaborative genetic mapping of the human genome. *Genomics* 6:575–577

[3] Feller W (1968) *An Introduction to Probability Theory and its Applications, Vol 1*, 3rd ed. Wiley, New York

[4] Gelfand AE, Smith AFM (1990) Sampling-based approaches to calculating marginal densities. *J Amer Stat Assoc* 85:398–409

[5] Geman S, Geman D (1984) Stochastic relaxation, Gibbs distributions and the Bayesian restoration of images. *IEEE Trans Pattern Anal Machine Intell* 6:721–741

[6] Geyer CJ (1991) Markov chain Monte Carlo maximum likelihood. *Computing Science and Statistics: Proceedings of the 23rd Symposium on the Interface,* Keramidas EM, editor, Interface Foundation, Fairfax, VA pp 156–163

[7] Gilks WR, Richardson S, Spiegelhalter DJ, editors, (1996) *Markov Chain Monte Carlo in Practice.* Chapman and Hall, London

[8] Grimmett GR, Stirzaker DR (1992) *Probability and Random Processes,* 2nd ed. Oxford University Press, Oxford

[9] Guo SW, Thompson EA (1992) A Monte Carlo method for combined segregation and linkage analysis. *Amer J Hum Genet* 51:1111–1126

[10] Hastings WK (1970) Monte Carlo sampling methods using Markov chains and their applications. *Biometrika* 57:97–109

[11] Jennison C (1993) Discussion on the meeting of the Gibbs sampler and other Markov chain Monte Carlo methods. *J Roy Stat Soc B* 55:54–56

[12] Kelly FP (1979) *Reversibility and Stochastic Networks.* Wiley, New York

[13] Kirkpatrick S, Gelatt CD, Vecchi MP (1983) Optimization by simulated annealing. *Science* 220:671–680

[14] Kruglyak L, Daly MJ, Reeve-Daly MP, Lander ES (1996) Parametric and nonparametric linkage analysis: A unified multipoint approach. *Amer J Hum Genet* 58:1347–1363

[15] Lange K, Matthysse S (1989) Simulation of pedigree genotypes by random walks. *Amer J Hum Genet* 45:959–970

[16] Lange K, Sobel E (1991) A random walk method for computing genetic location scores. *Amer J Hum Genet* 49:1320–1334

[17] Lange K, Weeks DE, Boehnke M (1988) Programs for pedigree analysis: MENDEL, FISHER, and dGENE. *Genet Epidemiol* 5:471–472

[18] Litt M, Kramer P, Browne D, Gancher S, Brunt ERP, Root D, Phromchotikul T, Dubay CJ, Nutt J (1994) A gene for Episodic Ataxia/Myokymia maps to chromosome 12p13. *Amer J Hum Genet* 55:702–709

[19] Metropolis N, Rosenbluth A, Rosenbluth M, Teller A, Teller E (1953) Equations of state calculations by fast computing machines. *J Chem Physics* 21:1087–1092

[20] Press WH, Teukolsky SA, Vetterling WT, Flannery BP (1992) *Numerical Recipes in Fortran: The Art of Scientific Computing,* 2nd ed. Cambridge University Press, Cambridge

[21] Sobel E, Lange K (1993) Metropolis sampling in pedigree analysis. *Stat Methods Med Res* 2:263–282

[22] Sobel E, Lange K, O'Connell J, Weeks DE (1996) Haplotyping algorithms. In *Genetic Mapping and DNA Sequencing, IMA Volume 81 in Mathematics and its Applications.* Speed TP, Waterman MS, editors, Springer-Verlag, New York, pp 89–110

[23] Swartz MN, Trautner TA, Kornberg A (1962) Enzymatic synthesis of deoxyribonucleic acid. XI. Further studies on the nearest neighbor base sequences in deoxyribonucleic acids. *J Biol Chem* 237:864–875

[24] Tanner MA (1993) *Tools for Statistical Inference: Methods for Exploration of Posterior Distributions and Likelihood Functions*, 2nd ed. Springer-Verlag, New York

[25] Thomas DC, Cortessis V (1992) A Gibbs sampling approach to linkage analysis. *Hum Hered* 42:63–76

[26] Thompson EA (1994) Monte Carlo likelihood in genetic mapping. *Stat Sci* 9:355–366

[27] Thompson EA (1996) Likelihood and linkage: From Fisher to the future. *Ann Stat* 24:449–465

[28] Thompson EA, Guo SW (1991) Evaluation of likelihood ratios on complex genetic models. *IMA J Math Appl Med Biol* 8:149–169

# 10
# Reconstruction of Evolutionary Trees

## 10.1   Introduction

Inferring the evolutionary relationships among related **taxa** (species, genera, families, or higher groupings) is one of the most fascinating problems of molecular genetics [11, 13, 14]. It is now relatively simple to sequence genes and to compare the results from several contemporary taxa. In the current chapter we will assume that the chore of aligning the DNA sequences from these taxa has been successfully accomplished. The taxa are then arranged in an **evolutionary tree** (or **phylogeny**) depicting how taxa diverge from common ancestors. A single ancestral taxon roots the binary tree describing the evolution of the contemporary taxa. The reconstruction problem can be briefly stated as finding the rooted evolutionary tree best fitting the current DNA data. Once the best tree is identified, it is also of interest to estimate the branch lengths of the tree. These tell us something about the pace of evolution. For the sake of brevity, we will focus on the problem of finding the best tree.

It is worth emphasizing that molecular phylogeny is an area of intense current research. The models described here are caricatures of reality. Besides the dubious assumption that alignment is perfect, the models fail to handle site-to-site variation in the rate of evolution, correlation in the evolution of neighboring sites, and sequence variation within a taxon. Nonetheless, evolutionary biologists take the attitude that it is necessary to start somewhere and that a failure to account for details will not distort overall patterns if the patterns are sufficiently obvious. Mathematical biology abounds with compromises of this sort.

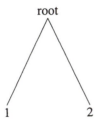

FIGURE 10.1. The Only Rooted Tree for Two Contemporary Taxa

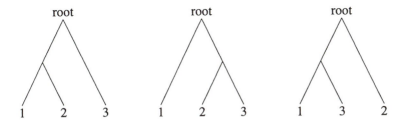

FIGURE 10.2. The Three Rooted Trees for Three Contemporary Taxa

## 10.2   Evolutionary Trees

An **evolutionary tree** is a directed graph showing the relationships between a group of contemporary taxa and their hypothetical common ancestors. The **root** of the tree is the common ancestor of all of the contemporary taxa. The other nodes are either the contemporary taxa at the **tips** of the tree or speciation events (**internal nodes**) from which two new taxa bifurcate.

The first theoretically interesting question about evolutionary trees is how many trees $T_n$ there are for $n$ contemporary taxa. For $n = 2$, obviously $T_n = 1$; the single tree with two contemporary taxa is depicted in Figure 10.1. The three possible trees for three contemporary taxa are depicted in Figure 10.2. In general, $T_n = \frac{(2n-3)!}{2^{n-2}(n-2)!}$. Thus, $T_4 = 15$, $T_5 = 105$, $T_6 = 945$, and so forth.

To verify the formula for $T_n$, first note that an evolutionary tree with $n$ tips has $2n - 1$ nodes and $2n - 2$ edges. This is certainly true for $n = 2$, and it follows inductively because every new tip to the tree adds two nodes and two edges. The formula for $T_n$ is proved in similar inductive fashion. $T_2 = (2 \cdot 2 - 3)! / (2^{2-2}0!) = 1$ obviously works. Given a tree with $n$ tips, tip $n + 1$ can be attached to any one of the existing $2n - 2$ edges or it can be attached directly to the root if the current bifurcation of the root is moved slightly forward in time. Thus,

$$T_{n+1} = (2n - 1)T_n$$

$$= \frac{(2n-1)(2n-3)!}{2^{n-2}(n-2)!}$$

$$= \frac{(2n-1)(2n-2)(2n-3)!}{2^{n-2}2(n-1)(n-2)!}$$

$$= \frac{(2n-1)!}{2^{n-1}(n-1)!}.$$

For instance with $n = 2$, the single rooted tree of Figure 10.1 is transformed into one of three rooted trees of Figure 10.2 by the addition of the tip 3.

## 10.3    Maximum Parsimony

The first step in constructing an evolutionary tree is to partially sequence the DNA of one representative member from each of several related taxa. A site-by-site comparison of the bases observed along some common stretch of DNA is then undertaken to ascertain which evolutionary tree best explains the relationships among the taxa. In the past, evolutionary biologists have also compared amino acid sequences deduced from one or more common proteins. DNA sequence data are now preferred because of their greater information content. As discussed in the Appendix, the four DNA bases are A (adenine), G (guanine), C (cytosine), and T (thymine). Of these, two are **purines** (A and G), and two are **pyrimidines** (C and T).

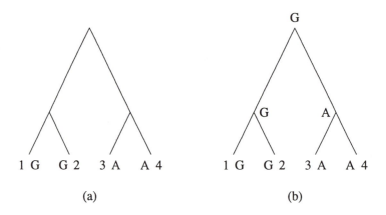

FIGURE 10.3. A Maximum Parsimony Assignment

The **maximum parsimony method** first devised by Eck and Dayhoff [2] and later modified by Fitch [5] provides a computationally fast technique for choosing a best evolutionary tree. Maximum parsimony assigns bases site by site to the internal nodes of an evolutionary tree so as to achieve the minimum number of

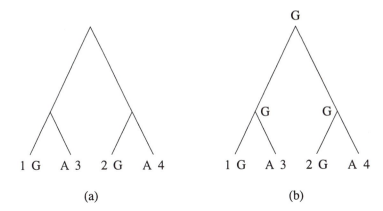

FIGURE 10.4. Another Maximum Parsimony Assignment

base changes as one passes from descendant nodes to ancestral nodes. For instance, Figure 10.3 (a) depicts the result of sequencing four contemporary taxa 1, 2, 3, and 4 at a particular DNA site; Figure 10.3 (b) represents an assignment of bases to the internal nodes that leads to only one disagreement between neighboring nodes in the given evolutionary tree. This assignment, which is not unique, is a maximum parsimony assignment. Figures 10.4 (a) and 10.4 (b) depict a maximum parsimony assignment to a different evolutionary tree. In this case, the minimum number of base changes is two. Hence, these two evolutionary trees can be distinguished given the bases observed on the four contemporary taxa. When many sites are considered, each rooted tree is assigned a maximum parsimony score at each site. These scores are then added over all sites to give a maximum parsimony criterion for a rooted tree.

One flaw in this scheme is that several rooted trees will possess the same maximum parsimony score. This fact can be appreciated by considering the role of the root. The root is unique in having exactly two neighbors. All other internal nodes have three neighbors, and the tips have one neighbor. If on one hand the bases at the two neighbors of the root agree, then the root will be assigned their shared base. If on the other hand the bases at the two neighbors of the root disagree, then the root will be assigned a base agreeing with one neighbor and disagreeing with the other neighbor. In either case, omitting the root leaves the maximum parsimony score assigned to the rooted tree unchanged. Thus, rooted trees that lead to the same **unrooted tree** are indistinguishable under maximum parsimony. Figure 10.5 illustrates how two different rooted trees can collapse to the same unrooted tree. The unrooted tree with the minimum maximum parsimony sum is declared the best unrooted tree.

Let us now demonstrate in detail how maximum parsimony operates [2, 5, 6]. Assignment of bases to nodes is done inductively starting with the tips, at which the bases are naturally fixed. Now suppose that we have solved the maximum

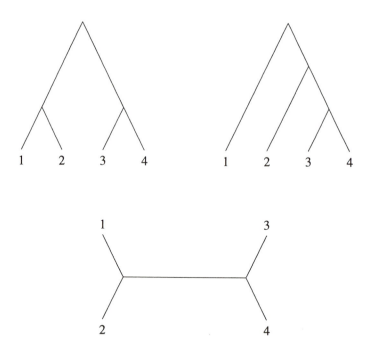

FIGURE 10.5. Two Rooted Trees Corresponding to the Same Unrooted Tree

parsimony problem for the two direct descendants $i$ and $j$ of an internal node $k$. By this we mean that we have constructed a maximum parsimony assignment to the subtree consisting of $i$ and all of its descendants under the condition that the base at $i$ is fixed at a particular value $b_i$. We likewise assume that we have constructed a maximum parsimony assignment to the subtree consisting of $j$ and all of its descendants under the condition that the base at $j$ is fixed at a particular value $b_j$. Let the corresponding maximum parsimony scores be $s_i(b_i)$ and $s_j(b_j)$. If node $i$ represents a contemporary taxon with observed base $b_i^{obs}$, then we take $s_i(b_i) = 1_{\{b_i = b_i^{obs}\}}$ and similarly for node $j$.

Suppose we now fix the base of the parent node $k$ at a particular value $b_k$. The value of the maximum parsimony score $s_k(b_k)$ assigned to $k$ under this condition is by definition

$$s_k(b_k) \quad = \quad \min_{(b_i, b_j)} [1_{\{b_k \neq b_i\}} + 1_{\{b_k \neq b_j\}} + s_i(b_i) + s_j(b_j)]. \qquad (10.1)$$

We now move inductively upward through the tree until reaching the root $l$. At that juncture, the maximum parsimony score for the whole tree is $s = \min_{b_l} s_l(b_l)$. If in equation (10.1) node $i$ is a contemporary taxon, then in view of the definition of $s_i(b_i)$, the intermediate score $s_k(b_k)$ reduces to

$$s_k(b_k) \quad = \quad \min_{b_j} [1_{\{b_k \neq b_i^{obs}\}} + 1_{\{b_k \neq b_j\}} + s_j(b_j)].$$

Analogous simplifications hold when node $j$ is a contemporary taxon and when nodes $i$ and $j$ are both contemporary taxa.

For three contemporary taxa, it is trivial to verify that the parsimony score $s$ takes one of the three values 0, 1, or 2. If the observed bases of the three taxa all agree, then $s = 0$; if two bases agree and one disagrees, then $s = 1$; and if all three disagree, then $s = 2$. Since the three possible rooted trees for three contemporary taxa collapse to the same unrooted tree, these numerical conclusions are consistent with the fact that maximum parsimony is actually a device for distinguishing unrooted trees. Problem 2 explores the more informative situation with four contemporary taxa.

In spite of maximum parsimony's speed and intuitive appeal, it can be misleading in extreme cases. Felsenstein [3] points out that maximum parsimony even fails the basic test of statistical consistency. In the remainder of this chapter, we focus on building a parametric model for the evolutionary changes at a single DNA site and implementing maximum likelihood estimation within the context of this model [4].

## 10.4    Review of Continuous-Time Markov Chains

As a prelude to the model-based approach, let us pause to review the theory of finite-state, continuous-time Markov chains. Just as in the discrete-time theory summarized in Chapter 9, the behavior of a Markov chain can be described by an indexed family $Z_t$ of random variables giving the state occupied by the chain at each time $t$. Of fundamental theoretical importance are the finite-time transition probabilities $p_{ij}(t) = \Pr(Z_t = j \mid Z_0 = i)$. For a chain having a finite number of states, these probabilities can be found by solving a matrix differential equation. To derive this equation, we use the short-time approximation

$$p_{ij}(t) \;=\; \lambda_{ij} t + o(t) \tag{10.2}$$

for $i \neq j$, where $\lambda_{ij}$ is the **transition rate** (or **infinitesimal transition probability**) from state $i$ to state $j$. Equation (10.2) implies the further short-time approximation

$$p_{ii}(t) \;=\; 1 - \lambda_i t + o(t), \tag{10.3}$$

where $\lambda_i = \sum_{j \neq i} \lambda_{ij}$.

Now consider the Chapman-Kolmogorov relation

$$p_{ij}(t + h) \;=\; p_{ij}(t) p_{jj}(h) + \sum_{k \neq j} p_{ik}(t) p_{kj}(h), \tag{10.4}$$

which simply says the process must pass through some intermediate state $k$ at time $t$ enroute to state $j$ at time $t + h$. Substituting the approximations (10.2) and (10.3) in (10.4) yields

$$p_{ij}(t + h) \;=\; p_{ij}(t)(1 - \lambda_j h) + \sum_{k \neq j} p_{ik}(t) \lambda_{kj} h + o(h).$$

Sending $h$ to 0 in the difference quotient

$$\frac{p_{ij}(t+h) - p_{ij}(t)}{h} = -p_{ij}(t)\lambda_j + \sum_{k \neq j} p_{ik}(t)\lambda_{kj} + \frac{o(h)}{h}$$

produces the forward differential equation

$$p'_{ij}(t) = -p_{ij}(t)\lambda_j + \sum_{k \neq j} p_{ik}(t)\lambda_{kj}. \tag{10.5}$$

The system of differential equations (10.5) can be summarized in matrix notation by introducing the matrices $P(t) = [p_{ij}(t)]$ and $\Lambda = (\Lambda_{ij})$, where $\Lambda_{ij} = \lambda_{ij}$ for $i \neq j$ and $\Lambda_{ii} = -\lambda_i$. The forward equations in this notation become

$$P'(t) = P(t)\Lambda \tag{10.6}$$
$$P(0) = I,$$

where $I$ is the identity matrix. It is easy to check that the solution of the initial value problem (10.6) is furnished by the **matrix exponential**

$$P(t) = e^{t\Lambda} = \sum_{k=0}^{\infty} \frac{t^k \Lambda^k}{k!}. \tag{10.7}$$

Probabilists call $\Lambda$ the **infinitesimal generator** or **infinitesimal transition matrix** of the process.

A probability distribution $\pi = (\pi_i)$ on the states of a Markov chain is a row vector whose components satisfy $\pi_i \geq 0$ for all $i$ and $\sum_i \pi_i = 1$. If

$$\pi P(t) = \pi \tag{10.8}$$

holds for all $t \geq 0$, then $\pi$ is said to be an **equilibrium distribution** for the chain. Written in components, the eigenvector equation (10.8) reduces to the balance condition $\sum_i \pi_i p_{ij}(t) = \pi_j$. Again, this is completely analogous to the discrete-time theory described in Chapter 9. For small $t$, equation (10.8) can be rewritten as

$$\pi(I + t\Lambda) + o(t) = \pi.$$

This approximate form makes it obvious that $\pi\Lambda = \mathbf{0}$ is a necessary condition for $\pi$ to be an equilibrium distribution. Multiplying (10.7) on the left by $\pi$ shows that $\pi\Lambda = \mathbf{0}$ is also a sufficient condition for $\pi$ to be an equilibrium distribution. In components this necessary and sufficient condition amounts to

$$\sum_{j \neq i} \pi_j \lambda_{ji} = \pi_i \sum_{j \neq i} \lambda_{ij} \tag{10.9}$$

for all $i$. If all the states of a Markov chain communicate, then there is one and only one equilibrium distribution $\pi$. Furthermore, each of the rows of $P(t)$ approaches $\pi$ as $t \to \infty$. Lamperti [10] provides a clear exposition of these facts.

Fortunately, the annoying feature of periodicity present in discrete-time theory disappears in the continuous-time theory. The definition and properties of reversible chains carry over directly from discrete time to continuous time provided we substitute infinitesimal transition probabilities for transition probabilities. For instance, the detailed balance condition becomes

$$\pi_i \lambda_{ij} \;=\; \pi_j \lambda_{ji} \tag{10.10}$$

for all pairs $i \neq j$. Kolmogorov's circulation criterion for reversibility continues to hold, and when it is true, the equilibrium distribution is constructed from the infinitesimal transition probabilities exactly as in discrete time.

## 10.5    A Nucleotide Substitution Model

Models for nucleotide substitution are of great importance in molecular evolution. Kimura [7], among others, views the changes occurring at a single position or site as a continuous-time Markov chain involving the four bases A, G, C, and T. The matrix $\Lambda$ below gives the transition rates under a generalization of Kimura's model of neutral evolution. In this matrix the rows and columns are labeled by the four states in the order A, G, C, and T from top to bottom and left to right.

$$\Lambda = \begin{array}{c} \\ A \\ G \\ C \\ T \end{array} \begin{pmatrix} -(\alpha + \gamma + \lambda) & \alpha & \gamma & \lambda \\ \epsilon & -(\epsilon + \gamma + \lambda) & \gamma & \lambda \\ \delta & \kappa & -(\delta + \kappa + \beta) & \beta \\ \delta & \kappa & \sigma & -(\delta + \kappa + \sigma) \end{pmatrix}.$$

Without further restrictions, this chain does not satisfy detailed balance. If we impose the additional constraints $\beta\gamma = \lambda\sigma$ and $\alpha\delta = \epsilon\kappa$, then the distribution

$$\pi_A = \frac{\delta}{\gamma + \delta + \kappa + \lambda}$$
$$\pi_G = \frac{\kappa}{\gamma + \delta + \kappa + \lambda}$$
$$\pi_C = \frac{\gamma}{\gamma + \delta + \kappa + \lambda} \tag{10.11}$$
$$\pi_T = \frac{\lambda}{\gamma + \delta + \kappa + \lambda}$$

satisfies detailed balance. To verify detailed balance, one must check six equalities of the type (10.10). For instance, $\pi_A \alpha = \pi_G \epsilon$ follows directly from the definitions of $\pi_A$ and $\pi_G$ and the constraint $\alpha\delta = \epsilon\kappa$. Kolmogorov's criterion indicates that the two stated constraints are necessary as well as sufficient for detailed balance.

In the Markov chain, two purines or two pyrimidines are said to differ by a **transition**. (This convention of the evolutionary biologists is confusing. All states differ by what a probabilist would call a single transition of the chain. However, we will defer to the biologists on this point.) A purine and a pyrimidine are said to differ by a **transversion**. The matrix $\Lambda$ displays a modest amount of symmetry in the sense that the two transversions leading to any given state always share the same transition rate.

In principle, it is possible to solve for the finite-time transition matrix $P(t)$ in this model by exponentiating the infinitesimal generator $\Lambda$. To avoid this rather cumbersome calculation, we generalize the arguments of Kimura [7] and exploit the symmetry inherent in the model. Define $q_{AY}(t)$ to be the probability that the chain is in either of the two pyrimidines C or T at time $t$ given that it starts in A at time 0. We will derive a system of coupled ordinary differential equations obeyed by $q_{AY}(t)$ and $p_{AG}(t)$. The entry $p_{AG}(t)$ of $P(t)$ is the probability that the chain is in G at time $t$ given that it starts in A at time 0. By the same reasoning that led to the forward equation (10.5), we have

$$
\begin{aligned}
q_{AY}(t+h) \ &= \ q_{AY}(t)(1 - \delta h - \kappa h) + p_{AG}(t)(\gamma + \lambda)h \\
&\quad + [1 - q_{AY}(t) - p_{AG}(t)](\gamma + \lambda)h + o(h),
\end{aligned}
$$

where $1 - q_{AY}(t) - p_{AG}(t)$ equals the probability $p_{AA}(t)$ of being in A at time $t$. Forming the obvious difference quotient and letting $h \to 0$ yields the differential equation

$$
q'_{AY}(t) \ = \ -c_1 q_{AY}(t) + c_2,
$$

where

$$
\begin{aligned}
c_1 \ &= \ \delta + \kappa + \gamma + \lambda \\
c_2 \ &= \ \gamma + \lambda.
\end{aligned}
$$

This equation can be solved by multiplying by the integrating factor $e^{c_1 t}$ and isolating the terms $[q_{AY}(t)e^{c_1 t}]'$ involving $q_{AY}(t)$ on the left side of the equation. These manipulations yield the solution

$$
q_{AY}(t) \ = \ \frac{c_2}{c_1}(1 - e^{-c_1 t}) \tag{10.12}
$$

satisfying the initial condition $q_{AY}(0) = 0$.

To solve for $p_{AG}(t)$, write the forward approximation

$$
\begin{aligned}
p_{AG}(t+h) \ &= \ p_{AG}(t)(1 - \epsilon h - \gamma h - \lambda h) + q_{AY}(t)\kappa h \\
&\quad + [1 - q_{AY}(t) - p_{AG}(t)]\alpha h + o(h).
\end{aligned}
$$

This leads to the differential equation

$$
p'_{AG}(t) \ = \ -c_3 p_{AG}(t) + c_4 q_{AY}(t) + \alpha,
$$

where

$$
\begin{aligned}
c_3 \ &= \ \epsilon + \alpha + \gamma + \lambda \\
c_4 \ &= \ \kappa - \alpha.
\end{aligned}
$$

Substituting the solution (10.12) for $q_{AY}(t)$, one can straightforwardly verify that this last differential equation has solution

$$
\begin{aligned}
p_{AG}(t) &= \frac{c_2 c_4 + \alpha c_1}{c_1 c_3} - \frac{c_2 c_4}{c_1 (c_3 - c_1)} e^{-c_1 t} \\
&\quad + \frac{c_2 c_4 - \alpha(c_3 - c_1)}{c_3 (c_3 - c_1)} e^{-c_3 t}
\end{aligned} \tag{10.13}
$$

satisfying $p_{AG}(0) = 0$.

This analysis has produced the probabilities $p_{AA}(t)$, $p_{AG}(t)$, and $q_{AY}(t)$ of being in A, G, or either pyrimidine at time $t$ starting from A at time 0. To decompose $q_{AY}(t)$ into its two constituent probabilities $p_{AC}(t)$ and $p_{AT}(t)$, define $q_{UU}(t)$ to be the probability that the chain is in either purine at time $t$ given that it starts in either purine at time 0. Likewise, define $q_{UC}(t)$ to be the probability that the chain is in the pyrimidine C at time $t$ given that it starts in either purine at time 0. Because of the symmetry of the transition rates, the probability $q_{UU}(t)$ makes sense, and $q_{UC}(t) = p_{AC}(t) = p_{GC}(t)$. From the probabilities $q_{AY}(t)$ and $p_{AC}(t)$, we calculate $p_{AT}(t) = q_{AY}(t) - p_{AC}(t)$.

To derive a differential equation for $q_{UU}(t)$, note the approximation

$$
\begin{aligned}
q_{UU}(t + h) &= q_{UU}(t)(1 - \gamma h - \lambda h) + q_{UC}(t)(\delta + \kappa)h \\
&\quad + [1 - q_{UU}(t) - q_{UC}(t)](\delta + \kappa)h + o(h),
\end{aligned}
$$

where $1 - q_{UU}(t) - q_{UC}(t)$ is the probability of being in T at time $t$. This approximation leads to

$$
q'_{UU}(t) = -c_1 q_{UU}(t) + c_5,
$$

where $c_1$ was defined previously and

$$
c_5 = \delta + \kappa.
$$

Again, the solution

$$
q_{UU}(t) = \frac{c_5 + (c_1 - c_5) e^{-c_1 t}}{c_1} \tag{10.14}
$$

satisfying $q_{UU}(0) = 1$ follows directly.

The approximation for $q_{UC}(t)$,

$$
\begin{aligned}
q_{UC}(t + h) &= q_{UC}(t)(1 - \delta h - \kappa h - \beta h) + q_{UU}(t)\gamma h \\
&\quad + [1 - q_{UU}(t) - q_{UC}(t)]\sigma h + o(h),
\end{aligned}
$$

yields the differential equation

$$
q'_{UC}(t) = -c_6 q_{UC}(t) + c_7 q_{UU}(t) + \sigma,
$$

where

$$c_6 = \delta + \kappa + \sigma + \beta$$
$$c_7 = \gamma - \sigma.$$

In view of equation (10.14) and the initial condition $q_{UC}(0) = 0$, the solution for $q_{UC}(t)$ is

$$q_{UC}(t) = \frac{c_5 c_7 + \sigma c_1}{c_1 c_6} + \frac{(c_1 - c_5)c_7}{c_1(c_6 - c_1)} e^{-c_1 t}$$
$$- \frac{c_7(c_6 - c_5) + \sigma(c_6 - c_1)}{c_6(c_6 - c_1)} e^{-c_6 t}. \qquad (10.15)$$

In summary, we have found the top row $[p_{AA}(t), p_{AG}(t), p_{AC}(t), p_{AT}(t)]$ of $P(t)$ corresponding to the nucleotide A. By symmetrical arguments, the other rows of $P(t)$ can also be calculated. In the limit $t \to \infty$, the rows of $P(t)$ all collapse to the equilibrium distribution $\pi$.

## 10.6   Maximum Likelihood Reconstruction

Maximum likelihood provides a second method of comparing evolutionary trees. As with maximum parsimony, DNA data are gathered at several different sites for several different contemporary taxa. A model is then posed for how differences evolve at the various sites. Most models involve the following assumptions:

(a) All sites evolve according to the same tree.

(b) All sites evolve independently.

(c) All sites evolve according to identically distributed stochastic processes.

(d) Conditional on the base at a given site of an internal node, evolution proceeds independently at the site along the two branches of the tree descending from the node.

Further assumptions about the detailed nature of evolution at a site can be imposed. For instance, we can adopt the generalized Kimura substitution model as just developed.

We now discuss how to compute the likelihood of the bases observed at the tips of an evolutionary tree for a particular site. According to assumptions (a), (b), and (c), we need merely multiply these site-specific likelihoods to recover the overall likelihood of a given tree. For a tree with $n$ tips, it is convenient to label the internal nodes $1, \ldots, n-1$ and the tips $n, \ldots, 2n-1$. Also, let $b_i$ be either one of the four possible bases at an internal node or the observed base at a tip. If the root is node

1, then designate the prior probability of base $b_1$ at this node by $q_{b_1}$. Assumption (d) now provides the likelihood expression

$$\sum_{b_1} \cdots \sum_{b_{n-1}} q_{b_1} \prod_{(i,j)} \Pr(b_j \mid b_i), \qquad (10.16)$$

where $(i, j)$ ranges over all pairs of ancestral nodes $i$ and direct descendant nodes $j$.

The sums-of-products expression (10.16) is analogous to our earlier representation of a pedigree likelihood. The factor $q_{b_1}$ corresponds to a prior, and the factor $\Pr(b_j \mid b_i)$ to a transmission probability. There is no analog of a penetrance function or of genotype elimination in this context. To evaluate expression (10.16), we carry out one summation at a time. It is most economical to choose an order for the iterated sums consistent with the graphical structure of the tree. In particular, the tree should be pruned working inward from the tips. Each node eliminated should be on the current periphery of the tree. Typically, the last node eliminated is the root, but this is not absolutely essential.

It is a curious feature of reversible Markov-chain models that the root can be eliminated and one of the two direct descendants of the root substituted for it. Suppose that nodes 2 and 3 are the direct descendants of the root. Those arrays of the likelihood (10.16) involving the index $b_1$ are $q_{b_1}$, $\Pr(b_2 \mid b_1)$, and $\Pr(b_3 \mid b_1)$. In the Markov-chain context, it is natural to take $q_{b_1}$ as the equilibrium distribution. Furthermore, if $t_2$ and $t_3$ are the times separating the root from nodes 2 and 3, respectively, then

$$\begin{aligned}
\Pr(b_2 \mid b_1) &= p_{b_1 b_2}(t_2) \\
\Pr(b_3 \mid b_1) &= p_{b_1 b_3}(t_3).
\end{aligned}$$

Isolating the sum over $b_1$ and invoking finite-time detailed balance now give

$$\begin{aligned}
\sum_{b_1} q_{b_1} p_{b_1 b_2}(t_2) p_{b_1 b_3}(t_3) &= \sum_{b_1} q_{b_2} p_{b_2 b_1}(t_2) p_{b_1 b_3}(t_3) \\
&= q_{b_2} \sum_{b_1} p_{b_2 b_1}(t_2) p_{b_1 b_3}(t_3) \\
&= q_{b_2} p_{b_2 b_3}(t_2 + t_3).
\end{aligned}$$

This is Felsenstein's **pulley principle**. The root can be eliminated and moved to either one of its direct descendants—in this case, node 2. Thus, if only reversible chains are considered, then maximum likelihood cannot distinguish two rooted trees that correspond to the same unrooted tree [4].

# 10.7    Origin of the Eukaryotes

**Eukaryotic** organisms differ from **prokaryotic** organisms in possessing a **nucleus**, a cellular organelle housing the chromosomes. The origin of eukaryotes from prokaryotic bacteria is one of the most intriguing questions in evolutionary biology. Bacteria can be subdivided into four broad groups. The **eubacteria** are common pathogens of eukaryotes, the **halobacteria** are found at high salt concentrations, the **eocytes** metabolize sulfur and are found at high pressures and temperatures, and the **methanogens** metabolize methane. Evolutionary biologists have traditionally classified the latter three groups in a single phylum, the archebacteria, leaving the eubacteria as the natural candidates for the ancestors of eukaryotes. In support of this view is the fact that mitochondria and chloroplasts, important organelles of eukaryotic cells, derive from eubacteria. Lake [9] upset this tidy classification by comparing **16s ribosomal** RNA sequences from a variety of representative eukaryotic and prokaryotic organisms. His analysis refutes the archebacterial grouping and supports the eocytes as the closest bacterial ancestor of the eukaryotes.

In this example we examine a small portion of Lake's original data. The relevant subset consists of 1,092 aligned bases from the rRNA of the organisms *A. salina* (a eukaryote), *B. subtilis* (a eubacterium), *H. morrhuae* (a halobacterium), and *D. mobilis* (an eocyte). These four taxa can be arranged in the three unrooted evolutionary trees depicted in Figure 10.6. Maximum parsimony favors the G tree with a score of 975 versus a score of 981 for each of the E and F trees. Although this result supports the archebacteria theory of the origin of the eukaryotes, the evidence is hardly decisive.

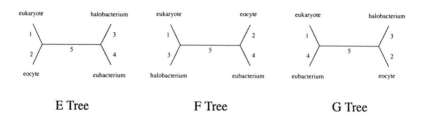

| E Tree | F Tree | G Tree |

FIGURE 10.6. Unrooted Trees for the Evolution of Eukaryotes

Maximum likelihood analysis of the same data contradicts the maximum parsimony ranking. Under the reversible version of the generalized Kimura model presented in Section 10.5, the E, F, and G trees have maximum loglikelihoods (base $e$) of $-4598.2$, $-4605.2$, and $-4606.6$, respectively. According to the pulley principle, we are justified in treating each of these unrooted trees as rooted at one node of branch 5. (See Figure 10.6 for the numbering of the branches.) Column 2 of Table 10.1 displays the parameter estimates and their standard errors for the favored E tree. In the table, certain entries are left blank. For instance, under reversibility the parameters $\epsilon$ and $\sigma$ are eliminated by the constraints $\epsilon = \alpha\delta/\kappa$ and $\sigma = \beta\gamma/\lambda$. The distribution at the root is specified as the stationary distribution (10.11). To avoid confounding branch lengths in the model with the infinitesimal rate parameters $\alpha$ through $\sigma$, we force the branch length of branch 4 to be 1.

A crude idea of the goodness of fit of the model can be gained by comparing it

TABLE 10.1. Parameter Estimates for the Best Eukaryotic Trees

| Parameter | Estimates for Best Unrooted Tree | Estimates for Best Rooted Tree |
|---|---|---|
| $\alpha$ | $.155 \pm .016$ | $.051 \pm .013$ |
| $\beta$ | $.174 \pm .019$ | $.217 \pm .023$ |
| $\gamma$ | $.084 \pm .007$ | $.078 \pm .009$ |
| $\delta$ | $.080 \pm .007$ | $.060 \pm .009$ |
| $\epsilon$ | | $.174 \pm .018$ |
| $\kappa$ | $.107 \pm .009$ | $.117 \pm .013$ |
| $\lambda$ | $.067 \pm .006$ | $.078 \pm .009$ |
| $\sigma$ | | $.134 \pm .025$ |
| $\pi_A$ | | $.193 \pm .012$ |
| $\pi_G$ | | $.373 \pm .015$ |
| $\pi_C$ | | $.297 \pm .015$ |
| $\pi_T$ | | $.137 \pm .011$ |
| Branch 1 | $1.642 \pm .153$ | $1.690 \pm .154$ |
| Branch 2 | $.234 \pm .047$ | $.110 \pm .044$ |
| Branch 3 | $.539 \pm .065$ | $.568 \pm .065$ |
| Branch 4 | $1 \pm 0$ | $1 \pm 0$ |
| Branch 5 | $.188 \pm .051$ | $.190 \pm .050$ |
| Branch 6 | | $.082 \pm .050$ |

to the unrestricted multinomial model with $4^4 = 256$ cells. Under the unrestricted model, the maximum loglikelihood of the data is $-4361.3$. The corresponding chi-square statistic of $473.8 = 2(-4361.3 + 4598.2)$ on 245 degrees of freedom is extremely significant. However, the multinomial data are sparse, and we should be cautious in applying large sample theory.

TABLE 10.2. Observed Base Proportions for the Four Contemporary Taxa

| Base | A. salina Eukaryote | D. mobilis Eocycte | H. morrhuae Halobacterium | B .subtilis Eubacterium |
|---|---|---|---|---|
| A | .250 | .232 | .202 | .255 |
| G | .319 | .328 | .367 | .279 |
| C | .230 | .263 | .290 | .234 |
| T | .202 | .178 | .141 | .232 |

Under the full version of the generalized Kimura model, all rooted trees are in principle distinguishable. Figure 10.7 depicts the best rooted tree, which, not surprisingly, collapses to the unrooted E tree. Column 3 of Table 10.1 provides maximum likelihood parameter estimates for this rooted tree. The corresponding

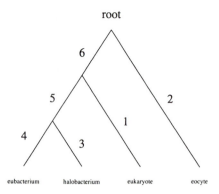

FIGURE 10.7. Best Rooted Tree for the Evolution of Eukaryotes

maximum loglikelihood ($-4536.2$) represents a substantial improvement over the maximum loglikelihood ($-4598.2$) of the E tree under the reversible version of the model. Most of this improvement occurs because imposing the stationary distribution on the root in the reversible model is incompatible with the wide variation in DNA base proportions in the contemporary species displayed in Table 10.2. Indeed, under the generalized Kimura model with stationarity imposed, the maximum loglikelihood of the best rooted tree falls to $-4588.7$. This best tree coincides with the tree depicted in Figure 10.7.

Without the assumption of stationarity, two other rooted trees reducing to the E tree have nearly the same loglikelihoods ($-4537.3$ and $-4537.4$) as the best rooted tree. Apparently, distinguishing rooted trees in practice is much harder than distinguishing them in theory. On the other hand, the best rooted tree corresponding to either an F or G tree has a much lower loglikelihood ($-4544.2$) than the best rooted tree.

## 10.8   Problems

1. Compute the number of unrooted evolutionary trees possible for $n$ contemporary taxa. (Hint: How does this relate to the number of rooted trees?)

2. Consider four contemporary taxa numbered 1, 2, 3, and 4. A total of $n$ shared DNA sites are sequenced for each taxon. Let $N_{wxyz}$ be the number of sites at which taxon 1 has base $w$, taxon 2 base $x$, and so forth. If we denote the three possible unrooted trees by E, F, and G, then we can define three statistics

$$N_E = \sum_{s \neq r} \sum_{r \in \{A,G,C,T\}} N_{rrss}$$

$$N_F = \sum_{s \neq r} \sum_{r \in \{A,G,C,T\}} N_{rsrs}$$

$$N_G = \sum_{s \neq r} \sum_{r \in \{A,G,C,T\}} N_{rssr}$$

for discriminating among the unrooted trees. Show that maximum parsimony selects the unrooted tree E, F, or G with largest statistic $N_E$, $N_F$, or $N_G$. Draw the unrooted tree corresponding to each statistic.

3. Let $u$ and $v$ be column vectors with the same number of components. Applying the definition of the matrix exponential (10.7), show that

$$e^{suv^t} = \begin{cases} I + suv^t & \text{if } v^t u = 0 \\ I + \frac{e^{sv^t u} - 1}{v^t u} uv^t & \text{otherwise.} \end{cases}$$

Using this, compute the $2 \times 2$ matrix exponential

$$\exp\left[ s \begin{pmatrix} -\alpha & \alpha \\ \beta & -\beta \end{pmatrix} \right],$$

and find its limit as $s \to \infty$.

4. Consider a continuous-time Markov chain with infinitesimal transition matrix $\Lambda = (\Lambda_{ij})$ and equilibrium distribution $\pi$. If the chain is at equilibrium at time 0, then show that it experiences $t \sum_i \pi_i \lambda_i$ transitions on average during the time interval $[0, t]$, where $\lambda_i = \sum_{j \neq i} \Lambda_{ij}$.

5. Let $\Lambda$ be the infinitesimal transition matrix and $\pi$ the equilibrium distribution of a reversible Markov chain with $n$ states. Define an inner product $\langle u, v \rangle_\pi$ on complex column vectors $u$ and $v$ with $n$ components by

$$\langle u, v \rangle_\pi = \sum_i u_i \pi_i v_i^*,$$

where * denotes complex conjugate. Verify that $\Lambda$ satisfies the self-adjointness condition

$$\langle \Lambda u, v \rangle_\pi = \langle u, \Lambda v \rangle_\pi.$$

Conclude by standard arguments that $\Lambda$ has only real eigenvalues.

6. For the nucleotide substitution model of Section 10.5, prove formally that $P(t)$ has the same pattern for equality of entries as $\Lambda$. For example, we have $p_{AC}(t) = p_{GC}(t)$. (Hint: Prove by induction that $\Lambda^k$ has the same pattern as $\Lambda$. Then note the matrix exponential definition (10.7).)

7. For the nucleotide substitution model of Section 10.5, show that $\Lambda$ has eigenvalues $0$, $-(\gamma + \lambda + \delta + \kappa)$, $-(\alpha + \epsilon + \gamma + \lambda)$, and $-(\delta + \kappa + \beta + \sigma)$ and corresponding right eigenvectors

$$\mathbf{1} = \left(1, 1, 1, 1\right)^t$$

$$u = \left(1, 1, -\frac{c_5}{c_2}, -\frac{c_5}{c_2}\right)^t$$

$$v = \left(\frac{\alpha(c_5 - c_3) + \kappa c_2}{\delta(c_3 - c_2) - \epsilon c_5}, \frac{-\epsilon(c_5 - c_3) - \delta c_2}{\delta(c_3 - c_2) - \epsilon c_5}, 1, 1\right)^t$$

$$w = \left(1, 1, \frac{\beta(c_2 - c_6) + \lambda c_5}{\gamma(c_6 - c_5) - \sigma c_2}, \frac{-\sigma(c_2 - c_6) - \gamma c_5}{\gamma(c_6 - c_5) - \sigma c_2}\right),$$

respectively, where the constants $c_1, \ldots, c_6$ are same ones defined in equations (10.12) through (10.15).

8. For the nucleotide substitution model of Section 10.5, verify in general that the equilibrium distribution is

$$\pi_A = \frac{\epsilon(\delta + \kappa) + \delta(\gamma + \lambda)}{(\alpha + \gamma + \epsilon + \lambda)(\gamma + \delta + \kappa + \lambda)}$$

$$\pi_G = \frac{\alpha(\delta + \kappa) + \kappa(\gamma + \lambda)}{(\alpha + \gamma + \epsilon + \lambda)(\gamma + \delta + \kappa + \lambda)}$$

$$\pi_C = \frac{\gamma(\delta + \kappa) + \sigma(\gamma + \lambda)}{(\beta + \delta + \kappa + \sigma)(\gamma + \delta + \kappa + \lambda)}$$

$$\pi_T = \frac{\lambda(\delta + \kappa) + \beta(\gamma + \lambda)}{(\beta + \delta + \kappa + \sigma)(\gamma + \delta + \kappa + \lambda)}.$$

9. Let $\Lambda$ be the infinitesimal transition matrix of a Markov chain, and suppose $\mu \geq \max_i \lambda_i$. If $R = I + \frac{1}{\mu}\Lambda$, prove that $R$ has nonnegative entries and that

$$S(t) = \sum_{i=0}^{\infty} e^{-\mu t} \frac{(\mu t)^i}{i!} R^i$$

coincides with $P(t)$. (Hint: Verify that $S(t)$ satisfies the same defining differential equation and the same initial condition as $P(t)$.)

10. There is an explicit formula for the equilibrium distribution of a continuous-time Markov chain in terms of weighted in-trees [12]. To describe this formula, we first define a directed graph on the states $1, \ldots, n$ of the chain. The vertices of the graph are the states of the chain, and the arcs of the graph are ordered pairs of states $(i, j)$ having transition rates $\lambda_{ij} > 0$. If it is possible to reach some designated state $k$ from every other state $i$, then a unique equilibrium distribution $\pi = (\pi_1, \ldots, \pi_n)$ exists for the chain. Note that this reachability condition is weaker than requiring that all states communicate.

The equilibrium distribution is characterized by defining certain subgraphs called **in-trees**. An in-tree $T_i$ to state $i$ is a subgraph having $n - 1$ arcs and connecting each vertex $j \neq i$ to $i$ by some directed path. Ignoring orientations, an in-tree is graphically a tree; observing orientations, all paths lead to $i$. The weight $w(T_i)$ associated with the in-tree $T_i$ is the product of the transition rates $\lambda_{jk}$ labeling the various arcs $(j, k)$ of the in-tree. For instance, in the nucleotide substitution chain, one in-tree to A has arcs (G,A), (C,A), and (T,C). Its associated weight is $\epsilon\delta\sigma$.

In general, the equilibrium distribution is given by

$$\pi_i = \frac{\sum_{T_i} w(T_i)}{\sum_j \sum_{T_j} w(T_j)}. \tag{10.17}$$

The reachability condition implies that in-trees to state $k$ exist and consequently that the denominator in (10.17) is positive. The value of the in-tree formula (10.17) is limited by the fact that in a Markov chain with $n$ states there can be as many as $n^{n-2}$ in-trees to a given state. Thus, in the nucleotide substitution model, there are $4^{4-2} = 16$ in-trees to each state and 64 in-trees in all. If you are undeterred by this revelation, then use formula (10.17) to find the equilibrium distribution of the nucleotide substitution chain of Section 10.5 when detailed balance is not assumed. Your answer should match the distribution appearing in Problem 8.

11. In his method of evolutionary parsimony, Lake [8] has highlighted the balanced transversion assumption. This assumption implies the constraints $\lambda_{AC} = \lambda_{AT}$, $\lambda_{GC} = \lambda_{GT}$, $\lambda_{CA} = \lambda_{CG}$, and $\lambda_{TA} = \lambda_{TG}$ in the nucleotide substitution model with general transition rates. Without further restrictions, infinitesimal balanced transversions do not imply finite-time balanced transversions. For example, the identity $p_{AC}(t) = p_{AT}(t)$ may not hold. Prove that finite-time balanced transversions follow if the additional closure assumptions

$$\begin{aligned} \lambda_{AG} - \lambda_{GA} &= \lambda_G - \lambda_A \\ \lambda_{CT} - \lambda_{TC} &= \lambda_T - \lambda_C \end{aligned}$$

are made [1]. (Hint: Show by induction that the matrices $\Lambda^k$ have the balanced transversion pattern for equality of entries.)

12. In Lake's balanced transversion model of the last problem, show that

$$\begin{aligned} \lambda_{AG}\lambda_{GT} &= \lambda_{AT}\lambda_{GA} \\ \lambda_{CT}\lambda_{TA} &= \lambda_{TC}\lambda_{CA} \end{aligned}$$

are necessary and sufficient conditions for the corresponding Markov chain to be reversible.

# References

[1] Cavender JA (1989) Mechanized derivation of linear invariants. *Mol Biol Evol* 6:301–316

[2] Eck RV, Dayhoff MO (1966) *Atlas of Protein Sequence and Structure*. National Biomedical Research Foundation, Silver Spring, MD

[3] Felsenstein J (1978) Cases in which parsimony or compatibility methods will be positively misleading. *Syst Zool* 27:401–410

[4] Felsenstein J (1981) Evolutionary trees from DNA sequences: A maximum likelihood approach. *J Mol Evol* 17:368–376

[5] Fitch WM (1971) Toward defining the course of evolution: Minimum change for a specific tree topology. *Syst Zool* 20:406–416

[6] Hartigan JA (1973) Minimum mutation fits to a given tree. *Biometrics* 29:53–65

[7] Kimura M (1980) A simple method for estimating evolutionary rates of base substitutions through comparative studies of nucleotide sequences. *J Mol Evol* 16:111–120

[8] Lake JA (1987) A rate-independent technique for analysis of nucleic acid sequences: evolutionary parsimony. *Mol Biol Evol* 4:167–191

[9] Lake JA (1988) Origin of the eukaryotic nucleus determined by rate-invariant analysis of rRNA sequences. *Nature* 331:184–186

[10] Lamperti J (1977) *Stochastic Processes. A Survey of the Mathematical Theory*. Springer-Verlag, New York

[11] Li W-H, Graur D (1991) *Fundamentals of Molecular Evolution*. Sinauer, Sunderland, MA

[12] Solberg JJ (1975) A graph theoretic formula for the steady state distribution of a finite Markov process. *Management Sci* 21:1040–1048

[13] Waterman MS (1995) *Introduction to Computational Biology: Maps, Sequences, and Genomes*. Chapman and Hall, London

[14] Weir BS (1990) *Genetic Data Analysis*. Sinauer, Sunderland, MA

# 11
# Radiation Hybrid Mapping

## 11.1 Introduction

In the 1970s Goss and Harris [12] developed a new method for mapping human chromosomes. This method was based on irradiating human cells, rescuing some of the irradiated cells by hybridization to rodent cells, and analyzing the hybrid cells for surviving fragments of a particular human chromosome. For various technical reasons, **radiation hybrid mapping** languished for nearly a decade and a half until revived by Cox et al. [10]. The current, more sophisticated and successful versions raise many fascinating statistical problems. We will first discuss the mathematically simpler case of haploid radiation hybrids. Once this case is thoroughly digested, we will turn to the mathematically subtler case of polyploid radiation hybrids.

In the haploid version of radiation hybrid mapping, an experiment starts with a human–rodent hybrid cell line [10]. This cell line incorporates a full rodent genome and a single copy of one of the human chromosomes. To fragment the human chromosome, the cell line is subjected to an intense dose of X-rays, which naturally also fragments the rodent chromosomes. The repair mechanisms of the cells rapidly heal chromosome breaks, and the human chromosome fragments are typically translocated or inserted into rodent chromosomes. However, the damage done by irradiation is lethal to the cell line unless further action is taken to rescue individual cells. The remedy is to fuse the irradiated cells with cells from a second unirradiated rodent cell line. The second cell line contains only rodent chromosomes, so no confusion about the source of the human chromosome fragments can arise for a new hybrid cell created by the fusion of two cells from the two different cell lines. The new hybrid cells have no particular growth advantage over the more numerous unfused cells of the second cell line. However, if cells from the second cell line lack an enzyme such as hypoxanthine phosphoribosyl transferase (HPRT) or thymidine kinase (TK), both the unfused and the hybrid cells can be grown in a selective medium that kills the unfused cells [10]. This selection process leaves a few hybrid cells, and each of the hybrid cells serves as a progenitor of a clone of identical cells.

Each clone can be assayed for the presence or absence of various human markers on the original human chromosome. Depending on the radiation dose and other experimental conditions, the cells of a clone generally contain from 20 to 60 percent of the human chromosome fragments generated by the irradiation of its ancestral human–rodent hybrid cell [8, 10]. The basic premise of radiation hybrid mapping is that the closer two loci are on the human chromosome, the less likely it is that irradiation will cause a break between them. Thus, close loci will tend to be concordantly retained or lost in the hybrid cells, while distant loci will tend to be independently retained or lost. The retention patterns from the various hybrid clones therefore give important clues for determining locus order and for estimating the distances between adjacent loci for a given order.

## 11.2    Models for Radiation Hybrids

The breakage phenomenon for a particular human chromosome can be reasonably modeled by a Poisson process. The preliminary evidence of Cox et al. [10] suggests that this Poisson breakage process is roughly homogeneous along the chromosome. For their data on human chromosome 21, Cox et al. [10] found that 8,000 rads of radiation produced on average about four breaks per cell. The intensity $\lambda$ characterizing the Poisson process is formally defined as the breakage probability per unit length. Assuming a length of $4 \times 10^4$ kilobases (kb) for chromosome 21, $\lambda \approx 4/(4 \times 10^4) = 10^{-4}$ breaks per kb when a cell is exposed to 8,000 rads [10].

For any two loci, the simple mapping function

$$1 - \theta \;=\; e^{-\lambda \delta} \tag{11.1}$$

relates the probability $\theta$ of at least one break between the loci to the physical distance $\delta$ between them. When $\lambda \delta$ is small, $\theta \approx \lambda \delta$. This is analogous to the approximate linear relationship between recombination fraction and map distance for small distances in genetic recombination experiments. Indeed, except for minor notational differences, equation (11.1) is Haldane's mapping function for recombination without chiasma interference.

In addition to breakage, fragment retention must be taken into account when analyzing radiation hybrid data. A reasonable assumption is that different fragments are retained independently. For the purposes of this exposition, we will make the further assumption that there is a common fragment retention probability $r$. Boehnke et al. [5] consider at length more complicated models for fragment retention. For instance, the fragment bearing the centromere of the chromosome may be retained more often than other fragments. This is biologically plausible because the centromere is involved in coordination of chromosome migration during cell division. However, these more complicated models appear to make little difference in ultimate conclusions.

In a radiation hybrid experiment, a certain number of clones are scored at several loci. For example, in the Cox et al. [1990] chromosome 21 data, 99 clones were scored at 14 loci. In some of the clones, only a subset of the loci was scored. One of their typical clones can be represented as $(0, 0, 0, 0, 0, 1, 0, 0, 0, 0, ?, 0, 0, 1)$. A "1" in a given position of this observation vector indicates that the corresponding human locus was present in the hybrid clone; a "0" indicates that the locus was absent; and a "?" indicates that the locus was untyped in the clone or gave conflicting typing results.

## 11.3    Minimum Obligate Breaks Criterion

Computing the minimum number of obligate breaks per order allows comparisons of different orders [3, 4, 5, 22]. If the order of the loci along the chromosome is the same as the scoring order, then the clone described in the last section requires three obligate breaks. These breaks occur whenever a run of 0's is broken by a 1 or vice versa; untyped loci are ignored in this accounting. The minimum number of obligate breaks for each clone can be summed over all clones to give a grand sum for a given order. This grand sum serves as a criterion for comparing orders. The minimum breaks criterion can be minimized over orders by a stepwise algorithm [5] or by standard combinatorial optimization techniques such as branch-and-bound [19] and simulated annealing [17].

The advantage of the minimum breaks criterion is that it depends on almost no assumptions about how breaks occur and fragments are retained. Given a common retention rate, this criterion is also strongly statistically consistent. Following Barrett [1] and Speed et al. [20], let us demonstrate this fact. Consider $m$ loci taken in their natural order $1, \ldots, m$ along a chromosome, and imagine an infinite number of independent, fully typed radiation hybrid clones at these loci. Let $B_i(\sigma)$ be the random number of obligate breaks occurring in the $i$th clone when the loci are ordered according to the permutation $\sigma$. In general, a permutation can be represented as an $m$-vector $(\sigma(1), \ldots, \sigma(m))$. Ambiguity about the left-to-right orientation of the loci can be avoided by confining our attention to permutations $\sigma$ with $\sigma(1) < \sigma(m)$. The correct order is given by the identity permutation $id$.

Given $n$ clones, the best order is identified by the permutation giving the smallest sum $S_n(\sigma) = \sum_{i=1}^{n} B_i(\sigma)$. Consistency requires that $S_n(id)$ be the smallest sum for $n$ large enough. Now the law of large numbers indicates that

$$\lim_{n \to \infty} \frac{1}{n} S_n(\sigma) = E[B_1(\sigma)]$$

with probability 1. Thus to demonstrate consistency, it suffices to show that the expected number of breaks $E[B_1(id)]$ under $id$ is strictly smaller than the expected number of breaks $E[B_1(\sigma)]$ under any other permutation $\sigma$.

To compute $E[B_1(id)]$, note that the interval separating loci $i$ and $i+1$ manifests an obligate break if and only if there is a break between the two loci and one locus is retained while the other locus is lost. This chain of events occurs with probability $2r(1-r)\theta_{i,i+1}$, where $r$ is the retention probability and $\theta_{i,i+1}$ is the breakage probability between the two loci. Thus,

$$E[B_1(id)] \quad = \quad 2r(1-r)\sum_{i=1}^{m-1}\theta_{i,i+1}. \tag{11.2}$$

The corresponding expression for an arbitrary permutation $\sigma$ is

$$E[B_1(\sigma)] \quad = \quad 2r(1-r)\sum_{i=1}^{m-1}\theta_{\sigma(i),\sigma(i+1)}. \tag{11.3}$$

The interval $I_{\sigma(i)}$ defined by a pair $\{\sigma(i), \sigma(i+1)\}$ is a union of adjacent intervals from the correct order $1, \ldots, m$. It is plausible to conjecture that we can match in a one-to-one fashion each interval $(k, k+1)$ against a union $I_{\sigma(i)}$ containing it. If this conjecture is true, then either $\theta_{k,k+1} = \theta_{\sigma(i),\sigma(i+1)}$ when the union $I_{\sigma(i)}$ contains a single interval, or $\theta_{k,k+1} < \theta_{\sigma(i),\sigma(i+1)}$ when the union $I_{\sigma(i)}$ contains several intervals. If the former case holds for all intervals $(k, k+1)$, then $\sigma = id$. The inequality $E[B_1(id)] < E[B_1(\sigma)]$ for $\sigma \neq id$ now follows by taking the indicated sums (11.2) and (11.3).

Thus, the crux of the proof reduces to showing that it is possible to match one to one each of the intervals $(k, k+1)$ against a union set $I_{\sigma(i)}$ that contains or covers it. This assertion is a special case of Hall's marriage theorem [6]. A simple direct proof avoiding appeal to Hall's theorem can be given by induction on $m$. The assertion is certainly true for $m = 2$. Suppose it is true for $m - 1 \geq 2$ and any permutation. There are two cases to consider.

In the first case, the last locus $m$ is internal to the given permutation $\sigma$ in the sense that $\sigma$ equals $(\sigma(1), \ldots, i, m, j, \ldots, \sigma(m))$. Omitting $m$ from $\sigma$ gives a permutation $\omega$ of $1, \ldots, m - 1$ for which the $m - 2$ intervals

$$(1, 2), \ldots, (m - 2, m - 1)$$

can be matched by induction. Assuming $j < i$, the pair $\{i, j\}$ in $\omega$ covers one of the intervals $(j, j+1), \ldots, (i-1, i)$ in this matching. In the permutation $\sigma$, match the pair $\{j, m\}$ to this covered interval. This is possible because $j < i$. To the pair $\{i, m\}$ in $\sigma$, match the interval $(m - 1, m)$. The full matching for $\sigma$ is constructed by appending these two matches to the matches for $\omega$ minus the match for the pair $\{i, j\}$. The situation with $i < j$ is handled similarly.

In the second case, $m$ is positioned at the end of $\sigma$. By our convention this means $\sigma = (\sigma(1), \ldots, \sigma(m-1), m)$. By induction, a matching can be constructed between $\omega = (\sigma(1), \ldots, \sigma(m - 1))$ and the intervals $(1, 2), \ldots, (m - 2, m - 1)$.

To this matching append the permitted match between the pair $\{\sigma(m-1), m\}$ and $(m-1, m)$. This completes the proof.

Clones with undetected typing errors can unduly influence the ranking of locus orders. A clone bearing a large number of obligate breaks probably should be retyped at the loci delimiting its obligate breaks. To identify outlier clones, one needs to compute the distribution of the number of obligate breaks under the true order and the true retention and breakage probabilities. This distribution can be computed recursively by defining $p_k(i, j)$ to be the joint probability that there are $j$ obligate breaks scored among the first $k$ loci of a clone and that the $k$th locus is present in the clone in $i$ copies. The index $k$ ranges from 1 to $m$, the index $j$ ranges from 0 to $k-1$, and the index $i$ equals 0 or 1. In this notation, the initial conditions

$$p_1(i, 0) \quad = \quad \begin{cases} 1 - r & \text{for } i = 0 \\ r & \text{for } i = 1 \end{cases}$$

are obvious. With no missing data and with $\theta_k$ now indicating the breakage probability between loci $k$ and $k + 1$, the appropriate recurrence relations for adding locus $k + 1$ are

$$p_{k+1}(0, j) \quad = \quad p_k(0, j)(1 - \theta_k r) + p_k(1, j - 1)\theta_k(1 - r)$$
$$p_{k+1}(1, j) \quad = \quad p_k(0, j - 1)\theta_k r + p_k(1, j)[1 - \theta_k(1 - r)].$$

In these recurrence relations, $p_k(i, j)$ is taken as 0 whenever $j < 0$. When the final locus $k = m$ is reached, the probabilities $p_m(i, j)$ can be summed on $i$ to produce the distribution of the number of obligate breaks. In practice, the best order identified and estimates of the retention and breakage probabilities under this order must be substituted in the above calculations. The next section addresses maximum likelihood estimation.

## 11.4    Maximum Likelihood Methods

The disadvantage of the minimum obligate breaks criterion is that it provides neither estimates of physical distances between loci nor comparisons of likelihoods for competing orders. Maximum likelihood obviously remedies the latter two defects, but does so at the expense of introducing some of the explicit assumptions mentioned earlier. We will now briefly discuss how likelihoods are computed and maximized for a given order. Different orders can be compared on the basis of their maximum likelihoods.

Because different clones are independent, it suffices to demonstrate how to compute the likelihood of a single clone. Let $X = (X_1, \ldots, X_m)$ be the observation vector for a clone potentially typed at $m$ loci. The component $X_i$ is defined as 0, 1, or ?, depending on what is observed at the $i$th locus. We can gain a feel for how to compute the likelihood of $X$ by considering two simple cases. If $m = 1$ and $X_1 \neq ?$, then $X_1$ follows the Bernoulli distribution

$$\Pr(X_1 = i) \quad = \quad r^i(1 - r)^{1-i} \tag{11.4}$$

for retention or nonretention. When $m = 2$ and both loci are typed, the likelihood must reflect breakage as well as retention. If $\theta$ is the probability of at least one break between the two loci, then

$$
\begin{aligned}
\Pr(X_1 = 0, X_2 = 0) &= (1 - r)(1 - \theta r) \\
\Pr(X_1 = 1, X_2 = 0) &= \Pr(X_1 = 0, X_2 = 1) \\
&= (1 - r)\theta r \\
\Pr(X_1 = 1, X_2 = 1) &= 1 - 2(1 - r)\theta r - (1 - r)(1 - \theta r) \\
&= (1 - \theta + \theta r)r.
\end{aligned}
\tag{11.5}
$$

Note that we parameterize in terms of the breakage probability $\theta$ between the two loci rather than the physical distance $\delta$ between them. Besides the obvious analytical simplification entailed in using $\theta$, only the product $\lambda\delta$ can be estimated anyway. The parameters $\lambda$ and $\delta$ cannot be separately identified.

As noted earlier, the probability of an obligate break between the two loci is $2r(1 - r)\theta$, in agreement with the calculated value

$$
\Pr(X_1 \neq X_2) = \Pr(X_1 = 1, X_2 = 0) + \Pr(X_1 = 0, X_2 = 1)
$$

from (11.5). It is natural to estimate $r$ and $\Pr(X_1 \neq X_2)$ by their empirical values. Given these estimates, one can then estimate $\theta$ via the identity

$$
\theta = \frac{\Pr(X_1 \neq X_2)}{2r(1 - r)}.
$$

Problems 2 and 3 elaborate on this point.

Generalization of the likelihood expressions in (11.5) to more loci involves two subtleties. First, the sheer number of terms accounting for all possible breakage and retention patterns quickly becomes unwieldy. Second, missing data can no longer be ignored. The key to efficient likelihood computation is to recognize that the likelihood splits into simple factors based on a hidden Markov property of the underlying model. To expose this factorization property, again assume that the loci $1, \ldots, m$ occur in numerical order along the chromosome. Let $\theta_i$ be the breakage probability on the interval connecting loci $i$ and $i + 1$, and suppose only loci $1 \leq t_1 < t_2 < \cdots < t_n \leq m$ are typed. If the typing result at locus $t_k$ is $x_{t_k}$, then

$$
\begin{aligned}
\Pr(X = x) = {}& \Pr(X_{t_1} = x_{t_1}) \\
& \times \prod_{i=2}^{n} \Pr(X_{t_i} = x_{t_i}) \mid X_{t_1} = x_{t_1}, \ldots, X_{t_{i-1}} = x_{t_{i-1}}).
\end{aligned}
\tag{11.6}
$$

Now $\Pr(X_{t_1} = x_{t_1})$ is immediately available from (11.4). In the degenerate case $n = 1$, the product in (11.6) is taken as 1. In general, the independence property of the governing Poisson process implies

$$
\Pr(X_{t_i} = x_{t_i} \mid X_{t_1}, \ldots, X_{t_{i-1}}) = \Pr(X_{t_i} = x_{t_i} \mid X_{t_{i-1}}).
$$

Indeed, when $X_{t_i} = X_{t_{i-1}}$,

$$\Pr(X_{t_i} = x_{t_i} \mid X_{t_1}, \ldots, X_{t_{i-1}}) = \left[1 - \prod_{j=t_{i-1}}^{t_i-1}(1 - \theta_j)\right] r^{x_{t_i}}(1 - r)^{1 - x_{t_i}}$$

$$+ \prod_{j=t_{i-1}}^{t_i-1}(1 - \theta_j). \tag{11.7}$$

The first term on the right of (11.7) involves conditioning on at least one break between loci $t_{i-1}$ and $t_i$. Here the retention fate of locus $t_i$ is no longer tied to that of locus $t_{i-1}$. The second term involves conditioning on the complementary event. When $X_{t_i} \neq X_{t_{i-1}}$, we have the simpler expression

$$\Pr(X_{t_i} = x_{t_i} \mid X_{t_1}, \ldots, X_{t_{i-1}}) = \left[1 - \prod_{j=t_{i-1}}^{t_i-1}(1 - \theta_j)\right] r^{x_{t_i}}(1 - r)^{1 - x_{t_i}}$$

since a break must occur somewhere between the two loci.

The EM algorithm provides an attractive avenue to maximum likelihood estimation of the $m$ parameters $\theta_1, \ldots, \theta_{m-1}$ and $r$. Collect these $m$ parameters into a vector $\gamma = (\theta_1, \ldots, \theta_{m-1}, r)^t$. Each of the entries $\gamma_i$ of $\gamma$ can be viewed as a success probability for a hidden binomial trial. As documented in Problem 9 of Chapter 2, the EM update for any one of these parameters takes either of the equivalent generic forms

$$\gamma_{\text{new},i} = \frac{E(\#\text{successes} \mid \text{obs}, \gamma_{\text{old}})}{E(\#\text{trials} \mid \text{obs}, \gamma_{\text{old}})}$$

$$= \gamma_{\text{old},i} + \frac{\gamma_{\text{old},i}(1 - \gamma_{\text{old},i})\frac{\partial L(\gamma_{\text{old}})}{\partial \gamma_i}}{E(\#\text{trials} \mid \text{obs}, \gamma_{\text{old}})}, \tag{11.8}$$

where obs denotes the observations $X$ over all clones, and $L$ is the loglikelihood function. The second form of the update (11.8) requires less thought to implement since only mechanical differentiations are involved in forming the score. If the number of clones is $H$, then $H$ is also the expected number of trials appearing in the denominator for both updates to $\theta_i$. The expected number of trials for $r$ coincides with the expected number of fragments. This expectation can be found by letting $N_i$ be the random number of breaks between loci $i$ and $i + 1$ over all clones. The first form of the update for $\theta_i$ shows that

$$\theta_{\text{new},i} = \frac{E(N_i \mid \text{obs}, \gamma_{\text{old}})}{H}.$$

It follows that the expected number of fragments over all $H$ clones is

$$H + \sum_{i=1}^{m-1} E(N_i \mid \text{obs}, \gamma_{\text{old}}) = H\left(1 + \sum_{i=1}^{m-1} \theta_{\text{new},i}\right).$$

## 11.5    Application to Haploid Data

TABLE 11.1. Best Locus Orders for Haploid Radiation Hybrid Data

| Orders | | | | | | | | | | | | | $\Delta L$ | Breaks |
|---|---|---|---|---|---|---|---|---|---|---|---|---|---|---|
| 1 | 2 | 3 | 4 | 5 | 6 | 7 | 8 | 9 | 10 | 11 | 12 | 13 | 0.00 | 123 |
| 1 | 2 | 3 | 4 | 5 | 6 | 7 | 8 | 10 | 9 | 11 | 12 | 13 | 1.49 | 125 |
| 1 | 2 | 3 | 4 | 5 | 13 | 12 | 11 | 10 | 9 | 8 | 7 | 6 | 1.79 | 126 |
| 5 | 4 | 3 | 2 | 1 | 6 | 7 | 8 | 9 | 10 | 11 | 12 | 13 | 1.84 | 128 |
| 1 | 2 | 3 | 4 | 5 | 7 | 6 | 8 | 9 | 10 | 11 | 12 | 13 | 1.93 | 127 |
| 6 | 7 | 1 | 2 | 3 | 4 | 5 | 8 | 9 | 10 | 11 | 12 | 13 | 2.26 | 127 |
| 6 | 7 | 5 | 4 | 3 | 2 | 1 | 8 | 9 | 10 | 11 | 12 | 13 | 2.43 | 128 |
| 1 | 2 | 3 | 4 | 5 | 6 | 7 | 8 | 11 | 10 | 9 | 12 | 13 | 3.22 | 127 |
| 1 | 2 | 3 | 4 | 5 | 6 | 7 | 8 | 11 | 9 | 10 | 12 | 13 | 3.23 | 127 |
| 1 | 2 | 3 | 4 | 5 | 13 | 12 | 11 | 9 | 10 | 8 | 7 | 6 | 3.28 | 128 |

The Cox et al. [10] data mentioned earlier involves 99 hybrids typed at 14 marker loci on chromosome 21. Examination of these data [5] shows that the markers D21S12 and D21S111 are always concordantly retained or lost. Since order cannot be resolved for these two loci, locus D21S111 is excluded from the analysis presented here. Table 11.1 presents the 10 best orders based on the maximum likelihood criterion. The difference in maximum loglikelihoods between the best order and the current order is given in the column labeled $\Delta L$. Logarithms here are to the base 10, so a difference of 3 corresponds to a likelihood ratio of 1,000. The minimum obligate breaks criterion is given in the column labeled Breaks. It is encouraging that the three best maximum likelihood orders are also the three best minimum obligate breaks orders. Evidently, some of the better orders involve complex rearrangements of the best order.

The diagram below gives the estimated distances between adjacent pairs of loci under the best order. These distances are expressed in the expected numbers of breaks × 100 between the two loci per chromosome. (One expected break is one **Ray**, so the appropriate units here are **centiRays**, abbreviated cR.) In Table 11.1 and the diagram, the loci D21S16, D21S48, D21S46, D21S4, D21S52, D21S11, D21S1, D21S18, D21S8, APP, D21S12, D21S47, and SOD1 are numbered 1 through 13, respectively.

**Interlocus Distances for the Best Order**

$$1 \overset{7.6}{-} 2 \overset{7.9}{-} 3 \overset{19.4}{-} 4 \overset{27.3}{-} 5 \overset{64.4}{-} 6 \overset{18.0}{-} 7 \overset{55.6}{-} 8 \overset{34.9}{-} 9 \overset{11.1}{-} 10 \overset{23.5}{-} 11 \overset{36.2}{-} 12 \overset{25.3}{-} 13$$

## 11.6    Polyploid Radiation Hybrids

Polyploid radiation hybrid samples can be constructed in several ways. For instance, one can pool different haploid clones and test each of the pools so constructed for the presence of the various markers to be mapped. If $c$ clones are pooled at a time, then a pool contains fragments generated by $c$ independently irradiated chromosomes. The overlapping nature of a pool obscures fragment retention patterns. Balanced against this information loss is the information gained by attaining a higher effective retention rate per locus per pool.

Another method of generating polyploid samples is to expose normal human diploid cells to a lethal dose of gamma irradiation. Some of the irradiated cells can again be rescued by hybridization to unirradiated cells from a rodent cell line. If the rodent cells are deficient in an enzyme such as HPRT, then only the hybrid cells will grow in a culture medium requiring the enzyme. Thus, the design of the diploid experiment is almost identical to the original haploid design. The diploid design carries with it the advantage that the same clones can be used to map any chromosome of interest. Although in principle one could employ heterozygous markers, geneticists forgo this temptation and score only the presence, and not the number of markers per locus in a diploid clone. Finally, just as with haploid clones, one can pool diploid clones to achieve sampling units with an arbitrary even number of chromosomes.

We now present methods for analyzing polyploid radiation hybrids with $c$ chromosomes per clone or sampling unit [15]. For the sake of brevity, we use the term "clone" to mean either a haploid clone, a diploid clone, or a fixed number of pooled haploid or diploid clones. Our analysis will assume that the breakage and fragment retention processes are independent among chromosomes and that typing can reveal only the presence and not the number of markers per locus in a clone.

## 11.7    Maximum Likelihood Under Polyploidy

Again let $X = (X_1, \ldots, X_m)$ denote the observation vector for a single clone. If no markers are observed at the $i$th locus, then $X_i = 0$. If one or more markers are observed, then $X_i = 1$. Because $(1 - r)^c$ is the probability that all $c$ copies of a given marker are lost, the single-locus polyploid likelihood reduces to the Bernoulli distribution

$$\Pr(X_1 = i) \quad = \quad [1 - (1 - r)^c]^i (1 - r)^{c(1-i)}.$$

The two-locus polyploid likelihoods

$$
\begin{aligned}
\Pr(X_1 = 0, X_2 = 0) &= [(1 - r)(1 - \theta r)]^c \\
\Pr(X_1 = 1, X_2 = 0) &= \Pr(X_1 = 0, X_2 = 1) \\
&= (1 - r)^c - [(1 - r)(1 - \theta r)]^c \qquad (11.9) \\
\Pr(X_1 = 1, X_2 = 1) &= 1 - 2(1 - r)^c + [(1 - r)(1 - \theta r)]^c
\end{aligned}
$$

generalize the two-locus haploid likelihoods (11.5). The expression for the first probability $\Pr(X_1 = 0, X_2 = 0)$ in (11.9) is a direct consequence of the independent fate of the $c$ chromosomes during fragmentation and retention. Considering a given chromosome, the marker at locus 1 is lost with probability $1 - r$. Conditional on this event, the marker at locus 2 must also be lost. This second event occurs with probability $1 - \theta r$ since its complementary event occurs only when there is a break between the two loci and the fragment bearing the second locus is retained. The expression for $\Pr(X_1 = 0, X_2 = 1)$ in (11.9) can be computed by subtracting $\Pr(X_1 = 0, X_2 = 0)$ from the probability $(1 - r)^c$ that all $c$ markers are lost at locus 1. Finally, $\Pr(X_1 = 1, X_2 = 1)$ is most easily computed by subtracting the three previous probabilities in (11.9) from 1 and simplifying.

Preliminary estimates of the parameters $r$ and $\theta$ can be derived from the empirically observed values of $\Pr(X_1 = 0, X_2 = 0)$ and $\Pr(X_1 = 1, X_2 = 1)$. In fact, the equation

$$\Pr(X_1 = 1, X_2 = 1) - \Pr(X_1 = 0, X_2 = 0) \quad = \quad 1 - 2(1 - r)^c$$

can be solved to give

$$r \quad = \quad 1 - \left[ \frac{1 - \Pr(X_1 = 1, X_2 = 1) + \Pr(X_1 = 0, X_2 = 0)}{2} \right]^{\frac{1}{c}}. \quad (11.10)$$

Once $r$ is known, $\theta$ is determined from the first equation in (11.9) as

$$\theta \quad = \quad \frac{1 - r - [\Pr(X_1 = 0, X_2 = 0)]^{\frac{1}{c}}}{r(1 - r)}. \quad (11.11)$$

Thus, the map $(\theta, r) \rightarrow (\Pr(X_1 = 1, X_2 = 1), \Pr(X_1 = 0, X_2 = 0))$ is one to one. Its range is not the entire set $\{(s, t) : s \geq 0, \ t \geq 0, \ s + t \leq 1\}$ since one can demonstrate that any image point of the map must in addition satisfy the inequality

$$\Pr(X_1 = 1, X_2 = 0)^2$$
$$\leq \quad \Pr(X_1 = 1, X_2 = 1) \Pr(X_1 = 0, X_2 = 0). \quad (11.12)$$

See Problem 8 for elaboration.

The observed values of $\Pr(X_1 = 1, X_2 = 1)$ and $\Pr(X_1 = 0, X_2 = 0)$ are maximum likelihood estimates for the simplified model in which the only constraints on the four probabilities displayed in (11.9) are nonnegativity, the symmetry condition $\Pr(X_1 = 1, X_2 = 0) = \Pr(X_1 = 0, X_2 = 1)$, and the requirement that the four probabilities sum to 1. This simplified model has in effect two parameters, which we can identify with $\Pr(X_1 = 1, X_2 = 1)$ and $\Pr(X_1 = 0, X_2 = 0)$ and estimate by their empirical values. These values are maximum likelihood estimates under the simplified model. If these estimates satisfy inequality (11.12), then they furnish maximum likelihood estimates of the radiation hybrid model as well. Since maximum likelihood estimates are preserved under reparameterization, the maximum likelihood estimates of $r$ and $\theta$ can then be computed by substituting estimated values for theoretical values in (11.10) and (11.11).

Under the polyploid model with many loci, likelihood calculation is hindered by the fact that likelihoods no longer factor. Nonetheless, it is possible to design a fast algorithm for likelihood calculation based on the theory of **hidden Markov chains** [18]. In the current context, there exists a Markov chain whose current state is the number of markers present in a clone at the current locus. As the chain progresses from one locus to the next, starting at the leftmost locus and ending at the rightmost locus, it counts the number of markers at each locus in the clone. These numbers are hidden from view because only the presence or absence of markers is directly observable. Suppose the chain is in state $i$ at locus $k$. The probability $t_{c,k}(i, j)$ of a transition from state $i$ at locus $k$ to state $j$ at locus $k + 1$ is of fundamental importance.

To compute $t_{c,k}(i, j)$, consider first a haploid clone. In this situation the chromosome copy number $c = 1$, and it is clear that

$$
\begin{aligned}
t_{1,k}(0, 0) &= 1 - \theta_k r \\
t_{1,k}(0, 1) &= \theta_k r \\
t_{1,k}(1, 0) &= \theta_k(1 - r) \\
t_{1,k}(1, 1) &= 1 - \theta_k(1 - r).
\end{aligned}
$$

Employing these simple transition probabilities, we can write the following general expression:

$$
\begin{aligned}
t_{c,k}(i, j) &= \sum_{l=\max\{0, i+j-c\}}^{\min\{i, j\}} \binom{i}{l}\binom{c-i}{j-l} \\
&\quad \times t_{1,k}(1, 1)^l t_{1,k}(1, 0)^{i-l} t_{1,k}(0, 1)^{j-l} t_{1,k}(0, 0)^{c-i-j+l}.
\end{aligned}
\tag{11.13}
$$

Formula (11.13) can be deduced by letting $l$ be the number of markers retained at locus $k$ that lead via the same original chromosomes to markers retained at locus $k + 1$. These $l$ markers can be chosen in $\binom{i}{l}$ ways. Among the $i$ markers retained at locus $k$, the fate of the $l$ markers retained at locus $k + 1$ and the remaining $i - l$ markers not retained at locus $k + 1$ is captured by the product $t_{1,k}(1, 1)^l t_{1,k}(1, 0)^{i-l}$ in (11.13). For $j$ total markers to be retained at locus $k + 1$, the $c - i$ markers not retained at locus $k$ must lead to $j - l$ markers retained at locus $k + 1$. These $j - l$ markers can be chosen in $\binom{c-i}{j-l}$ ways. The product $t_{1,k}(0, 1)^{j-l} t_{1,k}(0, 0)^{c-i-j+l}$ captures the fate of the $c - i$ markers not retained at locus $k$. Finally, the upper and lower bounds on the index of summation $l$ insure that none of the powers of the $t_{1,k}(u, v)$ appearing in (11.13) is negative.

In setting down the likelihood for the observations $(X_1, \ldots, X_m)$ from a single clone, it is helpful to define a set $O_i$ corresponding to each $X_i$. This set indicates the range of markers possible at locus $i$. Thus, let

$$
O_i = \begin{cases} \{0, 1, \ldots, c\} & \text{for } X_i \text{ missing} \\ \{0\} & \text{for } X_i = 0 \\ \{1, \ldots, c\} & \text{for } X_i = 1. \end{cases}
$$

The sets $O_1, \ldots, O_m$ encapsulate the same information as $(X_1, \ldots, X_m)$. Owing to the Markovian structure of the model, the likelihood of the observation vector $(X_1, \ldots, X_m)$ amounts to

$$P = \sum_{j_1 \in O_1} \cdots \sum_{j_m \in O_m} \binom{c}{j_1} r^{j_1}(1-r)^{c-j_1} \prod_{k=1}^{m-1} t_{c,k}(j_k, j_{k+1}). \tag{11.14}$$

The necessity of evaluating this sum of products lands us in familiar terrain. Here, however, there is more symmetry than in pedigree calculations. As suggested by Baum [2, 11], it is natural to evaluate the sum as an iterated sum in either the forward or reverse direction.

Suppose $Z_i$ is the unobserved number of markers at locus $i$. The only restriction on $Z_i$ is that $Z_i \in O_i$. Baum's **forward algorithm** is based on recursively evaluating the joint probabilities

$$\alpha_k(j) = \Pr(Z_1 \in O_1, \ldots, Z_{k-1} \in O_{k-1}, Z_k = j)$$

for $j \in O_k$. At the leftmost locus $\alpha_1(j) = \binom{c}{j}r^j(1-r)^{c-j}$, and the obvious update is

$$\alpha_{k+1}(j) = \sum_{i \in O_k} \alpha_k(i)t_{c,k}(i, j).$$

The likelihood (11.14) can be recovered by forming the sum $\sum_{j \in O_m} \alpha_m(j)$ at the rightmost locus.

In Baum's **backward algorithm** we recursively evaluate the conditional probabilities

$$\beta_k(i) = \Pr(Z_{k+1} \in O_{k+1} \ldots Z_m \in O_m \mid Z_k = i),$$

for $i \in O_k$, starting by convention at $\beta_m(j) = 1$ for $j \in O_m$. The required update is clearly

$$\beta_k(i) = \sum_{j \in O_{k+1}} t_{c,k}(i, j)\beta_{k+1}(j).$$

In this instance the likelihood (11.14) can be recovered at the leftmost locus by forming the sum $\sum_{i \in O_1} \alpha_1(i)\beta_1(i)$.

A quick search of the likelihood can be achieved if the partial derivatives of the likelihood can be computed analytically. Let us now indicate briefly how to do this based on the intermediate results of Baum's forward and backward algorithms. For instance, consider a partial derivative $\frac{\partial}{\partial \theta_i} P$ of $P$ with respect to a breakage probability. Inspection of equation (11.14) leads to the expression

$$\frac{\partial}{\partial \theta_i} P = \sum_{j_1 \in O_1} \cdots \sum_{j_m \in O_m} \binom{c}{j_1} r^{j_1}(1-r)^{c-j_1}$$

$$\times \left[ \frac{\partial}{\partial \theta_i} t_{c,i}(j_i, j_{i+1}) \right] \prod_{k \neq i} t_{c,k}(j_k, j_{k+1}). \tag{11.15}$$

Evidently, (11.15) can be evaluated as

$$\frac{\partial}{\partial \theta_i} P = \sum_{j_i \in O_i} \sum_{j_{i+1} \in O_{i+1}} \alpha_i(j_i) \left[ \frac{\partial}{\partial \theta_i} t_{c,i}(j_i, j_{i+1}) \right] \beta_{i+1}(j_{i+1}). \qquad (11.16)$$

Similar reasoning implies that

$$\frac{\partial}{\partial r} P = \sum_{j_1 \in O_1} \left[ \frac{\partial}{\partial r} \alpha_1(j_1) \right] \beta_1(j_1) \qquad (11.17)$$

$$+ \sum_{i=1}^{m-1} \sum_{j_i \in O_i} \sum_{j_{i+1} \in O_{i+1}} \alpha_i(j_i) \left[ \frac{\partial}{\partial r} t_{c,i}(j_i, j_{i+1}) \right] \beta_{i+1}(j_{i+1}).$$

The partial derivatives appearing on the right-hand sides of (11.16) and (11.17) are tedious but straightforward to evaluate. An efficient evaluation of $P$ and its partial derivatives can therefore be orchestrated by carrying out the backward algorithm first, followed by the forward algorithm performed simultaneously with the computation of all partial derivatives. Given a partial derivative $\frac{\partial}{\partial \gamma_i} P$ of the likelihood $P$, one forms the corresponding entry in the score vector by taking the quotient $(\frac{\partial}{\partial \gamma_i} P)/P$.

Finally, we note that the EM algorithm for maximum likelihood estimation generalizes easily to the polyploid case. The only differences are that now the expected number of trials for a breakage parameter is $cH$ and the expected number of trials for the common retention probability is $cH(1 + \sum_{i=1}^{m-1} \theta_{\mathrm{new},i})$, where $H$ is again the total number of clones.

## 11.8   Obligate Breaks Under Polyploidy

An obligate break is scored between two loci $i$ and $i+1$ of a clone whenever $X_i = 1$ and $X_{i+1} = 0$ or vice versa. According to equation (11.9), the probability of this event is

$$\Pr(X_1 = 1, X_2 = 0) + \Pr(X_1 = 0, X_2 = 1) = 2(1 - r)^c [1 - (1 - \theta_i r)^c].$$

Because the probability $2(1 - r)^c [1 - (1 - \theta_i r)^c]$ has a positive partial derivative $2cr(1 - r)^c (1 - \theta_i r)^{c-1}$ with respect to $\theta_i$, it is increasing as a function of $\theta_i$. Monotonicity of the obligate breakage probability was the only property used in establishing the statistical consistency of the minimum breaks criterion for ordering loci. Thus, the minimum breaks criterion is applicable to polyploid radiation hybrids and can form the basis of a quick method for ranking locus orders.

Outlier detection by counting obligate breaks is also feasible. With the probabilities $p_k(i, j)$ defined as in the haploid case, we can again compute the distribution of the number of obligate breaks per clone assuming no missing data. Note that now the index $i$ specifying the number of marker copies present at locus $k$ ranges from 0 to $c$ instead of from 0 to 1. The initial conditions are

$$p_1(i, 0) = \binom{c}{i} r^i (1 - r)^{c-i}.$$

Taking into account the defining condition for an obligate break leads to the recurrence relation

$$p_{k+1}(i, j) = \sum_{l \sim i} p_k(l, j) t_{c,k}(l, i) + \sum_{l \nsim i} p_k(l, j - 1) t_{c,k}(l, i)$$

for all $0 \le i \le c$, where $l \sim i$ indicates that $l$ and $i$ are simultaneously in either the set $\{0\}$ or the set $\{1, \ldots, c\}$, and where the transition probabilities $t_{c,k}(l, i)$ are defined in (11.13). As already noted in the haploid case, when the final locus $k = m$ is reached, the probabilities $p_m(i, j)$ can be summed on $i$ to produce the distribution of the number of obligate breaks.

## 11.9   Bayesian Methods

Bayesian methods offer an attractive alternative to maximum likelihood methods. To implement a Bayesian analysis of locus ordering, two technical hurdles must be overcome. First, an appropriate prior must be chosen. Once this choice is made, efficient numerical schemes for estimating parameters and posterior probabilities must be constructed.

It is more convenient to put a prior on the distances between the adjacent loci of an order than on the breakage probabilities determined by these distances. In designing a prior for interlocus distances, we can assume with impunity that the intensity of the breakage process satisfies $\lambda = 1$. It is also reasonable to assume that the $m$ loci to be mapped are sampled uniformly from a chromosome interval of known physical length. This length may be difficult to estimate in base pairs. Furthermore, physical distances measured in base pairs are less relevant than physical distances measured in expected number of breaks (Rays). We can circumvent the calibration problem of converting from one measure of physical distance to the other by using the results of a maximum likelihood analysis. Suppose that under the best maximum likelihood order, we estimate a total of $b$ expected breaks between the first and last loci. With $m$ uniformly distributed loci, adjacent pairs of loci should be separated by an average distance of $\frac{b}{m-1}$. This quantity should also approximate the average distance from the left end of the interval to the first locus and from the right end of the interval to the last locus. These considerations suggest that $d = \frac{(m+1)b}{m-1}$ would be a reasonable expected number of breaks to assign to the prior interval. In practice, this value of $d$ may be too confining, and it is probably prudent to inflate it somewhat.

Given a prior interval of length $d$, let $\delta_i$ be the distance separating the adjacent loci $i$ and $i + 1$ under a given order. To calculate the joint distribution of the vector of distances $(\delta_1, \ldots, \delta_{m-1})$, expand this vector to include the distance $\delta_0$ separating the left end of the interval from the first locus. These spacings are related to the positions $t_1, \ldots, t_m$ of the loci on the interval by the lower triangular transformation

$$\delta_i = \begin{cases} t_1 & i = 0 \\ t_{i+1} - t_i & 1 \le i \le m - 1. \end{cases} \tag{11.18}$$

Because the $t_i$ correspond to order statistics from the uniform distribution on $[0, d]$, the positions vector $(t_1, \ldots, t_m)$ has uniform density $m!/d^m$ on the set

$$\{(t_1, \ldots, t_m) : 0 \le t_1 \le \cdots \le t_m \le d\}.$$

The fact that the Jacobian of the transformation (11.18) is 1 implies that the spacings vector $(\delta_0, \ldots, \delta_{m-1})$ has uniform density $m!/d^m$ on the set

$$\{(\delta_0, \ldots, \delta_{m-1}) : 0 \le \delta_i, \ i = 0, \ldots, m - 1, \ \sum_{i=0}^{m-1} \delta_i \le d\}.$$

The marginal density of the subvector $(\delta_1, \ldots, \delta_{m-1})$ can now be recovered by the integration

$$\int_0^{d - \delta_1 - \cdots - \delta_{m-1}} \frac{m!}{d^m} d\delta_0 = \frac{m!(d - \delta_1 - \cdots - \delta_{m-1})}{d^m}.$$

This prior for the spacings $\delta_1, \ldots, \delta_{m-1}$ resides on the set

$$\{(\delta_1, \ldots, \delta_{m-1}) : 0 \le \delta_i, \ i = 1, \ldots, m - 1, \ \sum_{i=1}^{m-1} \delta_i \le d\}.$$

A uniform prior on $[0,1]$ is plausible for the retention probability $r$. This prior should be independent of the prior on the spacings. With the resulting product prior now fixed for the parameter vector $\gamma = (\delta_1, \ldots, \delta_{m-1}, r)^t$, we can estimate parameters by maximizing the log posterior $L(\gamma) + R(\gamma)$, where $L(\gamma)$ is the loglikelihood and

$$R(\gamma) = \ln(d - \delta_1 - \cdots - \delta_{m-1})$$

is the log prior. This yields the **posterior mode**. Because the M step is intractable, the EM algorithm no longer directly applies. However, intractability of the M step is no hindrance to the EM gradient algorithm [13]. If $Q(\gamma \mid \gamma_{\text{old}})$ is the standard $Q$ function produced by the E step of the EM algorithm, then the EM gradient algorithm updates $\gamma$ via

$$\gamma_{\text{new}} = \gamma_{\text{old}} - \left[ d^{20} Q(\gamma_{\text{old}} \mid \gamma_{\text{old}}) + d^2 R(\gamma_{\text{old}}) \right]^{-1} \tag{11.19}$$
$$\times \ [dL(\gamma_{\text{old}}) + dR(\gamma_{\text{old}})]^t,$$

where $dL$ and $dR$ denote the differentials of $L$ and $R$, $d^2 R$ is the second differential of $R$, and $d^{20} Q(\gamma \mid \gamma_{\text{old}})$ is the second differential of $Q$ relative to its left argument.

In effect, we update $\gamma$ by one step of Newton's method applied to the function $Q(\gamma \mid \gamma_{old}) + R(\gamma)$, keeping in mind the identity $d^{10}Q(\gamma_{old} \mid \gamma_{old}) = dL(\gamma_{old})$ proved in Problem 9 of Chapter 2.

All of the terms appearing in (11.19) are straightforward to evaluate. For instance, taking into account relation (11.1) with $\lambda = 1$, we have

$$
\begin{aligned}
\frac{\partial}{\partial \delta_i} L(\gamma) &= \frac{\partial}{\partial \theta_i} L(\gamma) \frac{d\theta_i}{d\delta_i} \\
&= \frac{\partial}{\partial \theta_i} L(\gamma)(1 - \theta_i).
\end{aligned}
$$

Differentiation of the log prior produces

$$
\begin{aligned}
\frac{\partial}{\partial \delta_i} R(\gamma) &= -\frac{1}{d - \delta_1 - \cdots - \delta_{m-1}} \\
\frac{\partial}{\partial r} R(\gamma) &= 0 \\
-d^2 R &= (dR)^t dR.
\end{aligned}
$$

Computation of the diagonal matrix $d^{20}Q(\gamma \mid \gamma)$ is more complicated. Let $N_i$ be the random number of chromosomes in the sample with breaks between loci $i$ and $i + 1$. As noted earlier, this random variable has a binomial distribution with success probability $\theta_i$ and $cH$ trials. Because of the nature of the complete data likelihood, it follows that modulo an irrelevant constant,

$$
\begin{aligned}
&Q(\gamma \mid \gamma_{old}) \\
&= E(N_i \mid obs, \gamma_{old}) \ln(\theta_i) + E([cH - N_i] \mid obs, \gamma_{old}) \ln(1 - \theta_i).
\end{aligned}
$$

Straightforward calculations show that

$$
\frac{\partial^2}{\partial \delta_i^2} Q(\gamma \mid \gamma_{old}) = -\frac{E(N_i \mid obs, \gamma_{old})(1 - \theta_i)}{\theta_i^2}.
$$

If $\tilde{\theta}_{new,i}$ is the EM update of $\theta_i$ ignoring the prior, then as remarked previously, $E(N_i \mid obs, \gamma_{old}) = cH\tilde{\theta}_{new,i}$ .

It is possible to simplify the EM gradient update (11.19) by applying the Sherman-Morrison formula discussed in Chapter 3. In the present context, we need to compute $(A + uu^t)^{-1}v$ for the diagonal matrix

$$
A = -d^{20}Q(\gamma_{old} \mid \gamma_{old})
$$

and the vectors $u = dR(\gamma_{old})$ and $v^t = dL(\gamma_{old}) + dR(\gamma_{old})$. Because $R(\gamma)$ does not depend on $r$, the partial derivative $\frac{\partial}{\partial r} R(\gamma)$ vanishes. Thus, the matrix $A + uu^t$ is block diagonal, and the EM gradient update for the parameter $r$ coincides with the EM gradient update for $r$ ignoring the prior. This suggests that the EM gradient update and the ordinary EM update for $r$ will be equally effective in finding the posterior mode.

Perhaps more important from the Bayesian perspective than finding the posterior mode is the possibility of computing posterior probabilities for the various locus orders. Under the natural assumption that all orders are a priori equally likely, the posterior probability of a given order $\alpha$ is

$$\frac{\int e^{L_\alpha(\gamma)+R_\alpha(\gamma)}d\gamma}{\sum_\beta \int e^{L_\beta(\gamma)+R_\beta(\gamma)}d\gamma}, \tag{11.20}$$

where the sum in the denominator ranges over all possible orders $\beta$ and $L_\beta$ and $R_\beta$ denote the loglikelihood and log prior appropriate to order $\beta$. Two ugly issues immediately rear their heads at this point. First, unless the number of loci $m$ is small, the number of possible orders can be astronomical. This problem can be finessed if the leading orders can be identified and the sum truncated to include only these orders. In many problems only a few orders contribute substantially to the denominator of the posterior probability (11.20).

The other issue is how to evaluate the integrals appearing in (11.20). Due to the complexity of the integrands, there is no obvious analytic method of carrying out the integrations. For haploid data, Lange and Boehnke [14] suggest two approximate methods. Both of these methods have their drawbacks and can be computationally demanding. Here we suggest an approximation based on Laplace's method from asymptotic analysis [7, 21]. The idea is to expand the logarithm of the integrand $e^{L_\alpha(\gamma)+R_\alpha(\gamma)}$ in a second-order Taylor's series around the posterior mode $\hat{\gamma}$. Recalling the well-known normalizing constant for the multivariate normal density and defining $F_\alpha(\gamma) = L_\alpha(\gamma) + R_\alpha(\gamma)$, this approximation yields

$$\begin{aligned}
\int e^{F_\alpha(\gamma)}d\gamma &\approx \int e^{F_\alpha(\hat{\gamma})+\frac{1}{2}(\gamma-\hat{\gamma})'d^2 F_\alpha(\hat{\gamma})(\gamma-\hat{\gamma})}d\gamma \\
&= e^{F_\alpha(\hat{\gamma})}(2\pi)^{\frac{m}{2}} \det(-d^2 F_\alpha(\hat{\gamma}))^{-\frac{1}{2}}.
\end{aligned} \tag{11.21}$$

The accuracy of Laplace's approximation increases as the log posterior function becomes more peaked around the posterior mode $\hat{\gamma}$. The quadratic form $d^2 F_\alpha(\hat{\gamma})$ measures the curvature of $F_\alpha(\gamma)$ at $\hat{\gamma}$.

## 11.10   Application to Diploid Data

Table 11.2 lists the 10 best orders identified for 6 loci on chromosome 4 from 83 diploid clones created at the Stanford Human Genome Center and distributed by Research Genetics of Huntsville, Alabama. These six **sequence tagged sites** constitute a small subset of a much more extensive set of chromosome 4 markers. The columns labeled Prob. 1, Prob. 2, and Prob. 3 are posterior probabilities calculated under various approximations of the integrals $\int e^{L_\beta(\gamma)+R_\beta(\gamma)}d\gamma$ in formula (11.20). The first approximation is

$$\int e^{L_\beta(\gamma)+R_\beta(\gamma)}d\gamma \quad \propto \quad e^{L_\beta(\hat{\gamma})+R_\beta(\hat{\gamma})}, \tag{11.22}$$

where $\hat{\gamma}$ is the maximum likelihood estimate and the log prior function $R_\beta(\gamma)$ is taken as 0. The second approximation uses the actual log prior function in (11.22) and replaces the maximum likelihood estimate by the posterior mode. The third approximation is just the Laplace approximation (11.21). For the numbers shown in Table 11.2, all $360 = \frac{6!}{2}$ orders were included in the denominator of (11.20).

The three posterior probabilities displayed in Table 11.2 evidently agree well. Except for two minor reversals for the Laplace approximation, the 10 listed orders have the same ranks. These posterior probability ranks are roughly similar to the ranks based on minimum obligate breaks.

TABLE 11.2. Best Locus Orders for Diploid Radiation Hybrid Data

| Orders | | | | | | Prob. 1 | Prob. 2 | Prob. 3 | Breaks |
|---|---|---|---|---|---|---|---|---|---|
| 1 | 2 | 3 | 4 | 5 | 6 | .36110 | .35688 | .34360 | 52 |
| 1 | 2 | 3 | 5 | 4 | 6 | .32010 | .33041 | .33398 | 51 |
| 2 | 3 | 4 | 5 | 6 | 1 | .16750 | .16307 | .16365 | 51 |
| 2 | 3 | 5 | 4 | 6 | 1 | .14562 | .14458 | .15322 | 51 |
| 1 | 2 | 3 | 6 | 5 | 4 | .00244 | .00222 | .00222 | 54 |
| 1 | 4 | 5 | 6 | 3 | 2 | .00136 | .00119 | .00120 | 54 |
| 1 | 3 | 2 | 4 | 5 | 6 | .00054 | .00045 | .00049 | 56 |
| 1 | 2 | 3 | 5 | 6 | 4 | .00038 | .00036 | .00053 | 54 |
| 1 | 3 | 2 | 5 | 4 | 6 | .00025 | .00022 | .00027 | 55 |
| 1 | 4 | 6 | 5 | 3 | 2 | .00021 | .00019 | .00028 | 54 |

The failure of Table 11.2 to identify a decisively best order reflects uncertainties in placing locus 1 to the right or to the left of the major cluster of loci and in reversing loci 4 and 5 in this cluster. The maximum likelihood odds for pair reversals under the best identified order are given in the diagram below. It is interesting that the odds for inverting loci 1 and 2 provide no hint of the overall ambiguity in ordering locus 1. Clearly, caution should be exercised in interpreting pairwise inversion odds.

**Pairwise Inversion Odds for the Best Order**

$$1 \ \overset{5.8\times10^9}{\rule{1cm}{0.4pt}} \ 2 \ \overset{6.7\times10^2}{\rule{1cm}{0.4pt}} \ 3 \ \overset{1.7\times10^{13}}{\rule{1cm}{0.4pt}} \ 4 \ \overset{1.1}{\rule{1cm}{0.4pt}} \ 5 \ \overset{1.5\times10^4}{\rule{1cm}{0.4pt}} \ 6$$

Maximum likelihood estimates of the interlocus distances under the best order are as follows:

**Interlocus Distances for the Best Order**

$$1 \ \overset{109.9}{\rule{1cm}{0.4pt}} \ 2 \ \overset{17.5}{\rule{1cm}{0.4pt}} \ 3 \ \overset{38.0}{\rule{1cm}{0.4pt}} \ 4 \ \overset{11.9}{\rule{1cm}{0.4pt}} \ 5 \ \overset{12.6}{\rule{1cm}{0.4pt}} \ 6$$

The total map length between locus 1 and locus 6 is $b = 189.8$ cR under this order. In the Bayesian analyses, $b$ was increased to 190.5 cR to determine a prior interval length of $\frac{7 \times 190.5}{5}$ = 266.7 cR. Interlocus distances based on the posterior mode then give a total map length of only 177.6 cR. Apparently, imposition of a tight prior tends to decrease estimated interlocus distances.

## 11.11    Problems

1. For $m$ loci in a haploid clone with no missing observations, the expected number of obligate breaks $E[B(id)]$ is given by expression (11.2).

   (a) Under the correct order, show [1] that

   $$\text{Var}[B(id)] = 2r(1-r)\{\sum_{i=1}^{m-1} \theta_{i,i+1} - 2r(1-r)\sum_{i=1}^{m-1} \theta_{i,i+1}^2$$
   $$+ (1-2r)^2 \sum_{i=1}^{m-2}\sum_{j=i+1}^{m-1} \theta_{i,i+1}\theta_{j,j+1}(1-\theta_{i+1,j})\},$$

   where the breakage probability $\theta_{i+1,j} = 0$ when $i+1 = j$. (Hint: Let $S_i$ be the indicator of whether a break has occurred between loci $i$ and $i+1$. Verify that

   $$E(S_i S_j) = r(1-r)\theta_{i,i+1}\theta_{j,j+1}[1 - \theta_{i+1,j}(1-2r)^2]$$

   by considering four possible cases consistent with $S_i S_j = 1$. The first case is characterized by retention at locus $i$, nonretention at locus $i+1$, retention at locus $j$, and nonretention at locus $j+1$.)

   (b) The above expression for $\text{Var}[B(id)]$ can be simplified in the Poisson model by noting that

   $$1 - \theta_{i+1,j} = \prod_{k=i+1}^{j-1}(1 - \theta_{k,k+1}).$$

   Using this last identity, argue by induction that

   $$\sum_{i=1}^{m-2}\sum_{j=i+1}^{m-1} \theta_{i,i+1}\theta_{j,j+1}(1-\theta_{i+1,j})$$
   $$= \sum_{i=1}^{m-1}\theta_{i,i+1} - [1 - \prod_{i=1}^{m-1}(1-\theta_{i,i+1})].$$

2. In a haploid radiation hybrid experiment with $m$ loci, let $X_{ij}$ be the observation at locus $j$ of clone $i$. Assuming independence of the clones and no missing data, show that

   $$\hat{r}_n = \frac{1}{n}\sum_{i=1}^{n}\frac{1}{m}\sum_{j=1}^{m} 1_{\{X_{ij}=1\}} \qquad (11.23)$$

is a strongly consistent sequence of estimators of $r$. Let $a_{jk}$ be the probability $\Pr(X_{ij} \neq X_{ik}) = 2r(1 - r)\theta_{jk}$. Show that

$$\hat{a}_{njk} = \frac{1}{n}\sum_{i=1}^{n} 1_{\{X_{ij} \neq X_{ik}\}}$$

is a strongly consistent sequence of estimators of $a_{jk}$. Finally, prove that

$$\hat{\theta}_{njk} = \frac{\hat{a}_{njk}}{2\hat{r}_n(1 - \hat{r}_n)} \tag{11.24}$$

is a strongly consistent sequence of estimators of $\theta_{jk}$, the breakage probability between loci $j$ and $k$.

3. In addition to the assumptions of the last problem, suppose that there are just $m = 2$ loci. Prove that the estimates (11.23) and (11.24) of $r$ and $\theta$ reduce to the maximum likelihood estimates described in Sections 11.4 and 11.7 when inequality (11.12) is satisfied empirically.

4. Let $\hat{\theta}_{njk}$ be any strongly consistent sequence of estimators of $\theta_{jk}$ for polyploid radiation hybrid data. Prove that minimizing the estimated total distance

$$D(\sigma) = -\sum_{i=1}^{m-1} \ln[1 - \hat{\theta}_{n,\sigma(i),\sigma(i+1)}]$$

between the first and last loci of an order $\sigma$ provides a strongly consistent criterion for choosing the true order.

5. Let $L(\gamma)$ be the loglikelihood for the data $X$ on a single, haploid clone fully typed at $m$ loci. Here $\gamma = (\theta_1, \ldots, \theta_{m-1}, r)^t$ is the parameter vector. The expected information matrix $J$ has entries

$$J_{\gamma_i\gamma_j} = E\left[-\frac{\partial^2}{\partial\gamma_i\gamma_j}L(\gamma)\right].$$

Show that [9, 14]

$$J_{\theta_i\theta_i} = \frac{r(1-r)(2-\theta_i)}{(1-\theta_i r)\theta_i(1-\theta_i+\theta_i r)}$$

$$J_{\theta_i r} = J_{r\theta_i}$$

$$= \frac{(1-2r)(1-\theta_i)}{(1-\theta r)(1-\theta+\theta r)}$$

$$J_{rr} = \frac{1}{r(1-r)} + \sum_{i=1}^{m-1}\theta_i\left[\frac{1-r}{r(1-\theta_i r)} + \frac{r}{(1-r)(1-\theta_i+\theta_i r)}\right].$$

Prove that all other entries of $J$ are 0. Hints: Use the factorization property of the likelihood for a haploid clone. In the case of two loci, denote the probability $\Pr(X_1 = i, X_2 = j)$ by $p_{ij}$ for brevity. Then a typical entry $J_{\alpha\beta}$ of $J$ is given by

$$
J_{\alpha\beta} \;=\; \sum_{i=0}^{1}\sum_{j=0}^{1} \frac{1}{p_{ij}} \left[\frac{\partial p_{ij}}{\partial\alpha}\right]\left[\frac{\partial p_{ij}}{\partial\beta}\right].
$$

6. Continuing the last problem, prove that $J_{\theta_i\theta_i}$ has a maximum at $r = \frac{1}{2}$ when $\theta_i$ is fixed. Use this fact to show that $J_{\theta_i\theta_i} \leq 1/[2\theta_i(1-\theta_i)]$. Given a known retention probability $r$, this inequality proves that the asymptotic standard error of the estimated $\theta_i$ will be at least $\sqrt{2}$ times greater than that calculated for a simple binomial experiment with success probability $\theta$.

7. Complete the calculation of the partial derivatives of the likelihood for a single clone under the polyploid model by specifying the partial derivatives $\frac{\partial}{\partial\theta_i}t_{c,i}$, $\frac{\partial}{\partial r}t_{c,i}$, and $\frac{\partial}{\partial r}\alpha_i(j_i)$ appearing in equations (11.16) and (11.17).

8. Under the polyploid model for two loci, consider the map

$$
\begin{aligned}
(\theta, r) &\;\to\; (q_{00}, q_{11}) \\
q_{00} &\;=\; \Pr(X_1 = 0, X_2 = 0) \\
q_{11} &\;=\; \Pr(X_1 = 1, X_2 = 1).
\end{aligned}
$$

Show that this map from $\{(\theta, r) : \theta \in [0, 1],\ r \in (0, 1)\}$ is one to one and onto the region

$$
Q \;=\; \{(q_{00}, q_{11}) : q_{00} \in (0, 1),\ q_{11} \in (0, 1),\ q_{00}q_{11} \geq q_{01}^2\},
$$

where $q_{01} = \Pr(X_1 = 0, X_2 = 1)$. Prove that $\theta = 0$ if and only if $q_{00}+q_{11} = 1$, and $\theta = 1$ if and only if $q_{00}q_{11} = q_{01}^2$. The upper boundary of $Q$ is formed by the line $q_{00} + q_{11} = 1$ and the lower boundary by the curve $q_{00}q_{11} = q_{01}^2$. Prove that the curve is generated by the function $q_{11} = 1 + q_{00} - 2\sqrt{q_{00}}$.

9. Under the polyploid model for two loci, show that the expected information for $\theta$ is

$$
J_{\theta\theta} \;=\; \frac{c^2 r^2 (1-\theta r)^{c-2}(1-r)^c[1 - 2(1-r)^c + (1-\theta r)^c]}{[1 - (1-\theta r)^c][1 - 2(1-r)^c + (1-r)^c(1-\theta r)^c]}.
$$

Argue that $J_{\theta\theta}$ has a maximum as a function of $r$ near $r = \frac{1}{c+1}$ when $\theta$ is near 0. (Hint: Be careful because $\lim_{\theta\to 0} J_{\theta\theta} = \infty$. The singularity at $\theta = 0$ is removable in the function $\frac{\partial}{\partial r}\ln J_{\theta\theta}$.)

This result suggests that the value $r = \frac{1}{c+1}$ is nearly optimal for small $\theta$ in the sense of providing the smallest standard error of the maximum likelihood estimate $\hat\theta$ of $\theta$. In this regard note that $J_{\theta r}$ and $J_{rr}$ have finite limits as $\theta \to 0$. Thus for small $\theta$, the approximate standard error of $\hat\theta$ is proportional to $\frac{1}{\sqrt{J_{\theta\theta}}}$ even when $r$ is jointly estimated with $\theta$.

10. In computing the distribution of the number of obligate breaks per clone in the polyploid model, how must the initial conditions and recurrences for the probabilities $p_k(i, j)$ be modified when some loci are untyped?

11. The construction of radiation hybrids always involves a selectable enzyme such as HPRT or TK. On the chromosome containing the selectable locus, at least one fragment containing the locus is necessary to form a viable clone. This requirement invalidates our model of fragment retention for the chromosome in question. The obvious amendment of the model is to condition on the event of retention of at least one fragment bearing the selectable locus. Discuss how this change of the model affects likelihood calculation [16].

# References

[1] Barrett JH (1992) Genetic mapping based on radiation hybrid data. *Genomics* 13:95–103

[2] Baum LE (1972) An inequality and associated maximization technique in statistical estimation for probabilistic functions of Markov processes. *Inequalities* 3:1–8

[3] Bishop DT, Crockford GP (1991) Comparisons of radiation hybrid mapping and linkage mapping. *Genetic Analysis Workshop 7: Issues in Gene Mapping and Detection of Major Genes*. MacCluer JW, Chakravarti A, Cox D, Bishop DT, Bale SJ, Skolnick MH, editors, *Cytogenetics and Cell Genetics*, S Karger, Basel

[4] Boehnke M (1991) Radiation hybrid mapping by minimization of the number of obligate chromosome breaks. *Genetic Analysis Workshop 7: Issues in Gene Mapping and Detection of Major Genes*. MacCluer JW, Chakravarti A, Cox D, Bishop DT, Bale SJ, Skolnick MH, editors, *Cytogenetics and Cell Genetics*, S Karger, Basel

[5] Boehnke M, Lange K, Cox DR (1991) Statistical methods for multipoint radiation hybrid mapping. *Amer J Hum Genet* 49:1174–1188

[6] Brualdi RA (1977) *Introductory Combinatorics*. North-Holland, New York

[7] de Bruijn, NG (1981) *Asymptotic Methods in Analysis*. Dover, New York

[8] Burmeister M, Kim S, Price ER, de Lange T, Tantravahi U, Myers RM, Cox DR (1991) A map of the distal region of the long arm of human chromosome 21 constructed by radiation hybrid mapping and pulsed-field gel electrophoresis. *Genomics* 9:19–30

[9] Chernoff H (1993) Kullback-Leibler information for ordering genes using sperm typing and radiation hybrid mapping. *Statistical Sciences and Data Analysis: Proceedings of the Third Pacific Area Statistical Conference*, Matsusita K, Puri ML, Hayakawa T, editors, VSP, Utrecht, The Netherlands

[10] Cox DR, Burmeister M, Price ER, Kim S, Myers RM (1990) Radiation hybrid mapping: A somatic cell genetic method for constructing high-resolution maps of mammalian chromosomes. *Science* 250:245–250

[11] Devijver PA (1985) Baum's forward-backward algorithm revisited. *Pattern Recognition Letters* 3:369–373

[12] Goss SJ, Harris H (1975) New method for mapping genes in human chromosomes. *Nature* 255:680–684

[13] Lange K (1995) A gradient algorithm locally equivalent to the EM algorithm. *J Roy Stat Soc B* 57:425–437

[14] Lange K, Boehnke M (1992) Bayesian methods and optimal experimental design for gene mapping by radiation hybrids. *Ann Hum Genet* 56:119–144

[15] Lange K, Boehnke M, Cox DR, Lunetta KL (1995) Statistical methods for polyploid radiation hybrid mapping. *Genome Res* 5:136–150

[16] Lunetta KL, Boehnke M, Lange K, Cox DR (1996) Selected locus and multiple panel models for radiation hybrid mapping. *Amer J Hum Genet* 59:717–725

[17] Press WH, Flannery BP, Teukolsky SA, Vetterling WT (1992) *Numerical Recipes. The Art of Scientific Computing*, 2nd ed. Cambridge University Press, Cambridge

[18] Rabiner L (1989) A tutorial on hidden Markov models and selected applications in speech recognition. *Proc IEEE* 77:257–285

[19] Reingold EM, Nievergelt J, Deo N (1977) *Combinatorial Algorithms: Theory and Practice*. Prentice-Hall, Englewood Cliffs, NJ

[20] Speed TP, McPeek MS, Evans SN (1992) Robustness of the no-interference model for ordering genetic loci. *Proc Natl Acad Sci* 89:3103–3106

[21] Tierney L, Kadane JB (1986) Accurate approximations for posterior moments and marginal densities. *J Amer Stat Assoc* 81:82–86

[22] Weeks DE, Lehner T, Ott J (1991) Preliminary ranking procedures for multilocus ordering based on radiation hybrid data. *Genetic Analysis Workshop 7: Issues in Gene Mapping and Detection of Major Genes*. MacCluer JW, Chakravarti A, Cox D, Bishop DT, Bale SJ, Skolnick MH, editors, *Cytogenetics and Cell Genetics*, S Karger, Basel

# 12

# Models of Recombination

## 12.1   Introduction

At meiosis, each member of a pair of homologous chromosomes replicates to form two **sister** chromosomes known as **chromatids**. The maternally and paternally derived sister pairs then perfectly align to form a bundle of four chromatids. Crossing-over occurs at points along the bundle known as **chiasmata**. At each **chiasma**, one sister chromatid from each pair is randomly selected and cut at the crossover point. The cell then rejoins the partial paternal chromatid above the cut to the partial maternal chromatid below the cut, and vice versa, to form two hybrid maternal–paternal chromatids. The preponderance of evidence suggests that the two chromatids participating in a chiasma are chosen nearly independently from chiasma to chiasma [30]. This independence property is termed lack of **chromatid interference**. After crossing-over has occurred, the recombined chromatids of a bundle are coordinately separated by two cell divisions so that each of the four resulting gametes receives exactly one chromatid.

The number and positions of the chiasmata along a chromatid bundle provide an example of a **stochastic point process** [5]. Most probabilists are familiar with point processes such as Poisson processes and renewal processes. This chapter considers point process models for the formation of chiasmata. Each such **chiasma process** induces correlated and identically distributed **crossover processes** on the four gametes created from a chromatid bundle. We can conceive of both chiasma and crossover processes as occurring on a fixed interval of the real line. When one makes, as we do in this chapter, the assumption of no chromatid interference, then each of the four crossover processes is created from the chiasma process by **random thinning** of chiasmata. If we characterize a gamete by the origin of one of its **telomeres** (chromosome ends), then it participates in half of the crossovers on average. Random thinning amounts to independently choosing for each chiasma whether the gamete does or does not participate in the underlying crossover event. Because of the symmetry of the model, these two choices are equally likely.

For any well-behaved set $A$ on the chromatid bundle, the random variable $N_A$ counts the number of random points in $A$. The chiasma process is determined by these random variables. Two important functions of $A$ are the **intensity measure** $E(N_A)$ and the **avoidance probability** $\Pr(N_A = 0)$. It is natural to assume that random points never coincide and that $E(N_{\{a\}}) = 0$ for every fixed point $a$. Beyond these assumptions, we seek models that correctly capture the phenomenon of **chiasma interference**. This second kind of interference involves the suppression of additional chiasmata in the vicinity of a chiasma already formed. Although a fair amount is known about the biochemistry and cytology of crossing-over, no one has suggested a mechanism that fully explains chiasma interference [27, 29]. Until such a mechanism appears, we must be content with purely phenomenological models for the relatively small number of chiasmata per chromatid bundle. Even if a satisfactory model is devised, calculation of gamete probabilities under it may be very cumbersome. The models considered in this chapter have the advantage of permitting exact calculation of multilocus gamete probabilities.

## 12.2    Mather's Formula and Its Generalization

Mather [18] discovered a lovely formula connecting the recombination fraction $\theta$ separating two loci at positions $a$ and $b$ to the random number of chiasmata $N_{[a,b]}$ occurring on the interval $[a, b]$ of the chromatid bundle. Mather's formula

$$
\begin{aligned}
\theta &= \frac{1}{2}\Pr(N_{[a,b]} > 0) \\
&= \frac{1}{2}[1 - \Pr(N_{[a,b]} = 0)]
\end{aligned}
\tag{12.1}
$$

makes it clear that $0 \le \theta \le \frac{1}{2}$ and that $\theta$ increases as $b$ increases for $a$ fixed. The genetic map distance $d$ separating $a$ and $b$ is defined as $\frac{1}{2}E(N_{[a,b]})$, the expected number of chiasmata on $[a, b]$ per gamete. The unit of distance is the Morgan, in honor of Thomas Hunt Morgan. For short intervals, $\theta \approx d$ holds because $E(N_{[a,b]}) \approx \Pr(N_{[a,b]} > 0)$ holds.

To prove (12.1), note first that a gamete is recombinant between two loci $a$ and $b$ if and only if an odd number of crossovers occurs on the gamete between $a$ and $b$. Let $r_n$ be the probability that the gamete is recombinant given that $n$ chiasmata occur on the chromatid bundle between $a$ and $b$. It is clear that $r_0 = 0$. For $n > 0$, we have the recurrence

$$
r_n = \frac{1}{2}r_{n-1} + \frac{1}{2}(1 - r_{n-1})
\tag{12.2}
$$

because a gamete is recombinant after $n$ crossovers if it is recombinant after $n-1$ crossovers and does not participate in crossover $n$, or if it is nonrecombinant after $n-1$ crossovers and does participate in crossover $n$. In view of recurrence relation (12.2), it follows that $r_n = \frac{1}{2}$ for all $n > 0$, and this fact proves Mather's formula (12.1).

As a simple application of Mather's formula, suppose that the number of chiasmata on the chromatid bundle between $a$ and $b$ follows a Poisson distribution with mean $\lambda$. Then a gamete is recombinant with probability $\frac{1}{2}(1 - e^{-\lambda})$ and nonrecombinant with probability $\frac{1}{2}(1 + e^{-\lambda})$. Haldane's model, which postulates that the chiasma process is Poisson, turns out to be crucial in generalizing Mather's formula [16, 23]. To derive this generalization, we employ a randomization technique pioneered by Schrödinger.

Consider a sequence of $k + 1$ loci along a chromosome. The $k + 1$ loci define $k$ adjacent intervals $I_1, \ldots, I_k$. A gamete can be recombinant on some of these intervals and nonrecombinant on others. For a subset $S \subset \{1, \ldots, k\}$, let $y_S$ denote the probability that the gamete is recombinant on each of the intervals $I_i$ indexed by $i \in S$ and nonrecombinant on each of the remaining intervals $I_i$ indexed by $i \notin S$. Under Haldane's model, the numbers of chiasmata falling on disjoint intervals are independent. Therefore,

$$
y_S = \prod_{i \in S} \frac{1}{2}(1 - e^{-\lambda_i}) \prod_{i \notin S} \frac{1}{2}(1 + e^{-\lambda_i}), \tag{12.3}
$$

where $\lambda_i$ denotes the expected number of chiasmata on interval $I_i$.

Next consider what happens when we fix the number $N_{I_i} = n_i$ of chiasmata occurring on each interval $I_i$. The probability of the recombination pattern dictated by $S$ now changes to the conditional probability $y_{\mathbf{n}S}$, where $\mathbf{n}$ is the multi-index $(n_1, \ldots, n_k)$. We recover $y_S$ via

$$
y_S = \sum_{\mathbf{n}} y_{\mathbf{n}S} \prod_{i=1}^{k} \frac{\lambda_i^{n_i}}{n_i!} e^{-\lambda_i}. \tag{12.4}
$$

Equating formulas (12.3) and (12.4) and multiplying by $\prod_{i=1}^{k} e^{\lambda_i}$, we deduce that

$$
\begin{aligned}
y_S \prod_{i=1}^{k} e^{\lambda_i} &= \sum_{\mathbf{n}} y_{\mathbf{n}S} \prod_{i=1}^{k} \frac{\lambda_i^{n_i}}{n_i!} \\
&= \left(\frac{1}{2}\right)^k \prod_{i \in S} (e^{\lambda_i} - 1) \prod_{i \notin S} (e^{\lambda_i} + 1). \tag{12.5}
\end{aligned}
$$

We extract the coefficient $y_{\mathbf{n}S}$ by evaluating the partial derivative

$$
\frac{\partial^{\sum_{i=1}^{k} n_i}}{\partial \lambda_1^{n_1} \cdots \partial \lambda_k^{n_k}} \left( y_S \prod_{i=1}^{k} e^{\lambda_i} \right)
$$

at $\lambda_1 = \cdots = \lambda_k = 0$. Equating the two results from identity (12.5) yields

$$
\begin{aligned}
y_{n\,S} &= \left(\frac{1}{2}\right)^k \prod_{i \in S}(1 - 1_{\{n_i = 0\}}) \prod_{i \notin S}(1 + 1_{\{n_i = 0\}}) \\
&= \left(\frac{1}{2}\right)^k \sum_T (-1)^{|S \cap T|} 1_{\{\sum_{i \in T} n_i = 0\}},
\end{aligned}
$$

where $T$ ranges over all subsets of $\{1, \ldots, k\}$ and $|S \cap T|$ indicates the number of elements in the intersection $S \cap T$. This last formula continues to hold when the fixed counts $n_i$ are replaced by the random counts $N_{I_i}$. Taking expectations therefore produces

$$
\begin{aligned}
y_S &= E(y_{N\,S}) \\
&= \left(\frac{1}{2}\right)^k \sum_T (-1)^{|S \cap T|} \Pr\left(\sum_{i \in T} N_{I_i} = 0\right) \qquad (12.6) \\
&= \left(\frac{1}{2}\right)^k \sum_{t_1 = 0}^1 \cdots \sum_{t_k = 0}^1 (-1)^{\sum_{i=1}^k s_i t_i} \Pr\left(\sum_{i=1}^k t_i N_{I_i} = 0\right),
\end{aligned}
$$

where $s_i = 1_{\{i \in S\}}$ and $t_i = 1_{\{i \in T\}}$ are the obvious indicator functions. This is the sought-after generalization of Mather's formula [23, 25]. It expresses a multilocus gamete probability in terms of the avoidance probabilities of the chiasma process.

Special cases of (12.6) are easy to construct. For instance, Mather's formula (12.1) can be restated as $y_{\{1\}} = \frac{1}{2}[1 - \Pr(N_{I_1} = 0)]$ for $k = 1$. The probability of nonrecombination is $y_\emptyset = \frac{1}{2}[1 + \Pr(N_{I_1} = 0)]$. When $k = 2$, two of the relevant gamete probabilities are

$$
\begin{aligned}
& y_{\{1\}} \\
&= \frac{1}{4}[1 - \Pr(N_{I_1} = 0) + \Pr(N_{I_2} = 0) - \Pr(N_{I_1} + N_{I_2} = 0) \\
& y_{\{1,2\}} \qquad\qquad\qquad\qquad\qquad\qquad\qquad\qquad (12.7) \\
&= \frac{1}{4}[1 - \Pr(N_{I_1} = 0) - \Pr(N_{I_2} = 0) + \Pr(N_{I_1} + N_{I_2} = 0)].
\end{aligned}
$$

## 12.3   Count-Location Model

The count-location model operates by first choosing the total number $N$ of chiasmata along the bundle of four chromatids [12, 22]. If we identify the bundle with the unit interval $[0, 1]$, then $N = N_{[0,1]}$. Let $q_n = \Pr(N = n)$ be the distribution of $N$. Once the number of chiasmata is chosen, the individual chiasmata are located independently along the bundle according to some common continuous distribution $F(t)$. If $\lambda = E(N)$ in this setting, then the map length of an interval $[a, b]$ reduces to $d = \frac{1}{2}\lambda[F(b) - F(a)]$. The recombination fraction $\theta$ of the interval can be expressed compactly via the generating function $Q(s) = \sum_{n=0}^{\infty} q_n s^n$ of $N$. Conditioning on the value of $N$, we find that

$$
\begin{aligned}
\theta &= \frac{1}{2} \Pr(N_{[a,b]} > 0) \\
&= \frac{1}{2} \sum_{n=0}^{\infty} q_n [1 - \Pr(N_{[a,b]} = 0 \mid N = n)] \\
&= \frac{1}{2} \sum_{n=0}^{\infty} q_n \{1 - [1 - F(b) + F(a)]^n\} \\
&= \frac{1}{2} - \frac{1}{2} Q(1 - 2\lambda^{-1} d).
\end{aligned}
$$

Thus, the count-location model yields a **map function** $\theta = M(d)$ giving the recombination fraction $\theta$ of an interval purely in terms of the corresponding map distance $d$.

Gamete probabilities are easily computed in the count-location model based on formula (12.6). Indeed, if $I_i = [a_i, b_i]$ and $w_i = F(b_i) - F(a_i)$, then the gamete probability $y_S$ can be expressed in either of the equivalent forms

$$
\begin{aligned}
y_S &= \left(\frac{1}{2}\right)^k \sum_T (-1)^{|S \cap T|} Q\left(1 - \sum_{i \in T} w_i\right) \\
&= \sum_{n=0}^{\infty} q_n \left(\frac{1}{2}\right)^k \sum_T (-1)^{|S \cap T|} \left(1 - \sum_{i \in T} w_i\right)^n.
\end{aligned}
$$

Further algebraic simplification of $y_S$ is possible but will not be pursued here [23].

**Example 12.1** *Haldane's Model*

Haldane's Poisson model has chiasma count distribution $q_n = \lambda^n e^{-\lambda}/n!$ with generating function $Q(s) = e^{-\lambda(1-s)}$ [10]. In this case, the conversion between map distance $d$ and recombination fraction $\theta$ is mediated by the pair of functions mentioned in equations (7) and (8) of Chapter 7. Although Haldane's model is widely used, it unrealistically entails no chiasma interference. This defect is partially compensated for by its computational simplicity. See, for instance, Problem 1 of this chapter and Trow's formula in Problem 1 of Chapter 7. ∎

Other simple choices for the chiasma count distribution include the binomial and truncated Poisson distributions with generating functions

$$
\begin{aligned}
Q(s) &= \left(\frac{1}{2} + \frac{s}{2}\right)^r \\
Q(s) &= \frac{e^{-\lambda(1-s)} - e^{-\lambda}}{1 - e^{-\lambda}},
\end{aligned}
$$

respectively [11, 28]. The latter choice is particularly useful because it incorporates the empirical observation that almost all chromatid bundles display at least one chiasma.

respectively [11, 28]. The latter choice is particularly useful because it incorporates the empirical observation that almost all chromatid bundles display at least one chiasma.

## 12.4    Stationary Renewal Models

**Renewal processes** provide another important class of recombination models. A renewal process is generated by the partial sums of a sequence $X_1, X_2, \ldots$ of nonnegative, i.i.d. random variables. The first random point on $[0, \infty)$ occurs at $X_1$, the second at $X_1 + X_2$, the third at $X_1 + X_2 + X_3$, and so forth [7, 14, 24]. If the $X_i$ have common continuous distribution function $F(x)$ with mean $\mu$, then for large $x$ the interval $(x, x + \Delta x)$ contains approximately $\frac{\Delta x}{\mu}$ random points. The expected number of random points $U(x)$ on the interval $(0, x]$ coincides with the distribution function $E(N_{[0,x]})$ of the intensity measure. In the renewal theory literature, $F(x)$ is said to be the **interarrival distribution**, and $U(x)$ is said to be the **renewal function**. If the $X_i$ follow an exponential distribution, then the renewal process collapses to a Poisson process with intensity $\lambda = \frac{1}{\mu}$ and renewal function $U(x) = \frac{x}{\mu}$.

The identity $U(x) = \frac{x}{\mu}$ characterizes a Poisson process. In general, we can achieve a uniform distribution for the intensity measure by passing to a **delayed renewal process** where $X_1$ follows a different distribution function $F_\infty(x)$ than the subsequent $X_i$. The appropriate choice of $F_\infty(x)$ turns out to have density $F'_\infty(x) = [1 - F(x)]/\mu$. Problem 3 sketches a proof of this fact. The delayed renewal process with density $[1 - F(x)]/\mu$ is a **stationary point process** in the sense that it exhibits the same stochastic behavior regardless of whether we begin observing it at 0 or at some subsequent positive point $y$. In particular, the waiting time until the next random point after $y$ also follows the **equilibrium** distribution $F_\infty(x)$.

For more than two generations, geneticists have proposed various map functions. The recent work of Zhao and Speed [31] clarifies which of these map functions legitimately arise from point process models. They show that any valid map function can be realized by constructing a stationary renewal process for the underlying chiasma process. Implicit in this finding is the fact that a map function does not uniquely define its chiasma process.

In exploring the map function problem, let us consider for the sake of simplicity a map function $\theta = M(d)$ that is twice differentiable and whose derivative satisfies the fundamental theorem of calculus in the form

$$M'(d_2) - M'(d_1) = \int_{d_1}^{d_2} M''(x)dx.$$

By virtue of Mather's formula (12.1), it is clear that (a) $M(0) = 0$, (b) $M'(d) \geq 0$, and (c) $M'(0) = 1$. If at least one chiasma is certain on a segment of infinite length, then it is also clear that (d) $\lim_{d \to \infty} M(d) = \frac{1}{2}$. The final property (e) $M''(d) \leq 0$ is true but more subtle.

Property (e) can be proved by noting that formula (12.6) implies on one hand that

$$
\begin{aligned}
& y_{\{1,2,3\}} - y_{\{1,3\}} \\
= \ & \frac{1}{8}\Big[ -2\Pr(N_{I_2} = 0) + 2\Pr(N_{I_1} = 0, N_{I_2} = 0) \\
& + 2\Pr(N_{I_2} = 0, N_{I_3} = 0) - 2\Pr(N_{I_1} = 0, N_{I_2} = 0, N_{I_3} = 0) \Big] \\
= \ & -\frac{1}{4}\,\mathrm{E}\Big[ (1 - 1_{\{N_{I_1}=0\}}) 1_{\{N_{I_2}=0\}} (1 - 1_{\{N_{I_3}=0\}}) \Big] \qquad (12.8) \\
\leq \ & 0.
\end{aligned}
$$

On the other hand, if the map distance assigned to interval $I_j$ is $d_j$, then one can invoke Mather's formula (12.1) and exchange avoidance probabilities for recombination fractions in the above computation. This gives

$$
\begin{aligned}
& y_{\{1,2,3\}} - y_{\{1,3\}} \\
= \ & \frac{1}{8}\Big\{ -2[1 - 2M(d_2)] + 2[1 - 2M(d_1 + d_2)] \qquad (12.9) \\
& + 2[1 - 2M(d_2 + d_3)] - 2[1 - 2M(d_1 + d_2 + d_3)] \Big\}.
\end{aligned}
$$

Equality (12.9) and inequality (12.8) together imply that

$$
\begin{aligned}
& M(d_1 + d_2 + d_3) - M(d_1 + d_2) - M(d_2 + d_3) + M(d_2) \\
= \ & \int_{d_2}^{d_2+d_3} [M'(d_1 + u) - M'(u)]du \\
\leq \ & 0.
\end{aligned}
$$

Because this holds for all positive $d_j$, the integrand $M'(d_1 + u) - M'(u) \leq 0$, and property (e) follows from the difference quotient definition of $M''(d)$ [26].

Now suppose a chiasma process is determined by a stationary renewal model with distribution function $F(x)$ having density $f(x) = F'(x)$. Because of stationarity, the map length of the interval $[a, a + b]$ is $d = \frac{b}{2\mu}$. In view of Mather's formula (12.1) and the form of the equilibrium density $F'_\infty(x)$, the corresponding recombination fraction is

$$
\begin{aligned}
M(d) \ & = \ \frac{1}{2}\Big\{ 1 - \frac{1}{\mu}\int_b^\infty [1 - F(x)]dx \Big\} \\
& = \ \frac{1}{2}\Big\{ 1 - \frac{1}{\mu}\int_{2\mu d}^\infty \int_x^\infty f(y)dy dx \Big\}.
\end{aligned}
$$

Differentiating this expression twice with respect to $d$ yields

$$
M''(d) \ = \ -2\mu f(2\mu d). \qquad \bullet
$$

Thus, $f(x)$ can be recovered via

$$f(x) = -\frac{1}{2\mu} M''\left(\frac{x}{2\mu}\right). \tag{12.10}$$

Without loss of generality, we can always rescale distances so that $\mu = \frac{1}{2}$ in this formula.

Conversely, if we postulate the existence of a map function $M(d)$ satisfying properties (a) through (e), then equation (12.10) defines a valid density function $f(x)$. Indeed, property (e) indicates that $f(x) \geq 0$, and properties (a) and (d) indicate that $\int_0^\infty M'(x)dx = \frac{1}{2}$. In view of properties (b) and (e), this last fact entails $\lim_{x\to\infty} M'(x) = 0$. Thus, the calculation

$$\int_0^\infty f(x)dx = M'(0) - \lim_{x\to\infty} M'(x)$$
$$= 1$$

verifies that $f(x)$ has total mass 1. Using $f(x)$ to construct a stationary renewal process yields a map function matching $M(d)$ and proves Zhao and Speed's converse.

**Example 12.2** *Kosambi's Map Function*

Kosambi's map function [15] $M(d) = \frac{1}{2}\tanh(2d)$ has first two derivatives

$$M'(d) = \frac{4}{(e^{2d} + e^{-2d})^2}$$
$$M''(d) = -16\frac{e^{2d} - e^{-2d}}{(e^{2d} + e^{-2d})^3}.$$

From these expressions it is clear that properties (a) through (e) are true. Taking $\mu = \frac{1}{2}$ in equation (12.10) yields

$$f(x) = 16\frac{e^{2x} - e^{-2x}}{(e^{2x} + e^{-2x})^3}.$$

## 12.5  Poisson-Skip Model

A particularly simple stationary renewal model is the **Poisson-skip process**. This model is generated by a Poisson process with intensity $\lambda$ and a skip distribution $s_n$ on the positive integers. Random Poisson points are divided into $o$ points and $\chi$ points; $o$ points are "skipped" to reach $\chi$ points, which naturally correspond to chiasmata. At each $\chi$ point, one independently chooses with probability $s_n$ to skip $n - 1$ $o$ points before encountering the next $\chi$ point. This recipe creates a renewal process with interarrival distribution

$$F(x) = \sum_{n=1}^\infty s_n \sum_{m=n}^\infty \frac{(\lambda x)^m}{m!} e^{-\lambda x}.$$

The Poisson tail probability $\sum_{m=n}^{\infty} \frac{(\lambda x)^m}{m!} e^{-\lambda x}$ appearing in this formula is the probability that the $n$th random point to the right of the current $\chi$ point lies within a distance $x$ of the current $\chi$ point. If we let $\omega = \sum_{n=1}^{\infty} n s_n$ be the mean number of points until the next $\chi$ point, then Wald's formula [14] shows that $F(x)$ has mean $\frac{\omega}{\lambda}$. The density of the equilibrium distribution is

$$\frac{\lambda}{\omega}[1 - F(x)] \;=\; \frac{\lambda}{\omega} \sum_{n=1}^{\infty} s_n \left[ 1 - \sum_{m=n}^{\infty} \frac{(\lambda x)^m}{m!} e^{-\lambda x} \right]$$

$$=\; \frac{\lambda}{\omega} \sum_{n=1}^{\infty} s_n \sum_{m=0}^{n-1} \frac{(\lambda x)^m}{m!} e^{-\lambda x}.$$

According to equation (12.1), the map function for the Poisson-skip model boils down to

$$\theta \;=\; \frac{1}{2}\left\{ 1 - \frac{\lambda}{\omega} \int_x^{\infty} [1 - F(y)]dy \right\}$$

$$=\; \frac{1}{2}\left\{ 1 - \frac{\lambda}{\omega} \int_x^{\infty} \sum_{n=1}^{\infty} s_n \sum_{m=0}^{n-1} \frac{(\lambda y)^m}{m!} e^{-\lambda y} dy \right\}.$$

Because successive integrations by parts yield

$$\int_x^{\infty} \frac{(\lambda y)^m}{m!} e^{-\lambda y} dy \;=\; \frac{1}{\lambda} \sum_{k=0}^{m} \frac{(\lambda x)^k}{k!} e^{-\lambda x},$$

it follows that

$$\theta \;=\; \frac{1}{2}\left\{ 1 - \frac{1}{\omega} \sum_{n=1}^{\infty} s_n \sum_{m=0}^{n-1} \sum_{k=0}^{m} \frac{(\lambda x)^k}{k!} e^{-\lambda x} \right\}$$

$$=\; \frac{1}{2}\left\{ 1 - \frac{e^{-\lambda x}}{\omega} \sum_{n=1}^{\infty} s_n \sum_{k=0}^{n-1} (n-k) \frac{(\lambda x)^k}{k!} \right\}.$$

To calculate gamete probabilities under the Poisson-skip model, it is helpful to consider two associated Markov chains. The state space for the first chain is $\{0, 1, 2, \ldots\}$. When the chain is in state 0, the most recent point encountered was a $\chi$ point. When it is in state $i > 0$, it is must pass exactly $i - 1$ $o$ points before encountering the next $\chi$ point. Thus, if the chain is currently in state 0, then it moves to state $n - 1$, $n > 1$, with transition rate $\lambda s_n$. This transition mechanism implies that the chain decides how many $o$ points to skip simultaneously with moving to the next point. When the chain decides to skip no $o$ points, it remains in state 0. If the chain is currently in state $n > 0$, then it falls back to state $n - 1$ with transition rate $\lambda$. These are the only moves possible. If at most $r - 1$ $o$ points can be skipped, then the motion of the chain on the reduced state space $\{0, 1, 2, \ldots, r-1\}$ is summarized by the infinitesimal generator

$$\Gamma \;=\; \begin{array}{c} \\ 0 \\ 1 \\ \vdots \\ r-1 \end{array}
\begin{array}{cccccc}
0 & 1 & \cdots & r-2 & r-1 \\
\left(\begin{array}{ccccc}
-\lambda(1-s_1) & \lambda s_2 & \cdots & \lambda s_{r-1} & \lambda s_r \\
\lambda & -\lambda & \cdots & 0 & 0 \\
\vdots & \vdots & \vdots & \vdots & \vdots \\
0 & 0 & \cdots & \lambda & -\lambda
\end{array}\right).
\end{array}$$

The equilibrium distribution for the chain has entries $\pi_m = \frac{1}{\omega}\sum_{n>m} s_n$. Indeed, the balance condition $\pi\Gamma = \mathbf{0}$ reduces to

$$-\frac{1}{\omega}\sum_{n>0} s_n\lambda(1-s_1) + \frac{1}{\omega}\sum_{n>1} s_n\lambda \;=\; 0$$

for row $m = 0$ and to

$$\frac{1}{\omega}\sum_{n>0} s_n\lambda s_{m+1} - \frac{1}{\omega}\sum_{n>m} s_n\lambda + \frac{1}{\omega}\sum_{n>m+1} s_n\lambda \;=\; 0$$

for row $m > 0$. These equations follow from the identity $\sum_{n>0} s_n = 1$.

The second Markov chain is identical to the first except that it has an absorbing state $0_{\mathrm{abs}}$. In state 0 the chain moves to state $0_{\mathrm{abs}}$ with transition rate $\lambda s_1$. In state 1 it moves to state $0_{\mathrm{abs}}$ instead of state 0 with transition rate $\lambda$. If at most $r - 1$ o points can be skipped, then this second chain has infinitesimal generator

$$\Delta \;=\; \begin{array}{c} \\ 0 \\ 1 \\ \vdots \\ r-1 \\ 0_{\mathrm{abs}} \end{array}
\begin{array}{c}
\begin{array}{cccccc}
0 & 1 & \cdots & r-2 & r-1 & 0_{\mathrm{abs}}
\end{array} \\
\left(\begin{array}{cccccc}
-\lambda & \lambda s_2 & \cdots & \lambda s_{r-1} & \lambda s_r & \lambda s_1 \\
0 & -\lambda & \cdots & 0 & 0 & \lambda \\
\vdots & \vdots & \vdots & \vdots & \vdots & \vdots \\
0 & 0 & \cdots & \lambda & -\lambda & 0 \\
0 & 0 & \cdots & 0 & 0 & 0
\end{array}\right).
\end{array}$$

As emphasized in Chapter 10, the entry $p_{ij}(t)$ of the matrix exponential $e^{t\Gamma}$ provides the probability that the Poisson-skip process moves from state $i$ of the first Markov chain at time 0 to state $j$ of the same chain at time $t$. The entry $q_{ij}(t)$ of the matrix exponential $e^{t\Delta}$ provides the probability that the Poisson-skip process moves from state $i$ of the first chain to state $j$ of the first chain without encountering a $\chi$ point during the interim. Because $\Delta$ has the partition structure

$$\Delta \;=\; \begin{pmatrix} \Phi & \upsilon \\ \mathbf{0} & 0 \end{pmatrix}$$

for $\Phi$ an $r \times r$ matrix and $\upsilon$ a $1 \times r$ column vector, one can easily demonstrate that $e^{t\Delta}$ has the corresponding partition structure with $e^{t\Phi}$ as its upper left block. More to the point, the entries of $q_{ij}(t)$ of $e^{t\Phi}$ can be explicitly evaluated as

$$q_{0j}(t) \;=\; \begin{cases} e^{-\lambda t} & j = 0 \\ \sum_{k>j} s_k \frac{(\lambda t)^{k-j}}{(k-j)!} e^{-\lambda t} & j > 0 \end{cases} \tag{12.11}$$

for $i = 0$, and as

$$q_{ij}(t) = \begin{cases} 0 & j > i \text{ or } j = 0 \\ \frac{(\lambda t)^{i-j}}{(i-j)!} e^{-\lambda t} & 0 < j \le i \end{cases} \tag{12.12}$$

for $i > 0$.

The solutions (12.11) and (12.12) can be established by path-counting arguments. For instance, the expression for $q_{00}(t)$ is based on the observation that the Poisson-skip process cannot leave state 0 and return to it without encountering a $\chi$ point. The process stays in state 0 with probability $e^{-\lambda t}$. On the other hand, the process can leave state 0 and end up in state $j > 0$ if the $k$th point to its right is the next $\chi$ point and if it encounters $k - j$ $o$ points during the time interval $[0, t]$. Conditioning on the value of $k$ gives the expression in (12.11) for $q_{0j}(t)$ when $j > 0$. Similar reasoning leads to the expressions (12.12) for $q_{ij}(t)$ when $i > 0$.

Although at first glance finding explicit solutions for the entries $p_{ij}(t)$ of $e^{t\Gamma}$ seems hopeless, some simplification can be achieved by considering the discrete renewal process corresponding to how many random points are skipped. Starting from a $\chi$ point, let $u_n$ be the probability that the $n$th point to the right of the current point is a $\chi$ point. By definition, $u_0 = 1$. Furthermore, the probabilities $u_n$ satisfy the classical recurrence relation

$$u_n = s_1 u_{n-1} + s_2 u_{n-2} + \cdots + s_{n-1} u_1 + s_n u_0,$$

which enables one to compute all of the $u_n$ beginning with $u_0$. This recurrence is derived by conditioning on the number of the next-to-last $\chi$ point.

Armed with these probabilities, we can now express

$$p_{ij}(t) = 1_{\{0 < j \le i\}} \frac{(\lambda t)^{i-j}}{(i-j)!} e^{-\lambda t} + 1_{\{j=0\}} \sum_{n=0}^{\infty} u_n \frac{(\lambda t)^{i+n}}{(i+n)!} e^{-\lambda t}$$
$$+ 1_{\{j>0\}} \sum_{n=0}^{\infty} u_n \sum_{k>j} s_k \frac{(\lambda t)^{i+n+k-j}}{(i+n+k-j)!} e^{-\lambda t}. \tag{12.13}$$

Indeed, the first term $(\lambda t)^{i-j} e^{-\lambda t}/(i-j)!$ of (12.13) expresses the probability of encountering $i - j$ $o$ points during $[0, t]$; this is relevant when there is a direct path from state $i$ to $j$ that does not pass through state 0. The term $u_n (\lambda t)^{i+n} e^{-\lambda t}/(i+n)!$ is the probability of passing through the $i - 1$ current $o$ points to the right, hitting the next $\chi$ point, and returning to a $\chi$ point after encountering $n$ further points. Finally, the term $u_n s_k (\lambda t)^{i+n+k-j} e^{-\lambda t}/(i+n+k-j)!$ is the probability of passing through the $i - 1$ current $o$ points to the right, hitting the next $\chi$ point, returning to a $\chi$ point after encountering $n$ further points, and then passing through $k - j$ remaining $o$ points enroute to a $\chi$ point $k$ points down the road.

We are now in a position to calculate gamete probabilities. According to formula (12.6), we must first calculate the avoidance probability

$$\Pr\left(\sum_{i \in T} N_{I_i} = 0\right)$$

for a subset $\{I_i : i \in T\}$ of the ordered, adjacent intervals $I_1, \ldots, I_k$. Suppose, for example, that $k = 3$ and $T = \{1, 3\}$. Let interval $I_i$ have length $x_i$. At the start of interval $I_1$, the Poisson-skip process is in state $r$ of the first Markov chain with equilibrium probability $\pi_r$. On the interval $I_1$, the process must not encounter a $\chi$ point. It successfully negotiates the interval and winds up at state $s$ with probability $q_{rs}(x_1)$. On interval $I_2$, there is no restriction on the process, so it moves from state $s$ at the start of the interval to state $t$ at the end of the interval with probability $p_{st}(x_2)$. On interval $I_3$, the process again must not encounter a $\chi$ point. Therefore, the process successfully ends in state $u$ with probability $q_{tu}(x_3)$. Summing over all possible states at the start and finish of each interval gives the avoidance probability

$$\Pr(N_{I_1} + N_{I_3} = 0) \;\;=\;\; \sum_r \sum_s \sum_t \sum_u \pi_r q_{rs}(x_1) p_{st}(x_2) q_{tu}(x_3).$$

In obvious matrix notation, this reduces to

$$\Pr(N_{I_1} + N_{I_3} = 0) \;\;=\;\; \pi e^{x_1 \Phi} e^{x_2 \Gamma} e^{x_3 \Phi} \mathbf{1}.$$

The general case is handled in exactly the same fashion.

Avoidance probabilities can be combined to give a compact formula for gamete probabilities by defining the two matrices

$$R(x) \;\;=\;\; \frac{1}{2}(e^{x\Gamma} - e^{x\Phi})$$

$$Z(x) \;\;=\;\; \frac{1}{2}(e^{x\Gamma} + e^{x\Phi}).$$

Returning to our special case with three intervals, suppose that we wish to calculate $y_{\{1,2\}}$. Applying the distributive law in the gamete probability formula (12.6), we easily deduce that

$$y_{\{1,2\}} \;\;=\;\; \pi R(x_1) R(x_2) Z(x_3) \mathbf{1}.$$

In effect, we choose on each interval whether to avoid $\chi$ points, and thus use matrix $e^{x\Phi}$, or whether to embrace both $\chi$ and $o$ points, and thus use matrix $e^{x\Gamma}$. The factors of $\frac{1}{2}$ in the definition of $R(x)$ and $Z(x)$ give the overall factor $(\frac{1}{2})^k$ in (12.6), and the factors of $+1$ and $-1$ make the sign $(-1)^{|S \cap T|}$ come out right. In general, if a set $S$ is characterized by the $k$-tuple of indicators $s_i = 1_{\{i \in S\}}$, then the gamete probability $y_S$ can be expressed as

$$y_S \;\;=\;\; \pi R(x_1)^{s_1} Z(x_1)^{1-s_1} \cdots R(x_k)^{s_k} Z(x_k)^{1-s_k} \mathbf{1}. \qquad (12.14)$$

Formula (12.14) can also be derived and implemented by defining an appropriate hidden Markov chain. Let $U_j$ be the unobserved state of the first Markov chain at locus $j$, and let $Y_j$ be the observed indicator random variable flagging whether recombination has occurred between loci $j - 1$ and $j$. To compute a gamete probability $\Pr(Y_2 = i_1, \ldots, Y_{k+1} = i_k)$, one can apply Baum's forward algorithm as in the radiation hybrid model of Chapter 11 [2, 6]. We begin the recursive computation of the joint probabilities

$$f_j(u_j) \quad = \quad \Pr(Y_2 = i_1, \ldots, Y_j = i_{j-1}, U_j = u_j)$$

by setting $f_1(u_1) = \pi_{u_1}$. At the final locus $k+1$, we recover the gamete probability $\Pr(Y_2 = i_1, \ldots, Y_{k+1} = i_k)$ from the identity

$$\Pr(Y_2 = i_1, \ldots, Y_{k+1} = i_k) \quad = \quad \sum_{u_{k+1}} f_{k+1}(u_{k+1}).$$

In view of Mather's formula (12.1), if $i_j = 1$, then

$$f_{j+1}(u_{j+1}) \quad = \quad \sum_{u_j} f_j(u_j) \frac{1}{2} \Big[ p_{u_j, u_{j+1}}(x_j) - q_{u_j, u_{j+1}}(x_j) \Big]$$

because $\frac{1}{2}[p_{u_j, u_{j+1}}(x_j) - q_{u_j, u_{j+1}}(x_j)]$ is the probability that the chain moves from state $U_j = u_j$ at locus $j$ to state $U_{j+1} = u_{j+1}$ at locus $j+1$ and that the chosen gamete is recombinant on the interval between the loci. On the other hand, if $i_j = 0$, then

$$f_{j+1}(u_{j+1}) \quad = \quad \sum_{u_j} f_j(u_j) \frac{1}{2} \Big[ p_{u_j, u_{j+1}}(x_j) + q_{u_j, u_{j+1}}(x_j) \Big]$$

because

$$q_{u_j, u_{j+1}}(x_j) + \frac{1}{2} \Big[ p_{u_j, u_{j+1}}(x_j) - q_{u_j, u_{j+1}}(x_j) \Big]$$
$$= \quad \frac{1}{2} \Big[ p_{u_j, u_{j+1}}(x_j) + q_{u_j, u_{j+1}}(x_j) \Big]$$

is the probability that the chain moves from state $U_j = u_j$ at locus $j$ to state $U_{j+1} = u_{j+1}$ at locus $j+1$ and that the chosen gamete is nonrecombinant on the interval between the loci. Hence, Baum's forward algorithm is simply a device for carrying out the vector times matrix multiplications implied by formula (12.14).

**Example 12.3** *Chi-Square Model*

In the chi-square model [1, 4, 9, 20, 21, 32], a fixed number of points is skipped. If every $r$th point is a $\chi$ point, then $s_i = 1_{\{i=r\}}$ and the equilibrium distribution $\pi$ is uniform on $\{0, 1, \ldots, r-1\}$. The discrete renewal density $u_n = 1_{\{n=0 \bmod r\}}$. The expressions for $q_{0j}(t)$ and $p_{ij}(t)$ simplify to

$$q_{0j}(t) \quad = \quad \begin{cases} e^{-\lambda t} & j = 0 \\ \frac{(\lambda t)^{r-j}}{(r-j)!} e^{-\lambda t} & j > 0 \end{cases}$$

and

$$p_{ij}(t) \quad = \quad 1_{\{j \leq i\}} \frac{(\lambda t)^{i-j}}{(i-j)!} e^{-\lambda t} + \sum_{m=1}^{\infty} \frac{(\lambda t)^{mr+i-j}}{(mr+i-j)!} e^{-\lambda t}.$$

This model tends to fit data well.                                    ∎

## 12.6   Chiasma Interference

Chiasma interference can be roughly divided into **count interference** and **position interference**. Count interference arises when the total number of chiasmata on a chromosome follows a non-Poisson distribution. Position interference arises when the formation of one chiasma actively discourages the formation of other chiasmata nearby. The count-location model exhibits count interference but not position interference. Stationary renewal models exhibit both types of interference and therefore are somewhat better equipped to capture the subtleties of recombination data.

Traditionally, geneticists have measured interference by coincidence coefficients. The coincidence coefficient $C(I_1, I_2)$ of two adjacent intervals $I_1$ and $I_2$ is defined as the ratio of the probability of recombination on both intervals to the product of their individual recombination fractions. Based on equations (12.1) and (12.7), this ratio is

$$C(I_1, I_2) = \frac{\frac{1}{4}[1 - \Pr(N_{I_1} = 0) - \Pr(N_{I_2} = 0) + \Pr(N_{I_1} + N_{I_2} = 0)]}{\frac{1}{2}[1 - \Pr(N_{I_1} = 0)]\frac{1}{2}[1 - \Pr(N_{I_2} = 0)]}.$$

The conditions $C(I_1, I_2) < 1$ and $C(I_1, I_2) > 1$ are referred to as **positive** and **negative interference**, respectively. Positive interference occurs when

$$\Pr(N_{I_1} + N_{I_2} = 0) \quad < \quad \Pr(N_{I_1} = 0)\Pr(N_{I_2} = 0), \qquad (12.15)$$

and negative interference occurs when the reverse inequality obtains. Haldane's model gives equality and exhibits no interference.

In the count-location model, inequality (12.15) is equivalent to

$$Q(1 - x_1 - x_2) \quad < \quad Q(1 - x_1)Q(1 - x_2), \qquad (12.16)$$

where $x_1 = \frac{2}{\lambda}d_1$ and $x_2 = \frac{2}{\lambda}d_2$ are the standardized map lengths of the intervals $I_1$ and $I_2$, and $Q(s)$ is the generating function of the total number of chiasmata on the chromatid bundle. It is of some interest to characterize the class $\mathcal{C}$ of discrete distributions for the chiasma count $N$ guaranteeing positive interference or at least noninterference. In general, we can say that $\mathcal{C}$

(a) is closed under convergence in distribution,

(b) is closed under convolution,

(c) contains all distributions whose generating functions $Q(s)$ are log-concave in the sense that $\frac{d^2}{ds^2} \ln Q(s) \le 0$,

(d) contains all distributions concentrated on the set $\{0, 1\}$ or on the set $\{1, 2, 3, 4\}$.

Properties (a) and (b) are trivial to deduce. Problems 7l and 7m address properties (c) and (d). Property (d) is particularly relevant because most chromatid bundles carry between one and four chiasmata. From these four properties, we can build up a list of specific members of $\mathcal{C}$. For instance, $\mathcal{C}$ contains all binomial distributions and all distributions concentrated on two adjacent integers. Compound Poisson distributions such as the negative binomial exhibit negative interference rather than positive interference.

For the stationary renewal model, we allow equality in inequality (12.15) and reexpress it for all $y, z \geq 0$ as

$$\bar{F}_\infty(y + z) \; \leq \; \bar{F}_\infty(y)\bar{F}_\infty(z), \tag{12.17}$$

where the right-tail probability

$$\begin{aligned}
\bar{F}_\infty(x) &= 1 - F_\infty(x) \\
&= \frac{1}{\mu} \int_0^\infty [1 - F(w + x)]dw. \tag{12.18}
\end{aligned}$$

Log-concavity of $\bar{F}_\infty(x)$ is a sufficient condition for the submultiplicative property (12.17). Indeed, the inequality

$$\begin{aligned}
\ln \bar{F}_\infty(y + z) &= \int_0^{y+z} \frac{d}{dx} \ln \bar{F}_\infty(x)dx \\
&\leq \int_0^y \frac{d}{dx} \ln \bar{F}_\infty(x)dx + \int_0^z \frac{d}{dx} \ln \bar{F}_\infty(x)dx \\
&= \ln \bar{F}_\infty(y) + \ln \bar{F}_\infty(z)
\end{aligned}$$

is equivalent to the inequality

$$0 \; \leq \; \int_0^y \left[\frac{d}{dx} \ln \bar{F}_\infty(x) - \frac{d}{dx} \ln \bar{F}_\infty(x + z)\right]dx,$$

which is certainly true if $\frac{d}{dx} \ln \bar{F}_\infty(x)$ is decreasing.

A sufficient condition in turn for

$$\begin{aligned}
\frac{d^2}{dx^2} \ln \bar{F}_\infty(x) &= -\frac{d}{dx} \frac{1 - F(x)}{\mu \bar{F}_\infty(x)} \\
&= \frac{f(x)}{\mu \bar{F}_\infty(x)} - \frac{[1 - F(x)]^2}{\mu^2 \bar{F}_\infty(x)^2} \tag{12.19} \\
&\leq 0
\end{aligned}$$

to hold is that the **hazard rate** $\frac{f(x)}{1 - F(x)}$ be increasing in $x$. If this is the case, then in view of equation (12.18), we can average the right-hand side of the inequality

$$\frac{f(x)}{1 - F(x)} \; \leq \; \frac{f(x + w)}{1 - F(x + w)}$$

with respect to the probability density

$$\frac{1 - F(x + w)}{\int_0^\infty [1 - F(x + v)]dv} = \frac{1 - F(x + w)}{\mu \bar{F}_\infty(x)}$$

to give the bound

$$\frac{f(x)}{1 - F(x)} \le \frac{1 - F(x)}{\mu \bar{F}_\infty(x)}.$$

This last bound implies the log-concavity condition (12.19).

In summary, increasing hazard rate leads to positive interference [13]. For the particular case of the Poisson-skip process, we can assert considerably more. Let $C$ now be the class of discrete skip distributions $\{s_n\}_{n=1}^\infty$ that guarantee positive interference or no interference. Then $C$ satisfies properties (a) and (b) enumerated for the count-location model. Furthermore, properties (c) and (d) are replaced by

(e) contains all distributions $\{s_n\}_{n=1}^\infty$ that are positive on some interval, 0 elsewhere, and log-concave in the sense that $s_n^2 \ge s_{n-1}s_{n+1}$ for all $n$,

(f) contains all distributions concentrated on a single integer or two adjacent integers, all binomial, negative binomial, Poisson, and uniform distributions, and all shifts of these distributions by positive integers.

Although the proofs of these assertions are not beyond us conceptually, we refer interested readers to [17] for details.

## 12.7 Application to *Drosophila* Data

As an application of the various recombination models, we now briefly discuss the classic *Drosophila* data of Morgan et al. [19]. These early geneticists phenotyped 16,136 flies at 9 loci covering almost the entire *Drosophila* X chromosome. Because of the nature of the genetic cross employed, a fly corresponds to a gamete scorable on each interlocus interval as recombinant or nonrecombinant. Table 12.1 presents the gamete counts $n$ recorded. Here the set $S$ denotes the recombinant intervals. For example, 6,607 flies were nonrecombinant on all intervals (the first category), and one fly was recombinant on intervals 5, 6, and 8 and nonrecombinant on all remaining intervals (the last category).

Table 12.2 summarizes the results presented in the papers [17, 22]. Haldane's model referred to in the first row of the table fits the data poorly. The count-location model yields an enormous improvement in the maximum loglikelihood displayed in the last column of the table. In the count-location model, the maximum likelihood estimates of the count probabilities are $(q_0, q_1, q_2, q_3) = (.06, .41, .48, .05)$; these estimates correct the slightly erroneous values given in [22]. The departure of the count probabilities from a Poisson distribution is one of the reasons Haldane's model fails so miserably. The chi-square and mixture models referred to in Table 12.2 are special cases of the Poisson-skip model. The best fitting chi-square and mixture models have skip distributions determined by $s_5 = 1$ and

TABLE 12.1. Gamete Counts in the Morgan et al. Data

| S | n | S | n | S | n | S | n |
|---|---|---|---|---|---|---|---|
| $\phi$ | 6607 | {2, 4} | 38 | {6, 7} | 21 | {2, 5, 6} | 3 |
| {1} | 506 | {2, 5} | 85 | {6, 8} | 30 | {2, 5, 7} | 4 |
| {2} | 1049 | {2, 6} | 237 | {7, 8} | 2 | {2, 5, 8} | 1 |
| {3} | 855 | {2, 7} | 123 | {1, 2, 3} | 1 | {2, 6, 7} | 2 |
| {4} | 1499 | {2, 8} | 70 | {1, 2, 6} | 1 | {2, 6, 8} | 3 |
| {5} | 937 | {3, 4} | 22 | {1, 3, 5} | 1 | {2, 7, 8} | 2 |
| {6} | 1647 | {3, 5} | 55 | {1, 4, 5} | 1 | {3, 4, 7} | 2 |
| {7} | 683 | {3, 6} | 177 | {1, 4, 6} | 1 | {3, 4, 8} | 1 |
| {8} | 379 | {3, 7} | 88 | {1, 4, 7} | 2 | {3, 5, 6} | 1 |
| {1, 2} | 3 | {3, 8} | 38 | {1, 4, 8} | 1 | {3, 5, 7} | 2 |
| {1, 3} | 6 | {4, 5} | 41 | {1, 5, 7} | 2 | {3, 5, 8} | 3 |
| {1, 4} | 41 | {4, 6} | 198 | {1, 5, 8} | 1 | {3, 6, 7} | 1 |
| {1, 5} | 55 | {4, 7} | 159 | {1, 6, 8} | 1 | {3, 6, 8} | 1 |
| {1, 6} | 118 | {4, 8} | 91 | {2, 3, 6} | 1 | {4, 5, 8} | 1 |
| {1, 7} | 54 | {5, 6} | 35 | {2, 4, 6} | 4 | {4, 6, 8} | 4 |
| {1, 8} | 34 | {5, 7} | 49 | {2, 4, 7} | 5 | {4, 7, 8} | 1 |
| {2, 3} | 3 | {5, 8} | 40 | {2, 4, 8} | 6 | {5, 6, 8} | 1 |

$(s_4, s_5) = (.06, .94)$, respectively. These two models yield a further improvement over the count-location model because they take into account position interference as well as count interference. In spite of the inadequacies of Haldane's model and the count-location model, all four models give roughly similar estimates of the map distances (in centiMorgans = $100 \times$ Morgans) between adjacent pairs of loci. Note that Haldane's model and the count-location model compensate for the reduced number of double crossovers on adjacent intervals by expanding map distances.

TABLE 12.2. Analysis of the Morgan et al. Data

| Model | Interval | | | | | | | | max ln $L$ |
|---|---|---|---|---|---|---|---|---|---|
| | 1 | 2 | 3 | 4 | 5 | 6 | 7 | 8 | |
| Haldane | 5.3 | 11.0 | 8.3 | 14.6 | 8.7 | 17.6 | 7.9 | 4.5 | $-43023.91$ |
| Count-Loc | 5.3 | 10.8 | 8.2 | 14.2 | 8.6 | 16.9 | 7.8 | 4.5 | $-37449.17$ |
| Chi-square | 5.1 | 9.8 | 7.5 | 13.3 | 8.4 | 15.6 | 7.5 | 4.4 | $-36986.87$ |
| Mixture | 5.1 | 9.8 | 7.5 | 13.3 | 8.4 | 15.5 | 7.5 | 4.4 | $-36986.34$ |

## 12.8   Problems

1. Prove that in Haldane's model the gamete probability formula (12.6) collapses to the obvious independence formula

$$y_S = \prod_{i=1}^{k} \theta_i^{s_i} (1 - \theta_i)^{1-s_i}.$$

2. Karlin's binomial count-location model [11] presupposes that the total number of chiasmata has binomial distribution with generating function $(\frac{1}{2} + \frac{s}{2})^r$. Compute the corresponding map function and its inverse.

3. Consider a delayed renewal process generated by the sequence of independent random variables $X_1, X_2, \ldots$ such that $X_1$ has distribution function $G(x)$ and $X_i$ has distribution function $F(x)$ for $i \geq 2$. If $G(x) = F(x)$, then show that the renewal function $U(x) = E(N_{[0,x]})$ satisfies the renewal equation

$$U(x) = F(x) + \int_0^x U(x - y) dF(y). \qquad (12.20)$$

If $G(x)$ is chosen to make the delayed renewal process stationary, then show that

$$\frac{x}{\mu} = G(x) + \int_0^x U(x - y) dG(y), \qquad (12.21)$$

where $\mu$ is the mean of $F(x)$. If $\widehat{dH}(\lambda) = \int_0^\infty e^{-\lambda x} dH(x)$ denotes the Laplace transform of the distribution function $H(x)$ defined on $(0, \infty)$, also verify the identity

$$\widehat{dG}(\lambda) = \frac{1 - \widehat{dF}(\lambda)}{\mu \lambda}.$$

Finally, prove that the Laplace transform of the density $\frac{1}{\mu}[1 - F(x)]$ matches $\widehat{dG}(\lambda)$.

4. Show that Felsenstein's [8] map function

$$\theta = \frac{1}{2} \frac{e^{2(2-\gamma)d} - 1}{e^{2(2-\gamma)d} - \gamma + 1} \qquad (12.22)$$

arises from a stationary renewal model when $0 \leq \gamma \leq 2$. Kosambi's map function is the special case $\gamma = 0$. Why does (12.22) fail to give a legal map function when $\gamma > 2$? Note that at $\gamma = 2$ we define $\theta = \frac{d}{2d+1}$ by l'Hôpital's rule.

5. Continuing Problem 4, prove that Felsenstein's map function has inverse

$$d = \frac{1}{2(\gamma - 2)} \ln \left[ \frac{1 - 2\theta}{1 - 2(\gamma - 1)\theta} \right].$$

6. The Carter and Falconer [3] map function has inverse

$$M^{-1}(\theta) = \frac{1}{4}[\tan^{-1}(2\theta) + \tanh^{-1}(2\theta)].$$

Prove that the map function satisfies the differential equation

$$M'(d) = 1 - 16M^4(d)$$

with initial condition $M(0) = 0$. Deduce from these facts that $M(d)$ arises from a stationary renewal model.

7. Fix a positive integer $m$, and let $w_m = e^{\frac{2\pi i}{m}}$ be the principal $m$th root of unity. For each integer $j$, define the **segmental function** $_m\alpha_j(x)$ of $x$ to be the finite Fourier transform

$$_m\alpha_j(x) = \frac{1}{m} \sum_{k=0}^{m-1} e^{xw_m^k} w_m^{-jk}.$$

These functions generalize the hyperbolic trig functions $\cosh(x)$ and $\sinh(x)$. Prove the following assertions:

(a) $_m\alpha_j(x) = {_m\alpha_{j+m}}(x)$.

(b) $_m\alpha_j(x + y) = \sum_{k=0}^{m-1} {_m\alpha_k}(x)_m\alpha_{j-k}(y)$.

(c) $_m\alpha_j(x) = \sum_{k=0}^{\infty} \frac{x^{j+km}}{(j+km)!}$ for $0 \leq j \leq m - 1$.

(d) $\frac{d}{dx}[_m\alpha_j(x)] = {_m\alpha_{j-1}}(x)$.

(e) Consider the differential equation $\frac{d^m}{dx^m} f(x) = kf(x)$ with initial conditions $\frac{d^j}{dx^j} f(0) = c_j$ for $0 \leq j \leq m - 1$, where $k$ and the $c_j$ are constants. Show that

$$f(x) = \sum_{j=0}^{m-1} c_j k^{-\frac{j}{m}} {_m\alpha_j}(k^{\frac{1}{m}} x).$$

(f) The differential equation $\frac{d^m}{dx^m} f(x) = kf(x) + g(x)$ with initial conditions $\frac{d^j}{dx^j} f(0) = c_j$ for $0 \leq j \leq m - 1$ has solution

$$f(x) = \int_0^x k^{-\frac{m-1}{m}} {_m\alpha_{m-1}}[k^{\frac{1}{m}} (x - y)]g(y)dy$$
$$+ \sum_{j=0}^{m-1} c_j k^{-\frac{j}{m}} {_m\alpha_j}(k^{\frac{1}{m}} x).$$

(g) $\lim_{x \to \infty} e^{-x} {}_m\alpha_j(x) = \frac{1}{m}$.

(h) In a Poisson process of intensity 1, $e^{-x} {}_m\alpha_j(x)$ is the probability that the number of random points on $[0, x]$ equals $j$ modulo $m$.

(i) Relative to this Poisson process, let $N_x$ count every $m$th random point on $[0, x]$. Then $N_x$ has probability generating function

$$P(s) = e^{-x} \sum_{j=0}^{m-1} s^{-\frac{j}{m}} {}_m\alpha_j(s^{\frac{1}{m}} x).$$

(j) Furthermore, $N_x$ has mean

$$E(N_x) = \frac{x}{m} - \frac{e^{-x}}{m} \sum_{j=0}^{m-1} j \, {}_m\alpha_j(x).$$

(k) $\lim_{x \to \infty} \left[ E(N_x) - \frac{x}{m} \right] = -\frac{m-1}{2m}$.

(l) In the count-location model, suppose that the count distribution has a log-concave generating function $Q(s)$. Prove that the model exhibits positive or no interference [22].

(m) In the count-location model, suppose that the count distribution is concentrated on the set $\{0, 1\}$ or on the set $\{1, 2, 3, 4\}$. Prove that the model exhibits positive or no interference [22]. (Hint: Check the log-concavity property by showing that $Q''(s)Q(s) - Q'(s)^2$ reduces to a sum of negative terms.)

(n) In the count-location model, suppose that the count distribution has generating function

$$Q(s) = \frac{8}{35}s + \frac{1}{5}s^4 + \frac{4}{7}s^5.$$

Prove that the opposite of inequality (12.16) holds for $x_1 = .4$ and $x_2 = .1$ and, therefore, that negative interference occurs [22].

(o) The Poisson-skip model with skip distribution $s_1 = p$ and $s_3 = 1 - p$ has interarrival density

$$f(x) = pe^{-x} + (1 - p)\frac{x^2}{2}e^{-x}.$$

Show that $f(x)$ has decreasing hazard rate for $x$ small and positive [17].

# References

[1] Bailey NTJ (1961) *Introduction to the Mathematical Theory of Genetic Linkage.* Oxford University Press, London

[2] Baum L (1972) An inequality and associated maximization technique in statistical estimation for probabilistic functions of Markov processes. *Inequalities* 3:1–8

[3] Carter TC, Falconer DS (1951) Stocks for detecting linkage in the mouse and the theory of their design. *J Genet* 50:307–323

[4] Carter TC, Robertson A (1952) A mathematical treatment of genetical recombination using a four-strand model. *Proc R Soc Lond B* 139:410–426.

[5] Cox DR, Isham V (1980) *Point Processes*. Chapman and Hall, New York

[6] Devijver PA (1985) Baum's forward-backward algorithm revisited. *Pattern Recognition Letters* 3:369–373

[7] Feller W (1971) *An Introduction to Probability Theory and its Applications, Vol 2*, 2nd ed. Wiley, New York

[8] Felsenstein J (1979) A mathematically tractable family of genetic mapping functions with different amounts of interference. *Genetics* 91:769–775

[9] Fisher, RA, Lyon MF, Owen ARG (1947) The sex chromosome in the house mouse. *Heredity* 1:335–365.

[10] Haldane JBS (1919) The combination of linkage values, and the calculation of distance between the loci of linked factors. *J Genet* 8:299–309

[11] Karlin S (1984) Theoretical aspects of genetic map functions in recombination processes. In *Human Population Genetics: The Pittsburgh Symposium*, Chakravarti A, editor, Van Nostrand Reinhold, New York, pp 209–228

[12] Karlin S, Liberman U (1979) A natural class of multilocus recombination processes and related measures of crossover interference. *Adv Appl Prob* 11:479–501

[13] Karlin S, Liberman U (1983) Measuring interference in the chiasma renewal formation process. *Adv Appl Prob* 15:471–487

[14] Karlin S, Taylor HM (1975) *A First Course in Stochastic Processes*, 2nd ed. Academic Press, New York

[15] Kosambi DD (1944) The estimation of map distance from recombination values. *Ann Eugen* 12:172–175

[16] Lange K, Risch N (1977) Comments on lack of interference in the four-strand model of crossingover. *J Math Biol* 5:55–59

[17] Lange K, Zhao H, Speed TP (1997) The Poisson-skip model of crossingover. *Ann Appl Prob* (in press)

[18] Mather K (1938) Crossing-over. *Biol Reviews Camb Phil Soc* 13:252–292

[19] Morgan TH, Bridges CB, Schultz J (1935) Constitution of the germinal material in relation to heredity. *Carnegie Inst Washington Yearbook* 34:284–291

[20] Owen ARG (1950) The theory of genetical recombination. *Adv Genet* 3:117–157

[21] Payne LC (1956) The theory of genetical recombination: A general formulation for a certain class of intercept length distributions appropriate to the discussion of multiple linkage. *Proc Roy Soc B* 144:528–544.

[22] Risch N, Lange K (1979) An alternative model of recombination and interference. *Ann Hum Genet* 43:61–70

[23] Risch N, Lange K (1983) Statistical analysis of multilocus recombination. *Biometrics* 39:949–963

[24] Ross SM (1983) *Stochastic Processes*. Wiley, New York

[25] Schnell FW (1961) Some general formulations of linkage effects in inbreeding. *Genetics* 46:947–957

[26] Speed TP (1996) What is a genetic map function? In *Genetic Mapping and DNA Sequencing, IMA Vol 81 In Mathematics and its Applications.* Speed TP, Waterman MS, editors, Springer-Verlag, New York, pp 65–88

[27] Stahl FW (1979) *Genetic Recombination: Thinking about it in Phage and Fungi*. WH Freeman, San Francisco

[28] Sturt E (1976) A mapping function for human chromosomes. *Ann Hum Genet* 40:147–163

[29] Whitehouse HLK (1982) *Genetic Recombination: Understanding the Mechanisms*. St. Martin's Press, New York

[30] Zhao H, McPeek MS, Speed TP (1995) Statistical analysis of chromatid interference. *Genetics* 139:1057–1065

[31] Zhao H, Speed TP (1996) On genetic map functions. *Genetics* 142:1369–1377

[32] Zhao H, Speed TP, McPeek MS (1995) Statistical analysis of crossover interference using the chi-square model. *Genetics* 139:1045–1056.

# 13

# Poisson Approximation

## 13.1 Introduction

In the past few years, mathematicians have developed a powerful technique known as the Chen-Stein method [2, 4] for approximating the distribution of a sum of weakly dependent Bernoulli random variables. In contrast to many asymptotic methods, this approximation carries with it explicit error bounds. Let $X_\alpha$ be a Bernoulli random variable with success probability $p_\alpha$, where $\alpha$ ranges over some finite index set $I$. It is natural to speculate that the sum $S = \sum_{\alpha \in I} X_\alpha$ is approximately Poisson with mean $\lambda = \sum_{\alpha \in I} p_\alpha$. The Chen-Stein method estimates the error in this approximation using the total variation distance between two integer-valued random variables $Y$ and $Z$. This distance is defined by

$$\|\mathcal{L}(Y) - \mathcal{L}(Z)\| = \sup_{A \subset \mathcal{N}} |\Pr(Y \in A) - \Pr(Z \in A)|,$$

where $\mathcal{L}$ denotes distribution, and $\mathcal{N}$ denotes the integers. Taking $A = \{0\}$ in this definition yields the useful inequality

$$|\Pr(Y = 0) - \Pr(Z = 0)| \leq \|\mathcal{L}(Y) - \mathcal{L}(W)\|.$$

The **coupling method** is one technique for explicitly bounding the total variation distance between $S = \sum_{\alpha \in I} X_\alpha$ and a Poisson random variable $Z$ with the same mean $\lambda$ [4, 13]. In many concrete examples, it is possible to construct for each $\alpha$ two random variables $U_\alpha$ and $V_\alpha$ on a common probability space in such a way that $V_\alpha$ is distributed as $S - 1$ conditional on the event $X_\alpha = 1$ and $U_\alpha$ is distributed as $S$ unconditionally. The bound

$$\|\mathcal{L}(S) - \mathcal{L}(Z)\| \leq \frac{1 - e^{-\lambda}}{\lambda} \sum_{\alpha \in I} p_\alpha \, \mathrm{E}(|U_\alpha - V_\alpha|) \qquad (13.1)$$

then applies. Because $U_\alpha$ and $V_\alpha$ live on the same probability space, they are said to be coupled. If $U_\alpha \geq V_\alpha$ for all $\alpha$, then the simplified bound

$$\|\mathcal{L}(S) - \mathcal{L}(Z)\| \leq \frac{1 - e^{-\lambda}}{\lambda} [\lambda - \mathrm{Var}(S)] \qquad (13.2)$$

holds. Inequality (13.2) shows that $\mathrm{Var}(S) \approx \mathrm{E}(S)$ is a sufficient as well as a necessary condition for $S$ to be approximately Poisson.

The **neighborhood method** of bounding the total variation distance exploits certain neighborhoods of dependency $B_\alpha$ associated with each $\alpha \in I$ [1]. Here $B_\alpha$ is a subset of $I$ containing $\alpha$. Usually $B_\alpha$ is chosen so that $X_\alpha$ is independent of those $X_\beta$ with $\beta$ outside $B_\alpha$. If this is the case, then define two constants

$$b_1 = \sum_{\alpha \in I} \sum_{\beta \in B_\alpha} p_\alpha p_\beta$$

$$b_2 = \sum_{\alpha \in I} \sum_{\beta \in B_\alpha \setminus \{\alpha\}} p_{\alpha\beta},$$

where

$$\begin{aligned} p_{\alpha\beta} &= \mathrm{E}(X_\alpha X_\beta) \\ &= \Pr(X_\alpha = 1, X_\beta = 1). \end{aligned}$$

In this context, $\lambda - \mathrm{Var}(S) = b_1 - b_2$, and the total variation distance between $S$ and its Poisson approximation $Z$ with mean $\lambda$ satisfies

$$\|\mathcal{L}(S) - \mathcal{L}(Z)\| \leq \frac{1 - e^{-\lambda}}{\lambda} (b_1 + b_2). \qquad (13.3)$$

Both Chen-Stein methods are well adapted to solving many problems arising in mathematical genetics. We will illustrate the main ideas through a sequence of examples. Readers interested in mastering the underlying theory are urged to consult the references [2, 4, 13].

## 13.2 Poisson Approximation to the $W_d$ Statistic

In Chapter 4 we studied the $W_d$ statistic for multinomial trials. Recall that $W_d$ denotes the number of categories with $d$ or more successes after $n$ trials. If we let $q_\alpha$ be the success rate per trial for category $\alpha \in I = \{1, \dots, m\}$, then this category accumulates $d$ or more successes with probability

$$p_\alpha = \sum_{k=d}^{n} \binom{n}{k} q_\alpha^k (1 - q_\alpha)^{n-k}.$$

The coupling method provides a bound on the total variation distance between $W_d$ and a Poisson random variable with mean $\lambda = \sum_{\alpha=1}^{m} p_\alpha$. Our argument will make it clear that we can even elect a different quota $d_\alpha$ for each category in defining the number of categories that meet or exceed their quotas.

To validate the coupling bound (13.1), we must construct the random variables $U_\alpha$ and $V_\alpha$ described in Section 13.1. For $U_\alpha$ we just imagine conducting the multinomial trials according to the usual rules and set $U_\alpha = S = W_d$. If the number of outcomes $Y_\alpha$ falling in category $\alpha$ satisfies $Y_\alpha \geq d$, then $X_\alpha = 1$, and we set $V_\alpha = \sum_{\beta \neq \alpha} X_\beta$. If $Y_\alpha < d$, then we resample from the conditional distribution of $Y_\alpha$ given the event $Y_\alpha \geq d$. This produces a random variable $Y_\alpha^* > Y_\alpha$, and we redefine the outcomes of the first $Y_\alpha^* - Y_\alpha$ trials falling outside category $\alpha$ so that they now fall in category $\alpha$. If we let $V_\alpha$ be the number of categories other than $\alpha$ that now exceed their quota $d$, it is obvious because of the redirection of outcomes that $S \geq V_\alpha$. Thus, the conditions for the Chen-Stein bound (13.2) apply. As pointed out in Problem 4 of Chapter 4, the sum $\sum_{\alpha=1}^{m} p_\alpha^2$ should be small for the Poisson approximation to have any chance of being accurate.

## 13.3    Construction of Somatic Cell Hybrid Panels

**Somatic cell hybrids** are routinely used to assign particular human genes to particular human chromosomes [5, 20]. In brief outline, somatic cell hybrids are constructed by fusing normal human cells with permanently transformed rodent cells. The resulting hybrid cells retain all of the rodent chromosomes while losing random subsets of the human chromosomes. A few generations after cell fusion, clones can be identified with stable subsets of the human chromosomes. All chromosomes, human and rodent, normally remain functional. With a broad enough collection of different hybrid clones, it is possible to establish a correspondence between the presence or absence of a given human gene and the presence or absence of each of the 24 distinct human chromosomes. From this pattern one can assign the gene to a particular chromosome.

For this program of gene assignment to be successful, certain major assumptions must be satisfied. First, the human gene should be present on a single human chromosome or on a single pair of homologous human chromosomes. Second, the human gene should be detectable when present in a clone and should be distinguishable from any rodent analog of the human gene in the clone. Genes are usually detected by electrophoresis of their protein products or by annealing of an appropriate DNA probe directly to part of the gene. Third, each of the 24 distinct human chromosomes should be either absent from a clone or cytologically or biochemically detectable in the clone. Chromosomes can be differentiated cytologically by size, by the position of their centromeres, and by their distinctive banding patterns under appropriate stains. It is also possible to distinguish chromosomes by in situ hybridization of large, fluorescent DNA probes or by isozyme assays that detect unique proteins produced by genes on the chromosomes.

In this application of the Chen-Stein method, we consider the information content of a panel of somatic cell hybrids [9]. Let $n$ denote the number of hybrid clones in a panel. Since the Y chromosome bears few genes of interest, hybrids are usually created from human female cells. This gives a total of 23 different chromosome types—22 autosomes and the X chromosome. Figure 13.1 depicts a hybrid panel with $n = 9$ clones. Each row of this panel corresponds to a particular clone. Each of the 23 columns corresponds to a particular chromosome. A 1 in row $i$ and column $j$ of the panel indicates the presence of chromosome $j$ in clone $i$. A 0 indicates the absence of a chromosome in a clone. An additional test column of 0's and 1's is constructed when each clone is assayed for the presence of a given human gene. Barring assay errors or failures of one of the major assumptions, the test column will uniquely match one of the columns of the panel. In this case the gene is assigned to the corresponding chromosome.

FIGURE 13.1. A Somatic Cell Hybrid Panel

```
0 1 0 1 0 0 0 1 0 0 0 0 0 0 1 0 1 1 0 1 1 1 1
1 0 1 0 1 1 0 0 1 0 0 0 0 1 0 0 1 0 1 0 1 1 1
0 1 1 1 1 0 1 0 0 0 0 0 1 0 0 1 1 0 1 1 0 1 1
1 1 1 0 0 1 1 0 0 1 0 1 0 0 0 1 1 1 0 0 1 0 1
0 0 0 1 1 1 1 0 0 0 1 1 1 1 0 1 0 0 0 1 1 0
0 1 1 1 1 1 1 1 1 1 0 0 0 0 0 1 0 0 0 0 0 0
0 0 1 0 1 0 1 1 0 1 1 1 0 0 0 0 1 1 1 1 1 0 0
0 0 0 1 0 1 1 1 0 0 0 1 0 1 1 1 1 0 1 0 1 0 1
1 0 0 0 1 1 0 0 0 1 0 1 1 0 1 0 1 0 1 1 0 0 1
```

If two columns of a panel are identical, then gene assignment becomes ambiguous for any gene residing on one of the two corresponding chromosomes. Fortunately, the columns of the panel in Figure 13.1 are unique. This panel has the unusual property that every pair of columns differs in at least three entries. This level of redundancy is useful. If a single assay error is made in creating a test column for a human gene, then the gene can still be successfully assigned to a particular human chromosome because it will differ from one column of the panel in one entry and from all other columns of the panel in at least two entries. This consideration suggests that built-in redundancy of a panel is desirable. In practice, the chromosome constitution of a clone cannot be predicted in advance, and the level of redundancy is random. Minimum Hamming distance is a natural measure of the redundancy of a panel. The **Hamming distance** $\rho(c_s, c_t)$ of two columns $c_s$ and $c_t$ counts the number of entries in which they differ. The minimum Hamming distance of a panel is obviously defined as $\min_{\{s,t\}} \rho(c_s, c_t)$, where $\{s, t\}$ ranges over all pairs of columns from the panel.

When somatic cell hybrid panels are randomly created, it is reasonable to make three assumptions. First, each human chromosome is lost or retained independently during the formation of a stable clone. Second, there is a common retention probability $p$ applying to all chromosome pairs. This means that at least one member of each pair of homologous chromosomes is retained with probability $p$. Rushton [15] estimates a range of $p$ from .07 to .75. The value $p = \frac{1}{2}$ simplifies our theory considerably. Third, different clones behave independently in their retention patterns.

Now denote column $s$ of a random panel of $n$ clones by $C_s^n$. For any two distinct columns $C_s^n$ and $C_t^n$, define $X_{\{s,t\}}^n$ to be the indicator of the event $\rho(C_s^n, C_t^n) < d$, where $d$ is some fixed Hamming distance. The random variable $Y_d^n = \sum_{\{s,t\}} X_{\{s,t\}}^n$ is 0 precisely when the minimum Hamming distance equals or exceeds $d$. There are $\binom{23}{2}$ pairs $\alpha = \{s, t\}$ in the index set $I$, and each of the associated $X_\alpha^n$ has the same mean

$$p_\alpha = \sum_{i=0}^{d-1} \binom{n}{i} q^i (1-q)^{n-i},$$

where $q = 2p(1-p)$ is the probability that $C_s^n$ and $C_t^n$ differ in any entry. This gives the mean of $Y_d^n$ as $\lambda = \binom{23}{2} p_\alpha$.

The Chen-Stein heuristic suggests estimating $\Pr(Y_d^n > 0)$ by the Poisson tail probability $1 - e^{-\lambda}$. The error bound (13.3) on this approximation can be computed by defining the neighborhoods $B_\alpha = \{\beta : |\beta| = 2, \ \beta \cap \alpha \neq \emptyset\}$, where vertical bars enclosing a set indicate the number of elements in the set. It is clear that $X_\alpha^n$ is independent of those $X_\beta^n$ with $\beta$ outside $B_\alpha$. The Chen-Stein constant $b_1$ reduces to $\binom{23}{2} |B_\alpha| p_\alpha^2$. An elementary counting argument shows that

$$|B_\alpha| = \binom{23}{2} - \binom{21}{2} = 43.$$

Since the joint probability $p_{\alpha\beta}$ does not depend on the particular pair $\beta \in B_\alpha \setminus \{\alpha\}$ chosen, the constant $b_2$ is $\binom{23}{2}(|B_\alpha| - 1) p_{\alpha\beta}$. Fortunately, $p_{\alpha\beta} = p_\alpha^2$ when $p = 1/2$. Indeed, by conditioning on the value of the common column shared by $\alpha$ and $\beta$, it is obvious in this special case that the events $X_\alpha^n = 1$ and $X_\beta^n = 1$ are independent and occur with constant probability $p_\alpha$. The case $p \neq 1/2$ is more subtle, and we defer the details of computing $p_{\alpha\beta}$ to Problem 9. Table 13.1 provides some representative estimates of the probabilities $\Pr(Y_d^n > 0)$ for $p = 1/2$. Because the Chen-Stein method also provides upper and lower bounds on the estimates, we can be confident that the estimates are accurate for large $n$. In two cases in Table 13.1, the Chen-Stein upper bound is truncated to the more realistic value 1.

## 13.4   Biggest Marker Gap

Spacings of uniformly distributed points are relevant to the question of saturating the human genome with randomly generated markers [12]. If we identify a chromosome with the unit interval $[0,1]$ and scatter $n$ markers randomly on it, then it is natural to ask for the distribution of the largest gap between two adjacent markers or between either endpoint and its nearest adjacent marker. We can attack this

TABLE 13.1. Chen-Stein Estimate of $\Pr(Y_d^n > 0)$

| $d$ | $n$ | Estimate | Lower Bound | Upper Bound |
|---|---|---|---|---|
| 1 | 10 | 0.2189 | 0.1999 | 0.2379 |
| 1 | 15 | 0.0077 | 0.0077 | 0.0077 |
| 1 | 20 | 0.0002 | 0.0002 | 0.0002 |
| 1 | 25 | 0.0000 | 0.0000 | 0.0000 |
| 2 | 10 | 0.9340 | 0.0410 | 1.0000 |
| 2 | 15 | 0.1162 | 0.1112 | 0.1213 |
| 2 | 20 | 0.0051 | 0.0050 | 0.0051 |
| 2 | 25 | 0.0002 | 0.0002 | 0.0002 |
| 3 | 10 | 1.0000 | 0.0410 | 1.0000 |
| 3 | 15 | 0.6071 | 0.4076 | 0.8066 |
| 3 | 20 | 0.0496 | 0.0487 | 0.0505 |
| 3 | 25 | 0.0025 | 0.0025 | 0.0025 |

problem by the coupling method of Chen-Stein approximation. Corresponding to the order statistics $W_1, \ldots, W_n$ of the $n$ points, define indicator random variables $X_1, \ldots, X_{n+1}$ such that $X_\alpha = 1$ when $W_\alpha - W_{\alpha-1} \geq d$. At the ends we take $W_0 = 0$ and $W_{n+1} = 1$. The sum $S = \sum_{\alpha=1}^{n+1} X_\alpha$ gives the number of gaps of length $d$ or greater.

Because we can circularize the interval, all gaps, including the first and the last, behave symmetrically. Just think of scattering $n+1$ points on the unit circle and then breaking the circle into an interval at the first random point. It therefore suffices in the coupling method to consider the first Bernoulli variable $X_1 = 1_{\{W_1 \geq d\}}$. Now scatter the $n$ points in the usual way, and let $U_1$ count the number of gaps that exceed $d$ in length. If $W_1 \geq d$, then define $V_1$ to be the number of gaps other than $W_1$ that exceed $d$. If, on the other hand, $W_1 < d$, then resample $W_1$ conditional on the event $W_1 \geq d$ to get $W_1^*$. For $\alpha > 1$, replace the gap $W_\alpha - W_{\alpha-1}$ by the gap $(W_\alpha - W_{\alpha-1})(1 - W_1^*)/(1 - W_1)$ so that the points to the right of $W_1$ are uniformly chosen from the interval $[W_1^*, 1]$ rather than from $[W_1, 1]$. This procedure narrows all remaining gaps but leaves them in the same proportion. If we now define $V_1$ as the number of remaining gaps that exceed $d$ in length, it is clear that $V_1$ has the same distribution as $S - 1$ conditional on $X_1 = 1$. Because $U_1 \geq V_1$, the Chen-Stein inequality (13.2) applies.

To calculate the mean $\lambda = E(S)$, we again focus on the first interval. The identity $\Pr(X_1 = 1) = \Pr(W_1 \geq d) = (1 - d)^n$ and symmetry then clearly imply that $\lambda = (n + 1)(1 - d)^n$. In similar fashion, we calculate

$$\begin{aligned} \text{Var}(S) &= (n + 1)\,\text{Var}(X_1) + (n + 1)n\,\text{Cov}(X_1, X_2) \\ &= (n + 1)(1 - d)^n - (n + 1)(1 - d)^{2n} \\ &\quad + (n + 1)n\,E(X_1 X_2) - (n + 1)n(1 - d)^{2n}. \end{aligned}$$

Because

$$
\begin{aligned}
E(X_1 X_2) &= \Pr(X_1 = 1,\ X_2 = 1) \\
&= \int_d^1 \int_{u_{(1)}+d}^1 n(n-1)[1 - u_{(2)}]^{n-2} du_{(2)} du_{(1)} \\
&= \int_d^{1-d} n[1 - d - u_{(1)}]^{n-1} du_{(1)} \\
&= (1 - 2d)^n
\end{aligned}
$$

for $2d < 1$, it follows that

$$
\begin{aligned}
\mathrm{Var}(S) &= (n+1)(1-d)^n - (n+1)(1-d)^{2n} \\
&\quad + (n+1)n(1-2d)^n - (n+1)n(1-d)^{2n}.
\end{aligned}
$$

If $d$ is small and $n$ is large, then one can demonstrate that $\mathrm{Var}(S) \approx E(S)$, and the Poisson approximation is good [4].

It is of some interest to estimate the average number of markers required to reduce the largest gap below $d$. From the Poisson approximation, the median $n$ should satisfy $e^{-(n+1)(1-d)^n} \approx \frac{1}{2}$. This approximate equality can be rewritten as

$$
n \approx \frac{-\ln(n+1) + \ln\ln 2}{\ln(1-d)} \tag{13.4}
$$

and used iteratively to approximate the median. If one chooses evenly spaced markers, it takes only $\frac{1}{d}$ markers to saturate the interval $[0, 1]$. For the crude guess $n = \frac{1}{d}$, substitution in (13.4) leads to the improved approximation

$$
\begin{aligned}
n &\approx \frac{-\ln(\frac{1}{d} + 1) + \ln\ln 2}{\ln(1-d)} \\
&\approx \frac{1}{d} \ln \frac{1}{d}.
\end{aligned}
$$

In fact, a detailed analysis shows that the average required number of markers is asymptotically similar to $\frac{1}{d} \ln \frac{1}{d}$ for $d$ small [6, 17]. The factor $\ln \frac{1}{d}$ is the penalty exacted for randomly selecting markers.

The tedium of filling the last few gaps also plagues other mapping endeavors such as covering a chromosome by random clones of fixed length $d$ [19]. If we let the center of each clone correspond to a marker, then except for edge effects, this problem is completely analogous to the marker coverage problem.

## 13.5  Randomness of Restriction Sites

**Restriction enzymes** are special bacterial proteins that snip DNA. The restriction sites where the cutting takes place vary from enzyme to enzyme. For instance, the restriction enzyme EcoRI recognizes the six-base sequence GAATTC and snips DNA wherever this sequence appears. The restriction enzyme NotI recognizes the rarer eight-base sequence GCGGCCGC and consequently tends to produce much longer fragments on average than EcoRI. To a good approximation, the restriction sites for a particular enzyme occur along a chromosome according to a homogeneous Poisson process. Clustering of restriction sites is a particularly interesting violation of the Poisson process assumptions.

If one visualizes $n$ restriction sites along a stretch of DNA as random points on the unit interval [0, 1], then under the Poisson process assumption, the $n$ points should constitute a random sample of size $n$ from the uniform distribution on [0, 1]. The distances between adjacent points are known as **spacings**, or **scans**. An $m$-spacing is the distance between the first and last point of $m + 1$ adjacent points. In Section 13.4, we approximated the distribution of the largest 1-spacing. Here we are interested in detecting clustering by examining the smallest $m$-spacing $S_m$ from a set of $n$ restriction sites. Values of $m > 1$ are important because very short DNA fragments are difficult to measure exactly. The Chen-Stein method provides a means of assessing the significance of an observed $m$-spacing $S_m = s$ [4, 11].

Consider the collection $I$ of subsets $\alpha$ of size $m + 1$ from the set of $n$ random points on [0, 1]. Let $X_\alpha$ be the indicator random variable of the event that the distance from the first point of $\alpha$ to the last point of $\alpha$ is less than or equal to $s$. There are $|I| = \binom{n}{m+1}$ such collections $\alpha$, and each $X_\alpha$ has the same expectation. Because the event $\{S_m \le s\}$ is equivalent to the event $S = \sum_\alpha X_\alpha > 0$, it suffices to compute the probability $S = 0$. The Chen-Stein approximation suggests that $\Pr(S = 0) \approx e^{-\lambda}$ with

$$\lambda = \binom{n}{m+1} E(X_\alpha) = \binom{n}{m+1}[(m+1)s^m - ms^{m+1}].$$

Karlin and Macken use this approximation with $m = 10$ to detect clustering of PstI restriction sites in the *E. coli* bacterial genome [11].

To verify the substitution $E(X_\alpha) = p_\alpha = (m+1)s^m - ms^{m+1}$, we proceed by conditioning on the position $u$ of the leftmost of the $m + 1$ points. Because the remaining $m$ points of $\alpha$ must lie within a distance $s$ to the right of $u$, it follows that

$$
\begin{aligned}
p_\alpha &= (m+1) \int_0^{1-s} s^m \, du + (m+1) \int_{1-s}^1 (1-u)^m \, du \\
&= (m+1)s^m(1-s) + s^{m+1} \\
&= (m+1)s^m - ms^{m+1}.
\end{aligned}
\tag{13.5}
$$

Thus, if $s$ is small,

$$
\begin{aligned}
\lambda &= \binom{n}{m+1}[(m+1)s^m - ms^{m+1}] \\
&\approx \frac{n(n-1)\cdots(n-m)}{m(m-1)\cdots 1} s^m.
\end{aligned}
$$

If $\lambda$ is to be bounded away from 0 and $\infty$, written $\lambda \asymp 1$, then $n^{m+1}s^m \asymp 1$. Here $n$ is taken as very large and $s$ as very small.

To compute the Chen-Stein bound (13.3), it is convenient to define the neighborhood $B_\alpha = \{\beta : |\beta| = m + 1, \ \beta \cap \alpha \neq \emptyset\}$. Again $X_\alpha$ is independent of those $X_\beta$ with $\beta$ outside $B_\alpha$. The Chen-Stein constant $b_1$ can be expressed as

$$
\begin{aligned}
b_1 &= |I||B_\alpha|p_\alpha^2 \\
&= \binom{n}{m+1}\left[\binom{n}{m+1} - \binom{n-m-1}{m+1}\right][(m+1)s^m - ms^{m+1}]^2 \\
&= \lambda^2\left[1 - \frac{\binom{n-m-1}{m+1}}{\binom{n}{m+1}}\right] \\
&= \lambda^2\left[1 - (1 - \frac{m+1}{n})\cdots(1 - \frac{m+1}{n-m})\right].
\end{aligned}
$$

Now for any $m+1$ numbers $a_1, \ldots, a_{m+1}$ from $[0, 1]$, standard inclusion–exclusion arguments imply that

$$
a_1 + \cdots + a_{m+1} - \sum_{1 \leq i < j \leq m+1} a_i a_j \quad \leq \quad 1 - (1 - a_1)\cdots(1 - a_{m+1})
$$

$$
\leq \quad a_1 + \cdots + a_{m+1}.
$$

It follows that $b_1 \asymp \frac{\lambda^2(m+1)^2}{n} \asymp \frac{1}{n}$.

Evaluation of the constant $b_2$ is more difficult. Consider $\beta \in B_\alpha$ such that $|\beta \cap \alpha| = k$ for $0 < k < m + 1$. Let $U_{(1)}$ and $U_{(k)}$ be the positions of the first and last of the $k$ common points shared by $\beta$ and $\alpha$. If we condition on the values $U_{(1)} = u_{(1)}$ and $U_{(k)} = u_{(k)}$, then the indicator variables $X_\alpha$ and $X_\beta$ are independent and identically distributed. Hence,

$$
\begin{aligned}
&\Pr(X_\alpha = 1, X_\beta = 1 \mid U_{(1)} = u_{(1)}, U_{(k)} = u_{(k)}) \\
&= \Pr(X_\alpha = 1 \mid U_{(1)} = u_{(1)}, U_{(k)} = u_{(k)})^2.
\end{aligned}
$$

In order that $X_\alpha = 1$, the $m + 1 - k$ remaining points in $\alpha$ must be within a distance $s$ to the left of $u_{(1)}$ and a distance $s$ to the right of $u_{(k)}$. It follows that $\Pr(X_\alpha = 1 \mid U_{(1)} = u_{(1)}, U_{(k)} = u_{(k)}) \leq (3s)^{m+1-k}$ for $u_{(k)} - u_{(1)} \leq s$.

This uniform bound yields a crude upper bound on $p_{\alpha\beta}$ if combined with the probability that $U_{(k)} - U_{(1)} \leq s$. This latter probability is

$$
\Pr(U_{(k)} - U_{(1)} \leq s) = ks^{k-1} - (k-1)s^k
$$

for exactly the same reasons that produced equality (13.5). Thus, we can assert that $p_{\alpha\beta} \leq q_k = [ks^{k-1} - (k-1)s^k](3s)^{2[m+1-k]}$ whenever $|\beta \cap \alpha| = k$. Since there are $\binom{m+1}{k}\binom{n-m-1}{m+1-k}$ such collections $\beta$ for every $\alpha$,

$$
b_2 \leq \sum_{k=1}^{m} \binom{n}{m+1}\binom{m+1}{k}\binom{n-m-1}{m+1-k}q_k. \tag{13.6}
$$

The upper bound (13.6) is hard to evaluate explicitly, but its dominant contribution occurs when $k = m$. Indeed, because of the asymptotic relations $n^{m+1}s^m \asymp 1$ and $q_k \asymp s^{2m-k+1}$, the $k$th term of the sum satisfies

$$\binom{n}{m+1}\binom{m+1}{k}\binom{n-m-1}{m+1-k}q_k \;\asymp\; n^{2m-k+2}s^{2m-k+1}$$

$$\asymp\; n^{2m-k+2}n^{-\frac{(2m-k+1)(m+1)}{m}}$$

$$=\; n^{\frac{k-m-1}{m}}.$$

Thus, the dominant term of the sum is of order $n^{-1/m}$, and for sufficiently large $n$ and sufficiently small $s$, the Poisson approximation $\Pr(S = 0) \approx e^{-\lambda}$ applies. The slow rate $O(n^{-1/m})$ of convergence of the total variation distance to 0 is rather disappointing in this example. The theoretical arguments and numerical evidence presented by Glaz [7] and Roos [14] suggest that a compound Poisson approximation performs better.

## 13.6   DNA Sequence Matching

A basic problem in DNA sequence analysis is to test whether two different sequences share significant similarities. Strong similarity or **homology** often indicates a common evolutionary origin of the two sequences. In many cases homology also indicates a common biochemical or structural function of the genes encoded by the sequences.

To pose the problem of sequence comparison statistically, consider two sequences with $m$ and $n$ bases, respectively. Now imagine sliding the first sequence along the second sequence. For some alignment, the two sequences share a region attaining the longest perfect match. Since ties can occur, there may be several such regions. Let $M_{mn}$ be the random number of base pairs involved in a longest perfect match. Under the null hypothesis that the two sequences are unrelated, one can compute the approximate distribution of $M_{mn}$. If the observed value of $M_{mn}$ is inordinately large according to this distribution, then significant homology can be claimed.

Computing the distribution of $M_{mn}$ is subtle. Fortunately, the Chen-Stein method is applicable. Assume first that the bases appearing at the various positions of either sequence are chosen independently from the set of nucleotides {A,C,T,G} with probabilities $q_A$, $q_C$, $q_T$, and $q_G$, respectively. The probability of a match between any two positions is

$$q \;=\; q_A^2 + q_C^2 + q_T^2 + q_G^2.$$

Define $W_{ij}$ to be the indicator random variable for the event of a match between position $i$ of the first sequence and position $j$ of the second sequence. The indicator random variable $X_{ij}$ of the event that a perfect match of length $t$ or longer begins at positions $i$ and $j$ of the two sequences is given by

$$X_{ij} \;=\; (1 - W_{i-1,j-1})\prod_{k=0}^{t-1} W_{i+k,j+k}.$$

This expression for $X_{ij}$ ignores end effects. For $m$ and $n$ large compared to $t$, end effects will be trivial. Alternatively, imagine the two sequences wrapped into circles, and interpret the subscript arithmetic involved in defining $X_{ij}$ as modulo $m$ and $n$.

According to the Chen-Stein approximation, the distribution function of the longest match $M_{mn}$ satisfies

$$\Pr(M_{mn} < t) \quad = \quad \Pr(\sum_{(i,j)} X_{ij} = 0)$$

$$\approx \quad e^{-\lambda},$$

where $\lambda = \sum_{(i,j)} \Pr(X_{ij} = 1)$. There are $mn$ pairs $(i, j)$ and each has probability $\Pr(X_{ij} = 1) = (1 - q)q^t$ of initiating a perfect match. It follows that the mean $\lambda = mn(1 - q)q^t$. Evaluating the error bound (13.3) for the Poisson approximation is possible, but too complicated to present here. See [19] for the full treatment.

The Chen-Stein approximation does provide considerable insight into the distribution of $M_{mn}$. For instance, the approximate median of $M_{mn}$ satisfies $e^{-\lambda} \approx 1/2$. This gives $mn(1 - q)q^t \approx \ln 2$, or $mn(1 - q)/\ln 2 \approx (1/q)^t$. Solving for $t$ yields $t \approx \log_{\frac{1}{q}}[mn(1 - q)/\ln 2]$. This suggests that $M_{mn}$ is of order $\log_{\frac{1}{q}}[mn(1 - q)]$. In fact, it is known [3, 18] that $M_{mn}$ has mean and variance

$$\mathrm{E}(M_{mn}) \quad \approx \quad \log_{\frac{1}{q}}[mn(1 - q)] + \gamma \log_{\frac{1}{q}} e - \frac{1}{2} \tag{13.7}$$

$$\mathrm{Var}(M_{mn}) \quad \approx \quad \frac{\left[\pi \log_{\frac{1}{q}} e\right]^2}{6} + \frac{1}{12}, \tag{13.8}$$

where $\gamma \approx .577$ is Euler's constant. Note that $\mathrm{E}(M_{nn})$ grows like $2\log_{\frac{1}{q}} n$ as $n$ grows; $\mathrm{Var}(M_{nn})$ stays virtually constant.

To gain some insight into formulas (13.7) and (13.8), it is instructive to consider the simpler problem of characterizing the limiting behavior of the maximum number of failures observed in $n$ independent realizations of a geometric waiting time with failure probability $q$ per trial. The sequence matching problem is more complicated because it involves the maximum of $mn$ dependent waiting times, with failure equated to matching and success to nonmatching.

In the simplified problem, we construct a waiting time that counts the number of failures before an ultimate success by taking the integer part $\lfloor X \rfloor$ of an appropriate exponential waiting time $X$. Now $X$ can be viewed as the time until the first random point of a Poisson process on $[0, \infty)$. In this setting the random variable $\lfloor X \rfloor = k$ if and only if there are no random points on the disjoint intervals $[0, 1), [1, 2), \ldots,$ $[k - 1, k)$ and at least one random point on the interval $[k, k + 1)$. If the intensity of the Poisson process is $\lambda$, then this event occurs with probability $(e^{-\lambda})^k(1 - e^{-\lambda})$. It follows that $\lfloor X \rfloor$ is geometrically distributed with failure probability $q = e^{-\lambda}$.

Now let $X_1, \ldots, X_n$ be $n$ independent, exponentially distributed waiting times with common intensity $\lambda$. The integer part of the maximum $M_n = \max_{1 \leq i \leq n} X_i$ satisfies $\lfloor M_n \rfloor = \max_{1 \leq i \leq n} \lfloor X_i \rfloor$. In view of the inequalities $0 \leq M_n - \lfloor M_n \rfloor < 1$, the moments of $\lfloor M_n \rfloor$ are approximately the same as the moments of $M_n$.

Let us next show that $\lambda M_n - \ln n$ converges in distribution to the extreme value statistic having density $e^{-e^{-u}} e^{-u}$. This assertion can be most easily demonstrated by considering the moment generating function of $\lambda M_n - \ln n$. Since $M_n$ has density $n(1 - e^{-\lambda x})^{n-1} \lambda e^{-\lambda x}$, it follows upon making the change of variables $u = \lambda x - \ln n$ that

$$
\begin{aligned}
E[e^{s(\lambda M_n - \ln n)}] &= \int_0^\infty e^{s(\lambda x - \ln n)} n(1 - e^{-\lambda x})^{n-1} \lambda e^{-\lambda x} dx \\
&= \int_{-\ln n}^\infty e^{su} (1 - \frac{1}{n} e^{-u})^{n-1} e^{-u} du.
\end{aligned}
$$

Taking limits and then making the second change of variables $w = e^{-u}$ yield

$$
\begin{aligned}
\lim_{n \to \infty} E[e^{s(\lambda M_n - \ln n)}] &= \int_{-\infty}^\infty e^{su} e^{-e^{-u}} e^{-u} du \\
&= \int_0^\infty w^{-s} e^{-w} dw \qquad (13.9) \\
&= \Gamma(1 - s),
\end{aligned}
$$

where $\Gamma(x)$ is the gamma function.

For $|s|$ sufficiently small, $\Gamma(1 - s)$ is an analytic function of the complex variable $s$, and the convergence in (13.9) is uniform. Thus, the moments of $\lambda M_n - \ln n$ converge to the moments of the extreme value density $e^{-e^{-u}} e^{-u}$. The mean and variance of this density can be found by differentiating its cumulant generating function $\ln \Gamma(1 - s)$. In the limit, the mean and variance of $\lambda M_n - \ln n$ therefore reduce to the familiar quantities [10]

$$
\begin{aligned}
\frac{d}{ds} \ln \Gamma(1 - s)|_{s=0} &= \gamma \\
\frac{d^2}{ds^2} \ln \Gamma(1 - s)|_{s=0} &= \sum_{k=1}^\infty \frac{1}{k^2} \\
&= \frac{\pi^2}{6}.
\end{aligned}
$$

It does not take too much of a leap of logic to assume that $\lambda \lfloor M_n \rfloor - \ln n$ has similar moments to $\lambda M_n - \ln n$. However, some caution should be exercised because $\lambda \lfloor M_n \rfloor - \ln n$ remains discretely distributed with spacing $\lambda$ between adjacent values no matter how large $n$ becomes.

As a simple approximation we can write $\lfloor M_n \rfloor \approx M_n - U$, where $U$ is uniform on $[0, 1]$ and independent of $M_n$. Because $\lambda = 1/\log_{\frac{1}{q}}(e)$ and $U$ has mean $1/2$ and variance $1/12$, we find that

$$
\begin{aligned}
E(\lfloor M_n \rfloor) &\approx \frac{1}{\lambda} E(\lambda M_n - \ln n + \ln n) - \frac{1}{2} \\
&= \log_{\frac{1}{q}} n + \gamma \log_{\frac{1}{q}} e - \frac{1}{2}
\end{aligned}
$$

$$\mathrm{Var}(\lfloor M_n \rfloor) \approx \frac{1}{\lambda^2} \mathrm{Var}(\lambda M_n - \ln n) + \frac{1}{12}$$

$$= \frac{\left[\pi \log_{\frac{1}{q}} e\right]^2}{6} + \frac{1}{12}.$$

## 13.7   Problems

1. Prove that the Chen-Stein bound (13.1) implies the bound (13.2) when the inequality $U_\alpha \geq V_\alpha$ holds for all $\alpha$.

2. Show that in the neighborhood method $\lambda - \mathrm{Var}(S) = b_1 - b_2$.

3. Let $X_\alpha$, $\alpha \in I$, be independent Bernoulli random variables with means $p_\alpha$. Show that the total variation distance between $S = \sum_\alpha X_\alpha$ and a Poisson random variable $Z$ with mean $\lambda = \sum_\alpha p_\alpha$ is bounded above by the quantity $\lambda^{-1}(1 - e^{-\lambda}) \sum_\alpha p_\alpha^2$.

4. For a random permutation $\sigma_1, \ldots, \sigma_n$ of $\{1, \ldots, n\}$, let $X_\alpha = 1_{\{\sigma_\alpha = \alpha\}}$ be the indicator of a match at position $\alpha$. Show that the total number of matches $S = \sum_{\alpha=1}^n X_\alpha$ satisfies the coupling bound

$$\|\mathcal{L}(S) - \mathcal{L}(Z)\| \leq \frac{2(1 - e^{-1})}{n},$$

where $Z$ follows a Poisson distribution with mean 1. (Hint: Use inequality (13.1) rather than inequality (13.2).)

5. In certain situations the hypergeometric distribution can be approximated by a Poisson distribution. Suppose that $w$ white balls and $b$ black balls occupy a box. If you extract $n < w + b$ balls at random, then the number of white balls $S$ extracted follows a hypergeometric distribution. Note that if we label the white balls $1, \ldots, w$, and let $X_\alpha$ be the random variable indicating whether white ball $\alpha$ is chosen, then $S = \sum_{\alpha=1}^w X_\alpha$. Show that you can construct a coupling by performing the sampling experiment in the usual way. If white ball $\alpha$ does not show up, then randomly take one of the balls extracted and exchange it for white ball $\alpha$. Calculate an explicit Chen-Stein bound, and give conditions under which the Poisson approximation to $S$ will be good.

6. Consider the $n$-dimensional unit cube $[0, 1]^n$. Suppose that each of its $n2^{n-1}$ edges is independently assigned one of two equally likely orientations. Let $S$ be the number of vertices at which all neighboring edges point toward the vertex. The Chen-Stein method implies that $S$ has an approximate Poisson distribution $Z$ with mean 1. Verify the estimate

$$\|\mathcal{L}(S) - \mathcal{L}(Z)\| \leq (n + 1)2^{-n}(1 - e^{-1}).$$

(Hint: Let $I$ be the set of all $2^n$ vertices, $X_\alpha$ the indicator that vertex $\alpha$ has all of its edges directed toward $\alpha$, and $B_\alpha = \{\beta : \|\beta - \alpha\| \leq 1\}$. Note that $X_\alpha$ is independent of those $X_\beta$ with $\|\beta - \alpha\| > 1$. Also, $b_2 = 0$ because $p_{\alpha\beta} = 0$ for $\|\beta - \alpha\| = 1$.)

7. A graph with $n$ nodes is created by randomly connecting some pairs of nodes by edges. If the connection probability per pair is $p$, then all pairs from a triple of nodes are connected with probability $p^3$. For $p$ small and $\lambda = \binom{n}{3}p^3$ moderate in size, the number of such triangles in the random graph is approximately Poisson with mean $\lambda$. Use the neighborhood method to estimate the total variation error in this approximation.

8. Suppose $n$ balls (people) are uniformly and independently distributed into $m$ boxes (days of the year). The birthday problem involves finding the approximate distribution of the number of boxes that receive $d$ or more balls for some fixed positive integer $d$. This is a special case of the $W_d$ Poisson approximation treated in the text by the coupling method. In this exercise we attack the birthday problem by the neighborhood method. To get started, let the index set $I$ be the collection of all sets of trials $\alpha \subset \{1, \ldots, n\}$ having $|\alpha| = d$ elements. Let $X_\alpha$ be the indicator of the event that the balls indexed by $\alpha$ all fall into the same box. Argue that the approximation $\Pr(W_d = 0) \approx e^{-\lambda}$ with

$$\lambda = \binom{n}{d}\frac{1}{m^{d-1}}$$

is plausible. Now define the neighborhoods $B_\alpha$ so that $X_\alpha$ is independent of those $X_\beta$ with $\beta$ outside $B_\alpha$. Prove that the Chen-Stein constants $b_1$ is bounded above by $b_2$ are

$$b_1 = \binom{n}{d}\left[\binom{n}{d} - \binom{n-d}{d}\right]\left(\frac{1}{m}\right)^{2d-2}$$

$$b_2 = \binom{n}{d}\sum_{i=1}^{d-1}\binom{d}{i}\binom{n-d}{d-i}\left(\frac{1}{m}\right)^{2d-i-1}.$$

When $d = 2$, compute the total variation bound

$$\frac{1 - e^{-\lambda}}{\lambda}(b_1 + b_2) = \frac{1 - e^{-\lambda}}{\lambda}\frac{\binom{n}{2}(4n - 7)}{m^2}.$$

9. In the somatic cell hybrid model, suppose that the retention probability $p \neq \frac{1}{2}$. Define $w_{n,d_{12},d_{13}} = \Pr[\rho(C_1^n, C_2^n) = d_{12}, \rho(C_1^n, C_3^n) = d_{13}]$ for a random panel with $n$ clones. Show that

$$p_{\alpha\beta} = \sum_{d_{12}=0}^{d-1}\sum_{d_{13}=0}^{d-1} w_{n,d_{12},d_{13}},$$

regardless of which $\beta \in B_\alpha\backslash\{\alpha\}$ is chosen [8]. Setting $r = p(1 - p)$, verify the recurrence relation

$$w_{n+1,d_{12},d_{13}} = r(w_{n,d_{12}-1,d_{13}} + w_{n,d_{12},d_{13}-1} + w_{n,d_{12}-1,d_{13}-1})$$
$$+ (1 - 3r)w_{n,d_{12},d_{13}}.$$

Under the natural initial conditions, $w_{0,d_{12},d_{13}}$ is 1 when $d_{12} = d_{13} = 0$ and 0 otherwise.

10. In the somatic cell hybrid model, suppose that one knows a priori that the number of assay errors does not exceed some positive integer $d$. Prove that assay error can be detected if the minimum Hamming distance of the panel is strictly greater than $d$. Prove that the locus can still be correctly assigned to a single chromosome if the minimum Hamming distance is strictly greater than $2d$.

11. Consider an infinite sequence $W_1, W_2, \ldots$ of independent, Bernoulli random variables with common success probability $p$. Let $X_\alpha$ be the indicator of the event that a success run of length $t$ or longer begins at position $\alpha$. Note that $X_1 = \prod_{k=1}^{t} W_k$ and

$$X_j = (1 - W_{j-1}) \prod_{k=j}^{j+t-1} W_k$$

for $j > 1$. The number of such success runs starting in the first $n$ positions is given by $S = \sum_{\alpha \in I} X_\alpha$, where the index set $I = \{1, \ldots, n\}$. The Poisson heuristic suggests the $S$ is approximately Poisson with mean $\lambda = p^t[(n - 1)(1 - p) + 1]$. Let $B_\alpha = \{\beta \in I : |\beta - \alpha| \leq t\}$. Show that $X_\alpha$ is independent of those $X_\beta$ with $\beta$ outside $B_\alpha$. In the Chen-Stein bound (13.3), prove that the constant $b_2 = 0$. Finally, show that the Chen-Stein constant $b_1$ is bounded above by $\lambda^2(2t + 1)/n + 2\lambda p^t$. (Hint:

$$\begin{aligned} b_1 &= p^{2t} + 2tp^{2t}(1 - p) \\ &\quad + [2nt - t^2 + n - 3t - 1]p^{2t}(1 - p)^2 \end{aligned}$$

exactly. Note that the pairs $\alpha$ and $\beta$ entering into the double sum for $b_1$ are drawn from the integer lattice points $\{(i, j) : 1 \leq i, j \leq n\}$. An upper left triangle and a lower right triangle of lattice points from this square do not qualify for the double sum defining $b_1$. The term $p^{2t}$ in $b_1$ corresponds to the lattice point $(1, 1)$.)

12. Consider a word of length $n$ chosen from the four-letter DNA alphabet {A,T,C,G}. If letters are chosen for each position independently with probabilities $q_A$, $q_T$, $q_C$, and $q_G$, respectively, then let $R_n^A$ be the length of the longest run of A's in the word. Show that the distribution of $R_n^A$ satisfies the recurrence relation

$$\Pr(R_n^A \leq m) = \sum_{i=1}^{m+1} q_A^{i-1}(1 - q_A) \Pr(R_{n-i}^A \leq m)$$

for $n > m$ and the initial condition $\Pr(R_n^A \leq m) = 1$ for $n \leq m$ [16]. Similar results obtain for runs of any other letter.

13. In the context of the last problem, let $S_n^A$ be the length of the longest run of any letter when the word starts with an A. Define $S_n^T$, $S_n^C$, and $S_n^G$ similarly. Argue that

$$\Pr(S_n^A \leq m) = \sum_{i=1}^{m} q_A^{i-1} \big[ q_T \Pr(S_{n-i}^T \leq m) \tag{13.10}$$
$$+ q_C \Pr(S_{n-i}^C \leq m) + q_G \Pr(S_{n-i}^G \leq m) \big]$$

for $n > m$ and that $\Pr(S_n^A \leq m) = 1$ for $n \leq m$. Show how recurrence (13.10) and similar recurrences for the distributions of $S_n^T$, $S_n^C$, and $S_n^G$ permit exact calculation of the distribution of the length of the maximal run $R_n$ of any letter type.

# References

[1] Arratia R, Goldstein L, Gordon L (1989) Two moments suffice for Poisson approximations: the Chen-Stein method. *Ann Prob* 17:9–25

[2] Arratia R, Goldstein L, Gordon L (1990) Poisson approximation and the Chen-Stein method. *Stat Sci* 5:403–434.

[3] Arratia R, Gordon L, Waterman MS (1986) An extreme value theory for sequence matching. *Ann Stat* 14:971–993

[4] Barbour AD, Holst L, Janson S (1992) *Poisson Approximation*. Oxford University Press, Oxford

[5] D'Eustachio P, Ruddle FH (1983) Somatic cell genetics and gene families. *Science* 220:919–924

[6] Flatto L, Konheim AG (1962) The random division of an interval and the random covering of a circle. *SIAM Review* 4:211–222

[7] Glaz J (1992) Extreme order statistics for a sequence of dependent random variables. *Stochastic Inequalities*, Shaked M, Tong YL, editors, IMS Lecture Notes – Monograph Series, Vol 22, Hayward, CA, pp 100–115

[8] Goradia TM (1991) *Stochastic Models for Human Gene Mapping*. Ph.D. Thesis, Division of Applied Sciences, Harvard University

[9] Goradia TM, Lange K (1988) Applications of coding theory to the design of somatic cell hybrid panels. *Math Biosciences* 91:201–219

[10] Hille E (1959) *Analytic Function Theory, Vol 1*. Blaisdell, New York

[11] Karlin S, Macken C (1991) Some statistical problems in the assessment of inhomogeneities of DNA sequence data. *J Amer Stat Assoc* 86:27–35

[12] Lange K, Boehnke M (1982) How many polymorphic genes will it take to span the human genome? *Amer J Hum Genet* 34:842–845

[13] Lindvall T (1992) *Lectures on the Coupling Method.* Wiley, New York

[14] Roos M (1993) Compound Poisson approximations for the numbers of extreme spacings. *Adv Appl Prob* 25:847–874

[15] Rushton AR (1976) Quantitative analysis of human chromosome segregation in man-mouse somatic cell hybrids. *Cytogenetics Cell Genet* 17:243–253

[16] Schilling MF (1990) The longest run of heads. *The College Mathematics Journal* 21:196–207

[17] Solomon H (1978) *Geometric Probability.* SIAM, Philadelphia

[18] Waterman MS (1989) Sequence alignments. *Mathematical Methods for DNA Sequences*, CRC Press, Boca Raton, FL, pp 53–92

[19] Waterman MS (1995) *Introduction to Computational Biology: Maps, Sequences, and Genomes.* Chapman and Hall, London

[20] Weiss M, Green H (1967) Human-mouse hybrid cell lines containing partial complements of human chromosomes and functioning human genes. *Proc Natl Acad Sci* 58:1104–1111

# Appendix: Molecular Genetics in Brief

## A.1   Genes and Chromosomes

All of life is ultimately based on biochemistry and all of genetics on the biochemistry of DNA and RNA. The famous DNA double helix discovered by Watson and Crick carries the information necessary for the development, maintenance, and reproduction of all organisms, from bacteria to humans. The genetic code consists of an alphabet of four letters (or **bases**) organized into words (or **codons**) of three letters each. Codons are further grouped into genes or parts of genes known as **exons**. Genes are **translated** as needed by a cell into **proteins**; these in turn catalyze the many reactions taking place in the cell and serve as structural components of cellular organelles and membranes.

The four bases of DNA are **adenine, guanine, cytosine**, and **thymine**, abbreviated A, G, C, and T, respectively. The first two of these bases are **purines**; the latter two are **pyrimidines**. In RNA, a sister compound to DNA, the pyrimidine **uracil** (U) is substituted for thymine. Both RNA and DNA are **polymers** constructed by linking identical sugar units—ribose in the case of RNA and deoxyribose in the case of DNA—by identical phosphate groups. These sugar/phosphate linkages along the backbone of the polymer (or **strand**) occur at carbon sites on the sugars designated by the abbreviations $3'$ and $5'$. One base is attached to each sugar; a single repeat unit consisting of a sugar, phosphate group, and base is known as a **nucleotide**. Codons are read in the $5'$ to $3'$ direction.

DNA is distinguished from RNA by its stronger tendency to form double helices of complementary **strands**. The two strands of DNA are held together by hydrogen bonds between the bases projecting into the center of the double helix from the backbones. The geometry of base pairing dictates that adenine is paired to thymine and cytosine to guanine. These hydrogen bonds are strong enough to stabilize the double helix but weak enough to permit unzipping of the DNA for **transcription** of a gene on one strand or replication of both strands when a cell undergoes division. Note that in transcribing a gene, the **antisense** DNA strand serves as the template rather than the **sense** strand so that the copied **messenger** RNA will carry sense rather than antisense codons.

A **chromosome** is more than just a naked double helix. To protect DNA from the occasionally harsh environment of the cell, to keep it from getting hopelessly tangled, and to control what genes are expressed when, the double helix is wrapped around protein structures with a central core of eight **histone** proteins. A complex of wrapped DNA and histone core is known as a **nucleosome**. Nucleosomes are further organized into **chromatid** fibers by supercoiling. Thus, the helical motif is repeated at several levels in the construction of a chromosome. At the highest level of chromosome organization are the **centromere** and two **telomeres**. The centromere is critical to proper division of duplicated chromosomes at mitosis and meiosis. The telomeres protect the ends of a chromosome from degradation and control the maximal number of divisions a cell line can undergo. In gene mapping, the centromere and telomeres serve as cytologically visible landmarks. The adjectives **proximal** and **distal** indicate centromeric and telomeric directions, respectively.

The 22 **autosomes** and the **X chromosome** of humans contain a total of $3 \times 10^9$ base pairs. Embedded within the human **genome** are between 50,000 and 100,000 genes, most of which range from 10,000 to 100,000 bases in length. Much of the genome consists of noncoding DNA whose function is poorly understood. However, one should be careful in dismissing the noncoding regions as "junk" DNA. All organisms must control the timing and level of transcription of their genes. Geneticists have identified regulatory regions such as **promoters**, **enhancers**, and **silencers** upstream and downstream from many genes. Other regions provide recognition sites for recombination enzymes and attachment sites for the machinery of chromosome segregation during meiosis and mitosis. Even the patently junk DNA of **pseudogenes** provides a fossil record of how genes duplicate, evolve, and are eventually discarded. The full significance of the human genome will become clear only after it is sequenced. Exploring and interpreting this treasure trove should occupy geneticists for many years to come.

## A.2    From Gene to Protein

Proteins are constructed from the 20 amino acids shown in Table A.1. With an alphabet of four letters, three base words could in principle code for $4^3 = 64$ different amino acids. Nature has elected to forgo this opportunity and instead opts for redundancy in the genetic code. This redundancy is very evident in Table A.2. Several other features of Table A.2 are worth noting. First, three codons serve as genetic stop signals for terminating a growing protein polymer. Second, the codon

TABLE A.1. Amino Acids

| Amino Acid | Abbreviation | Amino Acid | Abbreviation |
|---|---|---|---|
| Alanine | Ala | Leucine | Leu |
| Arginine | Arg | Lysine | Lys |
| Aspartic acid | Asp | Methionine | Met |
| Asparginine | Asn | Phenylalanine | Phe |
| Cysteine | Cys | Proline | Pro |
| Glutamic acid | Glu | Serine | Ser |
| Glutamine | Gln | Threonine | Thr |
| Glycine | Gly | Tryptophan | Trp |
| Histidine | His | Tyrosine | Tyr |
| Isoleucine | Ile | Valine | Val |

AUG for methionine also plays the role of a start signal provided it is preceded by purine-rich sequences such as AGGA. Third, the base U is substituted for the base T. This substitution occurs because the DNA specifying a gene is first **transcribed** into single-stranded **premessenger** RNA. After appropriate processing, premessenger RNA is turned into **messenger** RNA, which is then translated into protein by cell organelles known as **ribosomes**. On release from a ribosome, a protein folds into its characteristic shape. Depending on its ultimate function, a protein may undergo further processing such as cleavage or the addition of lipid or carbohydrate groups.

The processing of premessenger RNA involves several steps. In eukaryotes (organisms with a well-defined **nucleus** for housing chromosomes), the exons of a typical gene are interrupted by noncoding sequences. These intervening sequences (or **introns**) must be spliced out of the premessenger RNA. A cap involving a methylated guanine is also added to the 5′ end of the RNA, and a poly(A) tail involving about 200 adenines is added to the 3′ end. These additions assist in stabilizing the RNA and binding it to the ribosomes. After messenger RNA is transported to the exterior of the nucleus, it is threaded through a ribosome like a magnetic tape through the head of a tape player. **Transfer** RNA molecules bring the appropriate amino acids into place for addition to the growing chain of the protein encoded by the messenger RNA. Figure A.1 summarizes the flow of information

FIGURE A.1. Information Flow from Gene to Protein

TABLE A.2. The Genetic Code

| 1st Codon | 2nd Codon | | | | 3rd Codon |
|---|---|---|---|---|---|
| | U | C | A | G | |
| U | Phe | Ser | Tyr | Cys | U |
| | Phe | Ser | Tyr | Cys | C |
| | Leu | Ser | Stop | Stop | A |
| | Leu | Ser | Stop | Trp | G |
| C | Leu | Pro | His | Arg | U |
| | Leu | Pro | His | Arg | C |
| | Leu | Pro | Gln | Arg | A |
| | Leu | Pro | Gln | Arg | G |
| A | Ile | Thr | Asn | Ser | U |
| | Ile | Thr | Asn | Ser | C |
| | Ile | Thr | Lys | Arg | A |
| | Met | Thr | Lys | Arg | G |
| G | Val | Ala | Asp | Gly | U |
| | Val | Ala | Asp | Gly | C |
| | Val | Ala | Glu | Gly | A |
| | Val | Ala | Glu | Gly | G |

from gene to protein.

## A.3   Manipulating DNA

Geneticists manipulate DNA in many ways. For instance, they unzip (or **denature**) double-stranded DNA by heating it in solution. They rezip (or **anneal**) it by cooling. Because double helices are so energetically favored, complementary strands quickly find and bind to one another. Even small segments of one strand will locally anneal to a large segment of a complementary strand. Geneticists exploit this behavior by devising small radioactive or fluorescent **probes** to identify large segments. A single base mismatch between probe and strand leads to poor annealing. Probes as short as 20 bases can provide a perfect match to a unique part of the human genome.

Chromosomes are much too large to handle conveniently. To reduce DNA to more manageable size, geneticists cut it into fragments and measure the length of the fragments. **Restriction enzymes** function as geneticists' molecular scissors.

TABLE A.3. Commonly Used Restriction Enzymes

| Restriction Enzyme | Recognition Site | Average Fragment Length in Man |
|---|---|---|
| AluI | AGCT | 0.3 kb |
| HaeIII | GGCC | 0.6 kb |
| TaqI | TCGA | 1.4 kb |
| HpaI | CCGG | 3.1 kb |
| EcoRI | GAATTC | 3.1 kb |
| PstI | CTGCAG | 7 kb |
| NotI | GCGGCCGC | 9766 kb |

Table A.3 lists some commonly used restriction enzymes, each of which recognizes a specific base sequence and cuts DNA there. Recognition sites are scattered more or less randomly throughout the genome. **Restriction maps** characterize the number, order, and approximate separation of recognition sites on large DNA segments. These maps are laborious to prepare and involve digesting a segment with different combinations of restriction enzymes or with a single enzyme at less than optimal laboratory conditions. These latter **partial digests** randomly miss some recognition sites and therefore give a mixture of fragments defined by adjacent sites and fragments spanning blocks of adjacent sites. Waterman [7] discusses the interesting computational issues that arise in piecing together a restriction map.

   **Gel electrophoresis** and **Southern blotting** are geneticists' molecular yardsticks. In electrophoresis, a sample of DNA is placed at the top of a gel subject to an electric field. Under the influence of the field, DNA migrates down the gel. Large DNA fragments encounter more obstacles than small fragments and consequently travel more slowly, just as in a flowing stream, large stones travel more slowly than small ones. Once the DNA fragments are separated by size, a Southern blot can be made. This involves denaturing the fragments by the addition of alkali and transferring the separated strands to a nitrocellulose or nylon membrane. After the strands are fixed to the membrane by baking or chemical crossbinding, radioactive probes are introduced to the membrane and anneal with specific fragments. When the membrane is applied to an X-ray film, a sequence of bands develops on the film highlighting those DNA fragments bound to probes. Alternatively, if sufficient DNA is sampled, then fluorescent or chemiluminescent probes can be substituted for radioactive probes.

   Geneticists often work with minuscule amounts of DNA. For instance, in genotyping human sperm cells, geneticists encounter single-copy DNA. The **polymerase chain reaction** (PCR) permits enormous copy-number amplification of a short DNA sequence. The chromosome region surrounding the target sequence is first denatured by heating it in a solution containing the four DNA bases, two specially chosen **primers**, and a **polymerase**. As the solution cools, the two primers anneal to the two strand-specific 3' regions flanking the target sequence. The polymerase then extends each primer through the target sequence, creating a new strand that partially complements one of the original strands. This constitutes the first cycle of PCR and doubles the number of target sequences. Each subsequent cycle of denaturation, primer annealing, and polymerase extension similarly doubles the number of target sequences. Figure A.2 depicts the first cycle of the process. Here the primers are shorter than they would be in practice.

**Cloning** is a kind of in vivo DNA amplification. DNA fragments isolated by restriction enzymes are ligated into circular DNA molecules called **vectors** and inserted into bacteria or yeast cells. Once inside the host cells, vectors resemble viruses in their ability to harness the machinery of the cell to replicate independently of the host chromosomes. Vast **libraries** of random DNA clones can be maintained in this manner. These clone libraries furnish the raw material for DNA sequencing.

## A.4   Mapping Strategies

Linkage mapping is described in detail in earlier chapters. It is worth emphasizing here the nature of most modern markers. **Restriction fragment polymorphisms** (RFLPs) exploit individual differences in the presence or absence of restriction sites. Suppose a probe is constructed to straddle a polymorphic restriction site for a particular restriction enzyme. If nonpolymorphic restriction sites flank the probe region on its left and right, then Southern blots of appropriately digested DNA from random individuals fall into three patterns. Homozygotes for the absence of the site show a single band high on the gel. This band corresponds to the long fragment between the two flanking restriction sites. Homozygotes for the presence of the site show two separate bands lower down on the gel. These correspond to the two smaller fragments defined by the flanking sites and separated by the interior restriction site. Heterozygotes show all three bands. If the probe falls to one side of the polymorphic restriction site, then at most two bands appear, but it is still possible to distinguish all three genotypes.

The single base pair differences revealed by RFLPs must fall within a restriction site for some restriction enzyme. These sites are naturally rare, and it is convenient to exploit single base pair differences wherever they occur. This can be accomplished by sequencing a small segment of DNA around a polymorphic site and designing short probes (or **oligonucleotides**) that match the sequence of the segment except at the site. At the site, each probe matches one of the dominant bases appearing in the population. Because annealing of a short probe requires a perfect match, Southern blots with different probes detect different alleles. This technique is particularly useful in screening for common mutations in disease loci.

The biallelic markers generated by single base pair differences exhibit limited polymorphism. **Short tandem repeat** markers are often much more polymorphic. For instance, the dinucleotide CA is repeated a random number of times in many regions of the human genome. Repeat numbers in a repetitive sequence ...CACA-CACACA... often vary from person to person. If a probe closely flanks a repeat

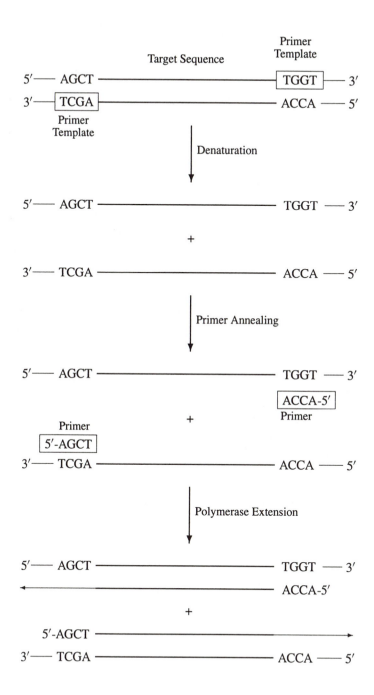

FIGURE A.2. DNA Amplification by PCR

252 Appendix: Molecular Genetics in Brief

region, then Southern blotting with the probe will reveal allelic differences in the number of repeat units as length differences in the fragments highlighted by the probe.

**Radiation hybrids** and **somatic cell hybrids** are physical mapping techniques covered in detail in Chapters 11 and 13, respectively. **Fluorescence in situ hybridization** (FISH) and **pulsed-field gel electrophoresis** are two other competing physical techniques with good resolution. In FISH, probes are directly annealed to chromosomes during the **interphase** period of cell division. Because chromosomes are less contracted during interphase, map resolution to within 100,000 bases is possible. In a recent variation of FISH, probes are annealed to DNA filaments stretched on a glass slide. In pulsed-field gel electrophoresis, the electric field applied to a gel is occasionally reversed. This permits large DNA fragments to untangle and slowly migrate down the gel without breaking. If two different probes anneal to the same large fragment, then presumably they coexist on the fragment. Using a sufficient number of fragments, closely spaced loci defined by well-defined probes can be ordered.

One advantage of physical mapping is that it does not require polymorphic loci. It is usually harder to find polymorphisms than it is to construct a probe from unique sequence DNA. For example, geneticists can easily identify **expressed sequence tags** by sequencing **complementary DNA** (cDNA). Because cDNA is synthesized from messenger RNA, an expressed sequence tag is guaranteed to be in the coding region of some gene. (Recall that all of the introns are spliced out of messenger RNA, leaving only the contributions from the exons.) Discovery of expressed sequence tags and classification of the genes they label have attracted the interest of the biotechnology industry. Because many genes are expressed only in certain tissues and only at certain times of development, systematic classification and mapping of expressed genes is apt to pay off in suggesting candidate genes for human diseases.

Finally, as a prelude to sequencing the human genome and the genomes of other species, a great deal of thought and effort has gone into ordering the clones present in clone libraries. Restriction maps of different clones often show sufficient similarity to suggest that two clones overlap. Alternatively, two clones may both harbor the same expressed sequence tag. The presence of such a chromosome anchor on both clones is proof of overlap. Chains of overlapping clones are referred to as **contigs**. Readers may consult Waterman [7] for a mathematical analysis of strategies for constructing contigs and closing the gaps between them.

# References

[1] Berg P, Singer M (1992) *Dealing with Genes: The Language of Heredity.* University Science Books, Mill Valley, CA

[2] Davies KE, Read AP (1992) *Molecular Basis of Inherited Disease*, 2nd ed. Oxford University Press, Oxford

[3] Gelehrter TD, Collins FS (1990) *Principles of Medical Genetics.* Williams and Wilkins, Baltimore

[4] Lewin B (1994) *Genes V.* Oxford University Press, Oxford

[5] Strachan T (1992) *The Human Genome.* Bios Scientific Publishers, Oxford

[6] Vogel F, Motulsky AG (1996) *Human Genetics: Problems and Approaches*, 3rd ed. Springer-Verlag, Berlin

[7] Waterman MS (1995) *Introduction to Computational Biology: Maps, Sequences, and Genomes.* Chapman and Hall, London

[8] Watson JD, Gilman M, Witkowski J, Zoller M (1992) *Recombinant DNA*, 2nd ed. Scientific American Books, New York

[9] Watson JD, Hopkins NH, Roberts JW, Steitz JA, Weiner AM (1987) *Molecular Biology of the Gene, Vols 1 and 2*, 4th ed. Benjamin/Cummings, Menlo Park, CA.

# Index